Networks and Communications Engineering

Networks and Communications Engineering

**Edited by
Bernhard Ekman**

www.willfordpress.com

Published by Willford Press,
118-35 Queens Blvd., Suite 400,
Forest Hills, NY 11375, USA

ISBN: 978-1-68285-347-4

Cataloging-in-Publication Data

Networks and communications engineering / edited by Bernhard Ekman.
p. cm.
Includes bibliographical references and index.
ISBN 978-1-68285-347-4
1. Computer networks. 2. Wireless communication systems. 3. Electronic data processing--Distributed processing.
4. Telecommunication. 5. Communication and technology. I. Ekman, Bernhard.
TK5105.5 .N48 2017
004.6--dc23

For information on all Willford Press publications
visit our website at www.willfordpress.com

Contents

Permissions

List of Contributors

Index

Preface

Communications engineering is the study of electrical and computer systems that improve telecommunications systems. This book on telecommunications engineering discusses topics related to the engineering and design of technology that optimizes and enhances networks and allows for data transfer at high speed. This book elucidates the concepts and innovative models around prospective developments with respect to communications engineering. Topics included herein delve into various applications of this field, as well as the rapid progress in technological innovations. Some of the diverse topics covered in this text address the varied branches that fall under this category. With state-of-the-art inputs by acclaimed experts of this field, this book targets students and professionals.

All of the data presented henceforth, was collaborated in the wake of recent advancements in the field. The aim of this book is to present the diversified developments from across the globe in a comprehensible manner. The opinions expressed in each chapter belong solely to the contributing authors. Their interpretations of the topics are the integral part of this book, which I have carefully compiled for a better understanding of the readers.

At the end, I would like to thank all those who dedicated their time and efforts for the successful completion of this book. I also wish to convey my gratitude towards my friends and family who supported me at every step.

Editor

A journal of real peak recognition of electrocardiogram (ecg) signals using neural network

Tarmizi Amani Izzah[1], Syed Sahal Nazli Alhady[1], Umi Kalthum Ngah, Wan Pauzi Ibrahim[2]

[1]School of Electrical Electronics Engineering, Universiti Sains Malaysia (Engineering Campus), Nibong Tebal, SPS, Penang
[2]School of Medical Sciences, Universiti Sains Malaysia (Health Campus), 16150, Kubang Kerian, Kelantan

Email address:
izzah.amani@ymail.com (I. A. Tarmizi), sahal@eng.usm.my (S. S. N. Alhady), eeumi@eng.usm.my (U. K. Ngah),
wpauzi@kb.usm.my (W. P. Ibrahim)

Abstract: This paper describes about the analysis of electrocardiogram (ECG) signals using neural network approach. Heart structure is a unique system that can generate ECG signals independently via heart contraction. Basically, an ECG signal consists of PQRST wave. All these waves are represented respective heart functions. Normal healthy heart can be simply recognized by normal ECG signal while heart disorder or arrhythmias signals contain differences in terms of features and morphological attributes in their corresponding ECG waveform. Some major important features will be extracted from ECG signals such as amplitude, duration, pre-gradient, post-gradient and so on. These features will then be fed as an input to neural network system. The target output represented real peaks of the signals is also being defined using a binary number. Result obtained showing that neural network pattern recognition is able to classify and recognize the real peaks accordingly with overall accuracy of 81.6% although there might be limitations and misclassification happened. Future recommendations have been highlighted to improve network's performance in order to get better and more accurate result.

Keywords: Heart, ECG Signal, Features Extraction, Neural Network And Matlab Simulation

1. State-of-the Art

Neural network nowadays has been applied extensively in wide areas including classification, detection, aerospace, forecasting, heart diagnosis and many more. This project applied neural network method in analysing cardiac rhythms. Recognition of real peaks of ECG signals is important to diagnose the heart diseases. Doctors obtained ECG data from Holter device that recorded patient's heart beat and they perform analysis manually based on the waveform characteristics. Currently, cardiologist or doctors identify the real peak based on their knowledge and previous experiences. Some ECG signals easily obtained the real peak by looking at the waveform pattern but there is also some signals which is very difficult to identify their real peak. This manual identification might contain inaccuracy and small percentage of error. In addition, it could be a tedious way especially when analyzing a very low frequency of ECG signals. Doctors normally need an adequate time to study and verify the ECG waveform before getting the correct result for every patient examined. This works done definitely time-consuming and not the efficient way.

Thus, this idea comes up that could be beneficial to cardiologists to recognize the real peak using new approach which is neural network. Neural network has the ability to memorize the pattern and directly gives the result accordingly. This eases the doctors and analysis could be done in way that is more efficient. Neural network also exhibits independent behavior as well as self-learning. It designed in such a way that exposed to enough training until it comes to a generalization state. Generalization means the network can memorize the data pattern and able to give result correctly based on the previous learning given. The system then tested to see its accuracy and performance.

It is then becoming a good momentum to exhort improvement in electrocardiography by offering a reliable as well as comprehensive solution for better ECG diagnosis [14]. Besides, by using neural network method, things will be much simpler, time-savings as well as reducing the needs of human efforts as machine has been trained to perform the desired workload.

2. Introduction

The needs of technology and computerized analysis usage has exhorted researchers, professionals, engineers and other expert people combining their efforts together in implementing quality diagnosis tools. The term quality has been interpreted as easier and faster analysis, lack maintenance, high efficient as well as low in the cost. Due to that, another approach to analyse ECG signals has been chosen by using neural network method via matlab software, as matlab is well known with multifunction and powerful computerized tool software. This project has applied neural network approach to analyze ECG signals focusing on real peaks recognition since it provides valuable information to doctors regarding heart diagnosis. Recognition of real peak correctly is absolutely essential as it indicated the condition of heart as well as reflects to its functionality.

2.1. Generation of ECG Signal

The corresponding part in the heart plays their respective roles. Sinoatrial (SA) node will excite the beats that caused heart muscles to contract. Below is shown the location of Atrioventricular (AV) node and SA node which are responsible for generating ECG signal in the human's heart.

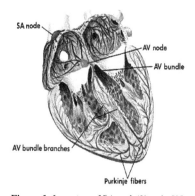

Figure 1. Location of SA and AV node [1].

The contraction of heart's muscles soon will be recorded as an electrical activity of the heart called ECG signal. Based on the pattern of ECG recording, heart status could be identified whether possessed of any cardiac arrhythmias or otherwise. As known, heart muscle possessed the characteristic of depolarization and repolarization. Depolarization is referred to the electrical potential activity excited by heart muscles while repolarization is a relaxation state when the heart changing back to its original position. P wave generated due to the atrial depolarization, QRS complex represented ventricular depolarization while T wave represented ventricular repolarization [2]. Figure below showed the corresponding part of heart function with respect to the ECG signal obtained.

Abnormalities happened in the respective waves segment will provide ideas to doctors and cardiologists at where the part of heart is having problems.

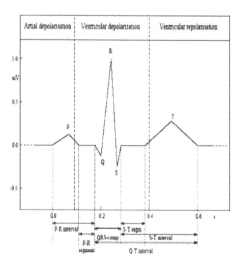

Figure 2. ECG signal based on heart function [3], [4].

2.2. Neural Network

A neural network is a type of computational model which is able to solve multi problems in various fields. It processes the information in a similar way as the human brain concept processing the information [5]. Basically, neural network consists of large processing elements called neurons working together to perform specific tasks. As in the human brain, there are thousands of dendrites which contain information signals. They transmitted the signals to the axon in the form of electrical spikes. The axon then sends the signals to another dendrites causing to a synapse. This synapse occurred when excitatory input is sufficiently large than the inhibitory input, and this concept of signal transmission also depicted on how neural network process inputs received. Figure below shown dendrites related structures for clearer understanding.

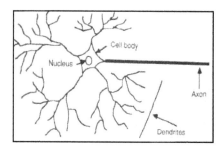

Figure 3. Dendrites [6].

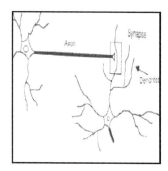

Figure 4. Synapses [6].

In neural network, dendrites carrying signals can be analogy as multiple inputs collected together for summation. Then, the combined inputs will be activated by activation function. Inputs that exceed threshold value will be feed to the output layer for final processing. During processing stage, inputs will be trained to produce desired target outputs until it come to a generalization stage. Generalization means at a condition where the network is able to recognize the inputs and the corresponding targets after undergone the training given. Then, the network will be tested by given new inputs signal to evaluate its performance and to see how accurate the output produced will be comparing to the target. Figure below depicted the analogy of human brain concept process the information to the neural network system.

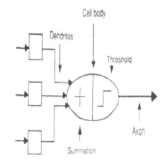

Figure 5. *Neuron model [6].*

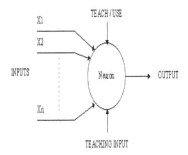

Figure 6. *Neural networks [6].*

Neural network consists of several architectures, from simple structure until the complicated ones.

2.2.1. Single-Layer Feed forward Networks

This is the simplest form of network architecture with only single layer output without any hidden layer. An input layer of source nodes will directly projects onto the output layer of neurons or computation nodes, just in one way butnot vice versa. Single layer is referring to the output layer which is just single output and not considered the input layer of source nodes since no computation performed there [7]. Figure 7 showed the corresponding figure including label.

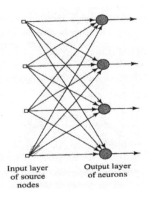

Figure 7. *Single layer feedforward networks [7].*

2.1.2. Multilayer Feed-forward Networks

This second layer is differ from above since it has one or more hidden layers. The computation also takes place in these hidden nodes. Hidden nodes are also used to intervene between the external input and the network output with respect to the network's manner. The structure of multilayer feedforward networks with one hidden layer as in figure 8.

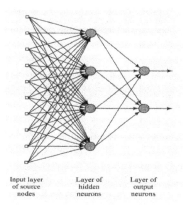

Figure 8. *Typical of multilayer feedforward networks [7].*

2.1.3. Recurrent Network with No-Self Feedback Loops and No Hidden Neurons

A recurrent neural network has at least one feedback loop. It may consist of a single layer of neurons with each neuron feeding its output back to all the input neurons as illustrated in figure 9. The feedback loops increased the learning capability of the network and on its performance. Besides, these feedback loops are also associated with unit delay elements (z^{-1}) which result in a nonlinear dynamical behavior in a condition when neural network contains nonlinear units.

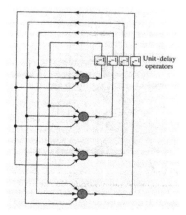

Figure 9. Typical of multilayer feedforward networks [7].

2.1.4. Recurrent Network With Hidden Neurons

This structure distinguishes itself from part (iii) with hidden neurons. The feedback connections originate from the hidden neurons and also from the output neurons. The structure illustrated in figure 10.

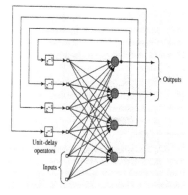

Figure 10. Typical of multilayer feedforward networks [7].

In neural networks, the input layer is passive while the hidden nodes and output layer are active and normally been activated by transfer function. These active nodes will modify weight and bias values to an optimum number where the network is best works at.There are many activation functions can be applied such as radial basis (radbas), competitive transfer function (compet), positive linear (poslin), saturating linear transfer function (satlin) and many more. The most commonly used is including hard-limit transfer function, linear transfer function and log-sigmoid transfer function. Neural networks need to be trained with suitable learning algorithm training functions corresponding to a network type. Among of the training functions are gradient descent backpropagation (traingd), Levenberg-Marquadt backpropagation (trainlm), gradient descent with adaptive learning rule backpropagation (traingda), random order incremental training with learning functions (trainr) and so on.

This experimental works used feed forward neural network with sigmoid hidden nodes and output neurons. It has been trained by scaled conjugate gradient backpropagation (trainscg) learning algorithm. Above are selected since they are most commonly used in various application, suitable for this project purpose and expected to be much efficient.

2.3. Methodology

Features extraction of ECG signals have been done to collect necessary data required for ECG analysis using neural network. In a truth fact, hundreds of input features could be extracted from the ECG signal. If all of the features are taken into consideration, identification information provided would be irrelevant and some do not give much significant to the network. Furthermore, the training duration also will be much longer. Neural networks also adaptive to a non-linear and irrelevant data with a degree of tolerance. Yet, their performance will be highly efficient when giving only appropriate and selected inputs [8].

First of all, several types of ECG signals obtained from healthy and unhealthy patients. Basically, ECG signal contains of PQRST peaks. In order to detect the peak correspondingly, an ECG signal has been divided into three segments. The first segment is P waveform, the second segment is QRS waveform and the last one is T waveform. U waveform is somehow exist in ECG signal, but it can be ignored as it does not really make any significant in cardiac diagnosis. The respective waveform after separating from the original signal is shown below.

Figure 11. P waveform.

Figure 12. QRS waveform.

Figure 13.T waveform.

The next stage is extracting important data features and characteristics of those waveforms using manual computation. The manual computation has been verified and approved by cardiologist, but it is more recommended to use LabVIEW software to perform automated extraction in order to get accurate values and precise computation. These features selected are amplitudes, durations, gradients and polarity. Then, all the values will be normalized within the range from 0 to 1 only. All of these features then fed into the neural network system as its inputs with certain desired target defined. The ECG data signals are collected and arranged in the form of numbers. There are 20 types of ECG signals involved in this experimental simulation with 49 samples (17 samples of P wave, 20 samples of QRS wave and 12 samples of T wave).Then, the simulation to get ECG waveform was performed by using Microsoft Excel for easier further analysis. Below is shown typical example of ECG signals and how features extraction is being accomplished. The same method of features extraction then been applied to the rest of ECG signals. This computation is quite time consuming and thus better system for extraction analysis is highly recommended to be implemented in the future. Referring to this figure below, real peak has been marked red

in colour. The process to identify real peak is via template matching and cardiologist verification. The examined signal is compared with the source or template image. Those waveforms possessed high similarity in their morphological attributes and features will be classified accordingly.

Some of the waveforms whichever not have source template will get verified by cardiologist after marking the expected real peak based on theoretical knowledge or peak criteria's.

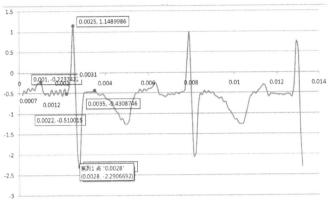

Figure 14. *Sample of ECG signal.*

Amplitude
P = -0.2237422
Q = -0.510015
R = -1.1489986
S = -2.2906692
T = -0.4308746
Duration
P wave = difference of x-axis values
0.0012 – 0.0007 = 0.0005
QRS complex (considered as one complete waveform)
Q (x-axis) = 0.0022
S (x-axis) = 0.0028
QRS duration: 0.0028 – 0.0022 = 0.0006
T wave = difference of x-axis values
0.0037 – 0.0031 = 0.0006
Gradient

$$m = \frac{y_2 - y_1}{x_2 - x_1}$$

$$Pre-gradP = \frac{-0.276.136 + .04123106}{0.0008 - 0.0007}$$

$$= \frac{0.13629}{0.0001}$$

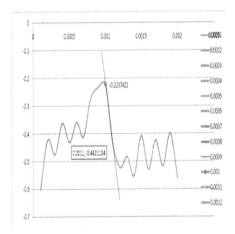

Figure 16. *Post-grad P.*

$$m = \frac{-0.2237422 + 0.4421104}{0.001 - 0.0011}$$

$$Post-gradP = \frac{0.2183682}{-0.0001}$$

$$= -2184$$

This computation to find pre-gradient and post-gradient of other waves applied the same method as above.
Polarity
Definition: If the waveform is positive, then correspond-

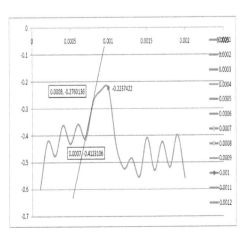

Figure 15. *Pre-grad P.*

ing peak is set as 1. If the waveform is negative, then the peak is set as 0. Positive is referred to same pattern as the normal waveform, while the shape oppose the corresponding normal waveform is denoted as negative. In this case, polarity for each waveform assigned as below

$$P = 1$$

$$QRS = 1$$

$$T = 1$$

Data then been arranged using Microsof Excel and after normalization, the data then be fed as an inputs to neural network pattern recognition system. This normalization is important to ensure the values of the data are in between the certain range. Thus, it is easier to train the neural networks efficiently as well as easier for the network system to learn. The network will learn input characteristics and attributes of the corresponding waveforms. If for example the characteristics of P wave are given to the network for training, the target of P wave will be put as 1 and the other is 0. Once the P wave is correctly detected, thus the real peak of P is easily obtained by the highest peak of that waveform. This is same goes to other waveforms.

3. Result

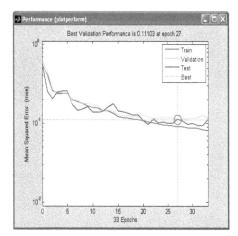

Figure 17. Network performance (MSE vs number of epochs).

Network's configuration has been detail identified with 80 hidden neurons, 2 layers feed forward, trained with scaled conjugate gradient back propagation (trainscg) with sigmoid type of hidden nodes and output neurons. 49 samples have been used with 5 input features, described as amplitude, duration, pre-gradient, post-gradient and polarity. The performance of the network is evaluated in mean squared error and confusion matrices. Network will be retrained to achieve as per desired target.

Simulation result showing that training and testing performance keep decreasing. In other words, it proved that the network is learning. During testing, it tried to achieved the target approximately as what has been trained. A part from that, it is seen that best validation performance, which

is the smallest difference of desired target with the network output is at 1.1103×10^{-1} after going through 27 iterations. At this point, the network possessed the ability to generalize very well after performance becomes minimized to the goal.

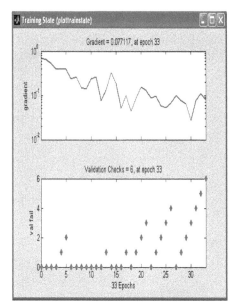

Figure 18. Network training state.

Result above presented training state of the network. The plot depicted the training state of the network from a training record. The minimum gradient reached at epochs 33 at a value of 7.77117×10^{-2}. Validation checks are at 6 also occurred during 33th iteration. The network has been well trained and learning to classify respective inputs to the corresponding target. It stops when minimum gradient has been reached to avoid network from overfitting and becomes uncontrolled. In neural network computing, validation is used to ensure the network able to generalize respective inputs mapping to the corresponding target. Validation will halt the training when generalization stops improving.

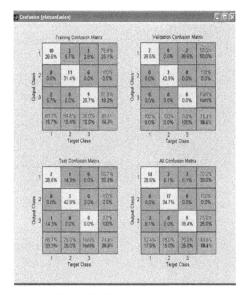

Figure 19. Confusion plot matrix.

A confusion matrix depicted clearly the actual versus predicted class values. It is also presented the classes which are correctly classified and misclassified. Thus, it is enabling us to see how well the model predicts the outcomes [9]. There are three waveforms that need to be recognized by neural networks, which is waveform of P, QRS and T. The real peak is detected by the highest peak of the waveform. This is different for QRS. Once the QRS waveform is correctly identified by neural network system, thus it is known that the highest peak should be the real peak of R, the peak before is Q and the peak after the R peak should be S peak. 35 dataset has been used for training purposes, 7 for validation and 7 for testing phase. Based on the result above, it showed that during training session, there are 30 set of data are correctly classified and only 5 data are misclassified. In validation, 5 data are correctly classified out of 7 while during testing phase, 5 data too are correctly classified accordingly while twice of misclassification happened. Several factors contribute to network's improvement will be included in a discussion part. The percentage accuracy of each process (training, validation and testing) is also can be seen in the confusion plot matrix above. Overall network performance showing that 40 data are correctly classified as desired while 9 data are misclassified. It performed good classification with total accuracy of 81.6%.

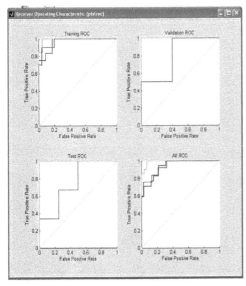

Figure 20. *Receiver Operating Characteristic (ROC).*

Above is shown the outcome of receiver operating characteristic (ROC) for this project. The ROC curve acts as fundamental tools for diagnostic evaluation for positive test which plotted true positive rate vs false positive rate [9], [10]. It reflects to a sensitivity and specificity of the network. True positive rate referred to real peak signals correctly identified as its real peak while false positive rate referred to non real peak of the signals incorrectly identified as real peak.

4. Discussion

The result for this project cannot achieve 100% or very high percentage of accuracy since small dataset was used. Thus, some recommendations could be suggested to improve network's performance in order to obtain approximately accurate outputs. There are including increase the numbers of hidden neurons and retrain the network several times. Furthermore, use larger data set so that the network will learn more and expose to enough trainings. Do try a different training algorithm as well as adjust the initial weighs and biases to new values. Then, train the networks again for several times until it reaches the desired target. In addition, it is recommended to perform automated features extraction using respective software such as LabVIEW or etc for accuracy and quick extraction purposes.

5. Conclusion

Neural network pattern recognition is suitable software with high ability to classify certain input patterns into a corresponding output target with overall accuracy of 81.6%. The training accuracy is 85.7%, validation accuracy is 71.4% while the testing accuracy is 71.4% too. It can be concluded that the real peak of ECG signals can be identified by training the network accordingly.

Acknowledgement

The appreciation goes to absolutely my main supervisor, Dr. Syed Sahal Nazli Alhady for provides endless help including motivation and guidance and also not forget to co-supervisor as well as field supervisor for some supported ideas directly or indirectly. My deepest gratitude then extends to Ministry of Science, Technology and Innovation, government of Malaysia for willing to sponsor me throughout my study, without the scholarship, this project might face difficulty and could not be accomplished in a comfortable manner. Thank you very much to those involved.

References

[1] A website on http://biology.about.com/library/organs/heart/blatrionode.htm, image courtesy of Carolina Biological Supply / Access Excellence.

[2] Chung, M.K, and Rich, M.W. Introduction to the cardiovascular system. Alcohol Health and Research World14 (4):269-276, 1990.

[3] Information, http://biology.about.com/od/humananatomybiology/ss/heart anatomy2.htm

[4] Mazhar B.Tayel, Mohamed E.El-Bouridy, ECG images classification using features extraction based on wavelet transformation and neural network, AIML 06 International Conference, 13-15 June 2006, Sharm El Sheikh, Egypt, 105-107.

[5] Rajesh Ghongade, Vishwakarma, Dr. Babasaheb, A brief

Performance Evaluation of ECG Feature Extraction Techniques for Artificial Neural Network Based Classification.

[6] Christos Stergiou and Dimitrios Siganos, A Report on Neural Networks.

[7] Simon Haykin, A book of Neural Networks A Comprehensive Foundation, Second Edition, 1999.

[8] International journals on WSEAS Transactions on Systems, Issue 1, Volume 4, January 2005, ISSN 1109-2777, paper titled P, Q, R, S, and T peaks recognition of ECG using MRBF with selected features, 137.

[9] M.P.S Chawla, Department of Electrical Engineering, Indian Institute of Technology, Roorkee, 247667 India, Parameterization and R-peak error estimations of ECG Signals using Independent Component Analysis, Computational and Mathematical Methods in Medicine, Vol. 8, No. 4, December 2007, 263-285.

[10] Exercise on Artificial Neural Networks, Information on rad.ihu.edu.gr/fileadmin/labsfiles/decision_support_systems /lessons/neural_nets/NNs.pdf.

2

Unified analytical models of parallel and distributed computing

Michal Hanuliak

Dubnica Technical Institute, Dubnica Nad Vahom, Slovakia

Email address:

michal.hanuliak@gmail.com

Abstract: The optimal resource allocation satisfies the needed capacity of the used resources. To such analysis we can use both analytical and simulation methods. Principally analytical methods (AM) belong to the preferred method in comparison to the simulation method, because of their potential ability of more general analysis and also of ability to analyze massive parallel computers. This article goes further in developing AM based on queuing theory results in relation to our published paper in [9]. The extensions are in extending derived AM to whole range of parallel computers and also to sum up public acceptance of our published paper. The article therefore describes deriving of correction factor of standard AM based on M/M/m and M/M/1queuing theory systems. In detail the paper describes derivation of a correction factor for standard AM to study more precise their performance. The paper contributions are in unified AM and in deriving correction factor in order to take into account real non-exponential nature of the inputs to the computing nodes and node's communication channels. The derived analytical results were compared with performed simulation results in order to estimate the magnitude of improvement. Likewise the corrected AM were tested under various ranges of parameters, which influence the architecture of the parallel computers and its communication networks too. These results are very important in practical use.

Keywords: Parallel Computer, Communication System, Correction Factor, Analytical Model, Performance, Queuing System, Overhead Latencies, Modeling

1. Introduction

In [9] we have characterized developing periods of parallel computers and their basic classification from the point of programmer as potential developers of parallel algorithms. In relation to it and also to other performed classifications [1, 8] we have divided them as following

• synchronous parallel computers. To this group belong actually dominated parallel computers based on multiply cores, processors or mix of them too (symmetrical multiprocessors - SMP) and most of realized massive parallel computers (classic supercomputers) [31]. The practical example of such synchronous parallel computer is illustrated at Figure 1.

• asynchronous parallel computers. According the mentioned characteristics this group consist of actually dominant distributed parallel computers based on NOW (Network of workstation) module. To this group belong mainly computer networks based on network of workstation (NOW) module [16, 30]. The example of typical asynchronous parallel computer illustrates Figure 2.

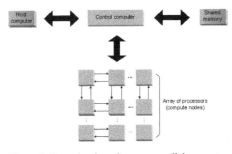

Figure 1. Example of synchronous parallel computer.

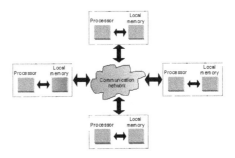

Figure 2. Example of asynchronous parallel computer.

Practical example of NOW module is representing at Figure 3. It represents also our outgoing architecture of laboratory parallel computer. On such modular parallel computer we have been able to study basic problems in parallel and distributed computing as load balancing, inter processor communication (IPC), modeling and optimization of parallel algorithms (effective PA) etc. [10, 11, 18]. The coupled computing nodes PC1, PC2, ... PCi (workstations) could be single extreme powerful personal computers (PC)| or SMP parallel computers. In this way parallel computing on networks of conventional PC workstations (single, multiprocessor, multicore) and Internet computing, suggest advantages of unifying parallel and distributed computing. Parallel computing and distributed computing have traditionally evolved as two separate research disciplines. Parallel computing has addressed problems of communication-intensive computation on highly - coupled processors [21, 22] while distributed computing has been concerned with coordination, availability, timeliness, etc., of more likely coupled computations [28, 34].

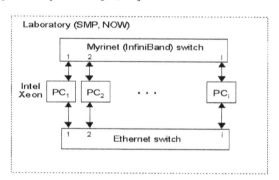

Figure 3. *Practical example of NOW module.*

Workstations could be connected using different network technologies such as off the shelf devices like Ethernet to specialized high speed communication networks (Infiniband, Quadrics, Myrinet) [35]. Such networks and the associated software and protocols introduce latency and throughput limitations thereby increasing the execution time of cluster - based computation [23, 26]. Researchers are engaged in designing algorithms and protocols to minimize the effect of these latencies [12, 13].

2. Abstract Models of Parallel Computer

For any realized parallel computer we can use one of the following two basic models according Figure 4 or Figure 5. The difference of both models technically consists in various types of memories (shared, distributed) because both of these memories are actually build from memory modules. On the other side applied using of various memory types of parallel computers is very different. But in abstract models existed differences are question of only defined technical parameters. Therefore recent trends point

to a convergence of research in parallel and distributed computing. Perhaps the most significant of these trends is architectural. Three architectural trends may be noted. First, increased communication bandwidth and reduced latency make geographical distribution of processing nodes less of a barrier to distributed computing.

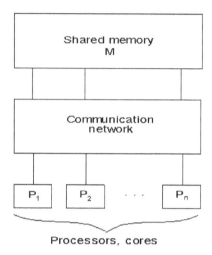

Figure 4. *Abstract model of the synchronous parallel computer.*

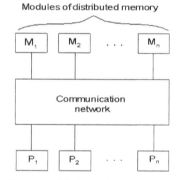

Figure 5. *Abstract model of asynchronous parallel computers.*

Second, the development of architecture transparent programming language, such as Java, provides a virtual computational environment in which nodes appear to be homogenous. Finally server machines in client/server (manager/worker) computing are increasingly adopting multiprocessor architecture, often multiple processors with a shared memory in a single workstations and symmetric multiprocessors (SMP) [36]. While such architectures are less scalable than computer networks (NOW, clusters), some concurrent programs with high communication traffic may execute on them more efficiently [19, 25].

3. Analytical Performance Evaluation of Parallel Computer in Queuing Theory

To the behavior analysis of parallel computer including their communication networks there have been developed various analytical models based on queuing theory results. Queuing theory is very good if you have to analyze a single independent computing node of sequential or parallel computers [6, 15]. But analysis of used dominant parallel computers (NOW, Grid - network of NOW modules) generally lead to multiple computing node case. The first problem, in comparison to a single node case, is existence of traffic dependency in any real networks. What comes out of one computing node feeds the second one and what comes out from second node could not be Poisson [3]. Typically we could get into trouble even in single SMP parallel computer based on multicore or multiprocessor platform. Suppose there is communication traffic entering the system M/M/1 (Poisson arrivals and exponential service times) as simplest applied queuing theory system. One must ask the question is what comes out of the same nature as at input traffic? It turns out that there is a fundamental theorem which says: only in the case when you have exponential service times (M/M/1, M/M/m) will you get something coming out which is also Poisson in nature [14].

There is a very important relationship for praxis which says: if you have Poisson arrivals and exponential service time what comes out has also Poisson characteristics. There is also a more general result which says: not only will the output be Poisson, if the input is Poisson, but we can consider also multiple service with defections, reneging, balking and in some cases finite queue storage. But it has been proved that the outgoing process will be exactly Poisson only on the assumption of unlimited length of queue in size.

If all the nodal traffic has the property that it is Poisson, then even in a complicated network we can do under some conditions network analysis on a node-by-node basis. In fact, however, that is not yet true, because in communication networks of parallel computers the time a communication message spends in one node is related to the time it spends in another node, because the service one is looking for is network communication. That is one very nasty problem, but there have been developed some solutions.

The second serious problem is blocking as consequence of always real limited technical resources. If one node is blocked, the node feeding could not enter more data into that node. Consider a communication network in which you are given the location of computing nodes and the required communication traffic between pairs of computing nodes. Then according mentioned theorem says that if you have Poisson traffic into an exponential server you get Poisson traffic out; but a message maintains its length as it passes through the network, so the service times are dependent as it goes along its path. Thus, one thing we want to do is to get rid of that dependence. We can do this by making an independence assumption; we just assume that the dependence does not exist. We manage this by allowing the communication message to change its length as it passes through the communication network. Every time it hits a new computing node, we are going to randomly choose the message length so that we come up with an exponential distribution again. With that assumption, we can then solve

the queuing problem of communication in parallel computers. Let us assume infinite storage at all points in the network of coupled computing nodes and refer to the problem M/M/1, where the question mark refers to the modified input process. We then run simulations, with and without the independence assumption for a variety of networks. The reason why it is good to do it is that a high degree of mixing takes place in a typical communication network; there are many ways into a node and many ways out of the node. The dependence between service times and between adjacent messages on a line need not be highly correlated even if there were originally. With that assumption, it is possible to use appropriate results of the queuing theory for modeling parallel computers including their communication systems. We can use the results from essentially single node queuing theory, since now we have independent arrival and services processes. Neither the queuing model nor the simulation hit the blocking effect; so blocking remains as nasty problem.

The assumption of independence permits us to break also the massive parallel computer into independent computing nodes, and allowed all node analysis to take place. The reason we had to make that assumption was because the communication message maintains the same length as they pass through the network. If we accept the independence assumption, it turns out that the queuing theory contains a number of results for cases where the service at a node is an independent random variable in an arbitrary network of queues. A basic theorem is due to Jackson [29, 32]. Jackson's result essentially gives us the probability distribution for various numbers of messages at each of the nodes in such a network. These nodes are essentially Markovian queues, and Markovian queuing theory is relatively simple in the nearest neighbor cases. There are more general cases, which one has to solve which involve solving a set of linear simultaneous equations for the probabilities of finding queue lengths at various points. Typically, in the complicated network, one looks for something called the product form solution. This term comes from Jackson's theorem, which expresses the joint probability of the numbers of customers at each queue being a particular combination is the product of their individual probabilities of having that number [27].

Product form networks have the property that they can be regarded as independently operating queues, where steady state can be expressed as both a set of global balance equations on each queue. Local flow balance says that the mean number of customers entering any queue from all others must equal the number leaving it to go to all others, including customers which leave and rejoin the same queue immediately.

3.1. Application of Queuing Theory

The basic premise behind the use of queuing models for computer systems analysis is that the components of a computer system can be represented by a network of servers (resources) and awaiting lines (queues). A server is defined as an entity that can affect, or even stop, the flow of jobs through the system. In a computer system, a server may be the CPU, I/O channel, memory, or a communication port. Awaiting line is just that: a place where jobs queue for service. To make a queuing model work, jobs or customers or communication message (blocks of data, packets) or anything else that requires the sort of processing provided by the server, are inserted into the network. A basic simple example could be the single server abstract model as single queuing theory system. In such model, jobs arrive at some rate, queue for service on a first-come first-served basis, receive service, and exit the system. This kind of model, with jobs entering and leaving the system, is called an open queuing system model [3, 33].

3.2. Background of Analytical Models

We will now turn our attention to some suitable network of queuing theory systems, the notation used to represent them, the performance quantities of interest, and the methods for calculating them. We have already introduced many notations for the quantities of interest for random variables and stochastic processes. For the analysis of parallel computers including their communication network communication networks we have been developed in [9] following analytical models

• standard analytical model on the basis of Jackson's theorem. In this case we suppose decomposition to the independent computing nodes on the basis of the M/M/m systems with infinite buffer. This together with the independence assumption reduces a very difficult problem to an open network of independent M/M/m queuing theory queues. This analytical model produced the worst results (upper limit)

• improved analytical model on the basis of the modern node's multiprocessor system [4], as computing node of parallel computer. This analytical model extends used analytical models considering the delays through the activities of communication processor and through awaiting these services with the more realistic M/D/m queuing theory systems. Satisfying conditions this improved analytical model will be produced the best results (lower limit)

• results of any other developed analytical model will be between results of these two analytical models (standard, improved).

Queuing theory systems are classified according to various characteristics, which are often summarized using Kendall's notation [7, 9]. The basic parameters of queuing theory systems are as following

• λ - arrival rate at entrance to a queue

• m - number of identical servers in the queuing system

• ρ - traffic intensity (dimensionless coefficient of utilization)

• q - random variable for the number of customers in a system at steady state

• w - random variable for the number of customers in a queue at steady state

• E (ts) - the expected (mean) service time of a server

• E (q) - the expected (mean) number of customers in a system at steady state

• E (w) - the expected (mean) number of customers in a queue at steady state

• E (tq) - the expected (mean) time spent in system (queue + servicing) at steady state

• E (tw) - the expected (mean) time spent in the queue at steady state.

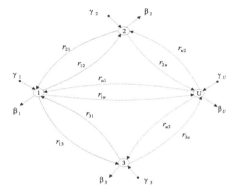

Figure 6. Parallel computer model including its communication network.

Communication demands (parallel processes, IPC data) arrive at random at a source node and follow a specific route in the communication networks towards their destination node. Data lengths of communicated IPC data units (for example in words) are considered to be random variables following distributions according Jackson theorem. Those data units are then sent independently through the communication network nodes towards the destination node. At each node a queue of incoming data units is served according to a first-come first-served (FCFS) discipline.

At Figure 6 we illustrate generalization of any parallel computer including their communication network as following

• computing nodes Ui (i=1, 2, 3, ... U) of any parallel computer are modeled as graph nodes

• network communication channels are modeled as graph edges rij (i≠j) representing communication intensities (relation probabilities).

The other used parameter of such abstract model are defined as following

• $\gamma1, \gamma2, ..., \gammaU$ represent the total intensity of input data stream to individual network computing nodes (the summary input stream from other connected computing nodes to the given i-th computing node. It is given as Poisson input stream with intensity λi demands in time unit

• rij are given as the relation probabilities from node i to the neighboring connected nodes j

• $\beta1, \beta2, .., \betaU$ correspond to the total extern output stream of data units from used nodes (the total output stream to the connected computing nodes of the given node).

The created abstract model according Figure 6 belongs in queuing theory to the class of open queuing theory systems (open queuing networks). Formally we can adjust abstract model adding virtual two nodes (node 0 and node U+1 according Figure 7 where

• virtual node 0 represent the sum of individual total extern input intensities $\gamma = \sum_{i=1}^{U} \gamma_i$ to computing nodes Ui

• virtual node U+1 represent the sum of individual total intern output intensities $\beta = \sum_{i=1}^{U} \beta_i$ from computing nodes ui.

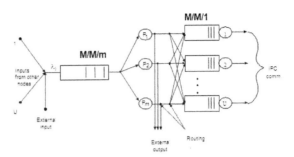

Figure 8. *Standard analytical model of i-th computing node.*

The intern input flow to i-th node is defined as the input from all other connected computing nodes. We can express it in two following ways

• through solving a system of linear equations in matrix form as $\bar{\lambda} = \gamma + \bar{\lambda} \cdot \bar{R}$

• using of two data structures in form of tables and that is the routing table (RT) and destination probability tables (DPT).

The used model were build on assumptions of modeling incoming demands to program queue as Poisson input stream and of the exponential inter arrival time between communication inputs to the communication channels. The idea of the previous models were the presumption of decomposition to the individual nondependent computing nodes together with the independence presumption of the demand length, that is the demand length is derived on the basis of the probability density function $pi = \mu\ e^{-\mu t}$ for $t > 0$ and $f(t) = 0$ for $t \le 0$ independent always at its input to the node. On this basis it was possible to model every used communication channel as the queuing theory system M/M/1 and derive the average value of delay individually for every communication channel. The whole delay was then simply the sum of the individual delays of the every used communication channel.

To improve the mentioned problems we suggested improved analytical model, which extends the used standard analytical model to more precise analytical model (improved analytical model) supposing that

• we consider to model computation activities in every node of NOW network as M/D/m system (assumption of inputted balanced parallel processes to every node)

• we consider an individual communication channels in i- th node as M/D/1 systems. In this way we can take into account also the influence of real non exponential nature of the inter arrival time of inputs to the communication channels (assumption of nearly equal IPC communication complexity).

Both published analytical models (standard, improved) are not fulfilled for every input load, for all parallel computer architectures and for the real character of computing node service time distributions. These changes may cause at some real cases imprecise results. Another survived problem of the used standard analytical model is assumption of the exponential inter arrival time between message inputs to the communication channels in case of

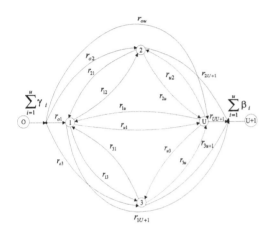

Figure 7. *Adjusted abstract model.*

3.3. Standard Analytical Model

In [9] we have presented open multi nodes module NOW consisting of U computation nodes including theirs communication network. Model of i-th computing node illustrates Figure 8. For the whole delay in NOW we have derived following final result based on M/M/m and M/M/1 queuing theory systems as

$$E(t_q)_{now} = \frac{1}{\gamma}\left[\sum_{i=1}^{U}\left(\lambda_i \cdot E(t_q)_i + \sum_{j=1}^{u_i}\lambda_{ij}\cdot E(t_q)_{ij}\right)\right]$$

where $\dfrac{\lambda_i \cdot E(t_q)_i}{\gamma}$ and $\dfrac{\lambda_{ij}\cdot E(t_q)_{ij}}{\gamma}$ defined individual contribution of computation queue delay (M/M/m) and communication channel delay (M/M/1) of every computing node to the whole delay. The meanings of used and not explained parameters are as following

• λi - the whole number of incoming demands to the i-th node, that is the sum both of external and internal inputs to the i-th node

• λij - the whole input flow to the j-th communication channel at i-th node

• E (tq)i - the average servicing time in the program queue (the waiting in a queue and servicing time) in the i-th node

• E (tq)ij - the average servicing time of the j-th queue of the communication channel (the queue waiting time and servicing time) at i-th node.

unbalanced communication complexity of parallel processes. To remove mentioned changes we derived a correction factor to standard analytical model.

4. Corrected Standard Analytical Model

The standard analytical model supposes that the inter arrival time to the node's communication channels has the exponential distribution. This assumption is not true mainly in the important cases of high communication utilization. The node servicing time of parallel processes (computation complexity) could vary from nearly deterministic (in case of balanced parallel processes) to exponential (in case of unbalanced ones). From this in case of node's high processors utilization the outputs from individual processor of node's multiprocessor may vary from the deterministic interval time distribution to exponential one. These facts violate the assumption of the random exponential distribution and could lead to erroneous value of whole node's delay calculation. Worst of all this error could the greater the higher is the node utilization. From these causes we have derived the correction factor which accounts the measure of violation for the exponential distribution assumption.

The inter arrival input time distribution to each node's communication channel depends on ρi, where ρi is the overall processor utilization at the node i. But because only the part λij from the total input rate λi for node i go to the node's communication channel j, it is necessary to weight the influence measure of the whole node's processors utilization trough the value $\lambda ij / \lambda i$ for channel j as $\rho i . (\lambda ij / \lambda i)$.

To clarify the node's processor utilization influence to the average delay of communication channel we have tested the 7-noded experimental parallel computer. The processing time was varied to develop the various workloads of node's processors. The achieved results are summarized at Table 1 for one of communication channels at the node 1. Graphical illustration of achieved results is at Figure 9.

Extensive testing have proved, that if we increase utilization of communication channel and that develops saturation of communication channel queue then average queue waiting time is less sensitive to the nature of inter arrival time distributions. This is due to the fact that the messages (communicating IPC data) wait longer in the queue what significantly influenced the increase of the average waiting time and the error influence of the non exponential inter arrival time distribution is decreased. To incorporate this knowledge for the correlation factor we investigated the influence of the weighting ρi ($\lambda ij / \lambda i$) through the value $(1 - \rho i)x$ for various values x. The performed experiments showed the best results for the value x = 1. Derived approximation of the average queue waiting time of the communication channel j at the node i, which eliminates violence of the exponential inter arrival time distribution is then given as

$$\frac{\rho_i \cdot (1 - \rho_{ij}) \cdot \lambda_{ij}}{\lambda_i}$$

The finally correction factor of the communication channel j at the node i, which we have named as cij is as following

$$c_{ij} = 1 - \frac{\rho_i \cdot (1 - \rho_{ij}) \cdot \lambda_{ij}}{\lambda_i}$$

With the derived correction factor cij we can define now the corrected average queue waiting time as:

$$W_{ij}'(LQ) = c_{ij} \cdot W_{ij}(LQ)$$

The standard analytical model we can simply correct in such a way that instead of Wij(LQ) we will consider its corrected value Wij'(LQ). In this way derived improved standard analytical model we have defined as corrected standard analytical model. From the performed tests it is also remarkable that decreasing of the node's processors workload the assumption of the exponential inter arrival message time distribution to the communication channel is more effective.

The average delay values of the node's communication channel achieved through simulation are compared with the results of the standard analytical model (exponential inter arrival time distribution) and with the results of the corrected standard model. Comparison of the relative errors is illustrated in the Figure 10.

Table 1. Achieved results for correction factor.

Processor utilization at node 1	Average channel delay at node 1 - simulation [msec]	Standard analytical model		Correct analytical model	
		Average channel delay [msec]	Relative error [%]	Average channel delay [msec]	Relative error [%]
0,6	21,97	22,27	1,4	22,03	0,3
0,7	21,72	22,27	2,5	21,92	0,9
0,8	21,43	22,27	3,9	21,70	1,3
0,9	21,05	22,27	5,8	21,45	1,9
0,95	21,91	22,20	6,5	21,31	1,9

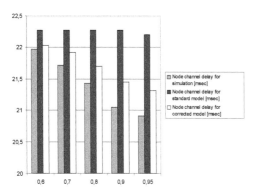

Figure 9. The influence of the exponential time distribution and its correction.

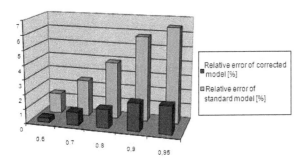

Figure 10. Comparison of relative errors.

At Table 2 there are results of the channel utilization influence to the average waiting time for the communication channel of 7-noded communication network. For this case the channel utilization was influenced through communication speed changes.

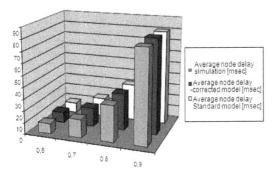

Figure 11. The channel utilization influence to the total node delay.

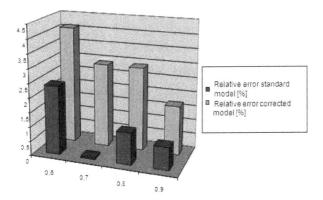

Figure 12. Influence of channel utilization to the accuracy of analytical model.

Table 2. The results of the channel influence.

Processor utilization at node 1	Average channel delay at node 1 - simulation [msec]	Standard analytical model		Correct analytical model	
		Average channel delay [msec]	Relative error [%]	Average channel delay [msec]	Relative error [%]
0,6	8,89	9,25	4,1	8,68	2,4
0,7	15,92	16,38	2,9	15,91	0,06
0,8	31,04	31,94	2,9	31,39	1,1
0,9	79,76	81,08	1,7	80,38	0,8

The achieved results in Table 2 are illustrated at Figure 11. The relative errors are incorporated in the Table 2. The influence of communication channel utilization to the result accuracy of the analytical models is at the Figure 12. From these achieved results follow that decreasing of the node's communication channel utilization the difference between simulated results and the standard analytical model increases.

5. The Achieved Results

The developed corrected model was intensive tested and compared with simulation results. We varied various range of parameters, which influence architecture of parallel computer – number of computing nodes, topology of communication network, communication network load, communication speed of node's communication channels etc. in order to verify stability and accuracy of the corrected model under various conditions. The following tables and graphs present essential parts of achieved results of the done experiments. In each tested case we compared percentages errors between corrected model and simulation results. In Table 3 are results for 5-noded communication network, in which all nodes are fully interconnected (every node with all other nodes). This rather small communication network has its causes through limitations of the simulation method than analytical one.

Table 3. The comparison of used methods

Processor number	Processor utilization p	The whole delay for simulation [msec]	Correct analytical model		Average message queue delay [msec]
			The whole delay [msec]	Relative error [%]	
4	0,24	25,44	24,34	4,52	0,01
3	0,31	25,45	24,39	4,35	0,05
2	0,47	25,94	24,83	4,47	0,30
1	0,94	51,91	54,79	5,55	15,60

For each node of tested communication network the number of node's processors and communication channels were the same. The processors utilization was varied through various numbers of processors in each node. Because the average queue waiting time is directed proportional to node's processor utilization, decreasing the processing utilization decrease also the delays in node's queue. This in turn influences the whole node's delay of communication channel. It has been remarkable that high node's processor utilization influence considerable node's communication delay through channel queue waiting time. The comparison of the whole delay for simulation and corrected analytical model is at the Figure 13. The comparison of their relative errors is at the Figure 14.

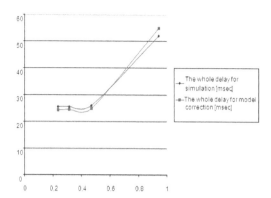

Figure 13. The accuracy of corrected model.

The relative error of the corrected model illustrates the Figure 14.

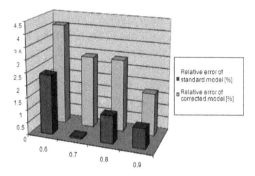

Figure 14. Comparison of relative errors.

The Tables 4 and 5 represent the results for the total message delay in alternatively analyzed 5-noded test model. Again all computing nodes have the same processors and the same communication channels to have the possibility to achieve uniform results for processor and channel utilization in individual nodes. For this tested net they were used always on processor in each node and the utilization of this processor was changed through the variation of message servicing time. Table 4 presents the results for the case of communication channel speeds of the used communication channels to 50%.

Table 4. Results for 50% communication channel utilization

Processor utilization	The whole delay for simulation [msec]	Correct analytical model		Average message queue delay [msec]
		The whole delay [msec]	Relative error [%]	
0,6	21,57	21,10	2,22	0,9
0,7	23,76	23,39	1,60	1,8
0,8	28,22	28,18	0,14	3,8
0,9	39,30	40,25	2,40	9,0
0,95	58,22	61,83	6,20	19,83

The comparison of the total message delays through simulation and corrected model for the 50% communication channel utilization is at the Figure 15.

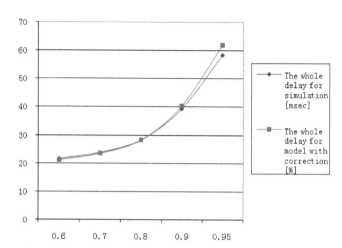

Figure 15. The comparison for 50% communication channel utilization.

Relative errors of corrected model for 50% communication channel utilization illustrate Figure 16.

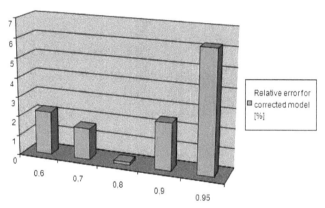

Figure 16. The relative errors for 50% channel utilization.

The portion of communication message queue delay to total communication message delay illustrates for 50% channel utilization Figure 17.

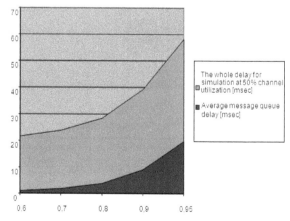

Figure 17. The portion of message queue delay to total message delay.

The Table 5 represents the results for the case of communication speed reduction to achieve the increased channel utilization to 80%.

Processor utilization	The whole delay for simulation [msec]	Correct analytical model		Average message queue delay [msec]
		The whole delay [msec]	Relative error [%]	
0,6	72,33	68,87	5,02	0,9
0,7	74,46	71,03	4,82	1,8
0,8	78,76	75,62	4,20	3,8
0,9	90,09	87,48	3,00	9,0
0,95	104,35	108,93	4,39	19,83

The comparison of the total delays for simulation and developed analytical model at 80% utilization is at the Figure 18.

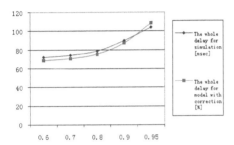

Figure 18. The comparison at 80% channel utilization.

The relative error of the corrected analytical model at 80% channel utilization illustrates Figure 18.

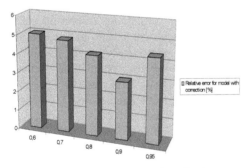

Figure 19. Relative errors at 80% channel utilization.

The influence of channel utilization to the total node influence of the utilization to whole communication channel delay illustrates Figure 20.

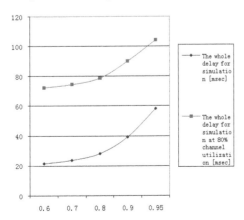

Figure 20. Influence of the utilization to the total delay.

In both considered cases the decreasing of processor utilization causes the decreasing of average node's channel communication delay. Therefore the communicating data are waiting in the node's channel queue shorter time and the total communication channel delay is lower. In contrary the decreasing of communication channel speed increase the channel utilization and then the communicating data have to wait longer in the communication channel queues and increase the total message node delay. These facts are clear also from the achieved results because all achieved results in Table 5 are greater than in Table 4.

The corrected analytical model allows modeling of delay influences through computation complexity of performed parallel processes influences of node's communication activities and also to correct violation influence of exponential inter arrival time distribution. The tested results has proved, that the corrected analytical model provides very precision results in the whole range of input workload of processor utilization, communication channels and network topologies with relative error, which does not exceed 6% and in most cases were in the range up to 5%.

To prove the effectiveness of corrected analytical model it is not satisfactory to document its accuracy but also his improvement in relation to standard analytical model. It is necessary to emphasize, that the effect of considering the delay influence caused through the processors communication activities to the total message delay intensify mainly under two following conditions

• if the average number of passed nodes at the message transport is high, the message queue delay has inconsiderable effect to the total message delay

• if the processor utilization is higher the average message queue waiting time is increased what influence the whole delay.

Table 6 represents the results and relative error for the average value of the total message delay in the 7 – nodes network through corrected analytical model and standard model.

To vary the processor utilization we modified the extern input flow in the same manner for each used node. For both analytical models (standard, corrected) are in Table 6 represented the whole delays (end-to-end) and relative errors in relation to simulated results. Comparison of whole delays illustrates Figure 21.

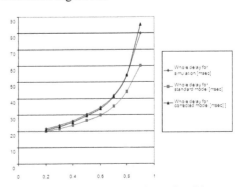

Figure 21. Comparison of the used models.

Table 6. Comparison of considered models.

Processor utilization	The whole delay for simulation [msec]	Standard analytical model		Correct analytical model	
		The whole delay [msec]	Relative error [%]	End to end delay [msec]	Relative error [%]
0,2	21,45	20,06	6,48	20,83	2,89
0,3	23,53	21,58	8,29	22,85	2,89
0,4	26,24	23,49	10,48	25,51	2,78
0,5	30,16	26,51	12,10	29,44	2,39
0,6	34,69	29,79	14,12	33,92	2,22
0,7	41,67	35,19	15,55	41,38	0,70
0,8	54,25	44,08	18,75	54,43	0,33
0,9	80,01	60,38	24,53	85,47	6,82

The relative errors comparison of both analytical models to simulation results according Figure 22 shows clearly the improvement of corrected analytical model. If the processor utilization varied from 20 to 90% the relative error of standard model changes from 6 to 24%. This is due the influences of communication message queue delays, the nature of inter arrival input to communication channels in the case of high processor utilization. In contrary the corrected analytical model in all cases has the relative number not greater than 7%. The achieved results in Table 6 indicate also other important critical fact. The derived corrected model produces more precise results in the whole range of node's processor utilization including the range of their higher utilization (in range 0,5 – 0,9) which are the most interesting to practical use.

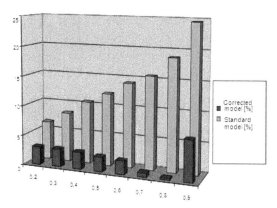

Figure 22. The comparison of relative errors.

We also point out, that accuracy contribution of corrected analytical model was achieved without the increasing the computation time in comparison to standard analytical model. It is also remarkable to emphasize increasing influence of the simulation complexity for the analysis of real massive parallel computers including their communication networks. The simulation models require three orders of magnitude more computation time for testing such complex parallel systems.

6. Conclusions

Performance evaluation of existed computers (sequential, parallel) used to be a very hard problem from their birthday.

This involves the investigation of the control and data flows within and between components of computers including their communication networks. The aim is to understand the behavior of the systems, which are sensitive from a performance point of view. It was, and still remains, not easy to apply typical analytical method (queuing theory) to performance evaluation of used computers because of their high number of not predictable parameters and for sequential computers the existence of only one control stream. Using of actual parallel computers (SMP -multiprocessor, multicore, NOV, Grid) open more possibilities to apply a queuing theory results to analyze more precise their performance. This imply existence of many inputs streams (control, data), which are inputs to modeled queuing theory systems and which are generated at various used resources by chance (theoretic assumption for good approximation of Poisson distribution). Therefore we could model computing nodes of parallel computers as M/D/m or M/M/m and their communication channels as M/D/1 or M/M/1 queuing theory systems in any existed parallel computer.

In relation to it this paper describes the deriving and testing of a correction factor of used standard analytical models in order to extend behavior analysis of parallel computers with another more precise analytical model. The developed corrected standard model was extensively tested and the results were compared with both standard analytical model and simulation results too. The results clearly show that the correction factor contributes to better results with a negligible increase in processing time. Its advantage, in comparison to possible used simulation method, is its ability to analyze also large existed communication networks of massive parallel computers (MPC). In this way we hope that more precise corrected analytical model would help in effective resource projecting of suggested parallel computers and parallel algorithms (effective PA) too. In this way parallel computing on networks of conventional personal workstations (single, multiprocessor) and Internet computing, awaits to unify advantages of parallel and distributed computing at their performance modeling too. In summary according input technical parameters of parallel computer we can apply to performance modeling of developed analytical models in actually typical following cases

• NOW based on workstations (single, multiprocessors or multicores, mix of them)
• Grid (network of NOW networks)
• mixed parallel computers based on SMP, NOW and Grid
• metacomputers (massive Grid etc.).

From a point of user application of any analytical method is to be preferred in comparison with other possible methods (simulation), because of universal and transparent character using of achieved results in form of used parameters. We can apply developed more precise analytical models to performance modeling of any parallel computer or parallel algorithm too. To practical applied

using we would like to advise following
- running of unbalanced parallel processes where λ is a parameter for incoming parallel processes with their exponential service time distribution as $E(ts) = 1/\mu$ (corrected standard model)
  - in case of potential considering incoming units of parallel processes (data block, packet etc.) at using model based on M/M/m and M/M1 queuing theory systems it would be necessary to recalculate at entrance incoming parallel processes to wanted data units. The way how to recalculate them to such units at first node entrance we would like to refer in next paper
- running of parallel processes (λ parameter for incoming parallel processes with their deterministic service time $E(ts) = 1/\mu = $ constant). The deterministic servicing times are a very good approximation of balanced parallel processes (M/D/m) with nearly equal amount of communication data blocks for every parallel process (M/D/1)
  - in case of using analytical model using M/D/m and M/D/1 we can consider λ parameter also for incoming units of parallel processes (data block, packet etc.) with their average service time for considered unit ti , where $E(ts) = 1/\mu = $ ti=constant.

Using developed more precise analytical models (in [9] and in this paper too) we are able to apply them so to SMP parallel computers (parallel computing) as to dominant distributed computers (NOW, Grid, metacomputer). In developed unified parallel computer models it is possible to study load balancing [2], inter-process communication (IPC) mechanisms, transport protocols, performance prediction etc. We would also like to analyze
- blocking problem (consequence of exhausted used resources)
- extending waiting times caused by limited technical resources (blocking consequence)
- the role of various routing algorithms
- to prove, or to indicate experimentally, the role of the supposed independence assumption, if you are looking for higher moments of delay [20]
- to verify the suggested model also for node limited buffer capacity and for other servicing algorithms than assumed FIFO (First In First Out)
- necessary unified decomposition strategies for parallel and distributed computing
- intensive testing [17], measurement and observation in order to obtain estimates of such system variables [5, 24].

Acknowledgements

This work was done within the project "Modeling, optimization and prediction of parallel computers and algorithms" at University of Zilina, Slovakia. The author gratefully acknowledges help of project supervisor Prof. Ing. Ivan Hanuliak, PhD.

References

[1] Abderazek A. B., Multicore systems on chip - Practical Software/Hardware design, Imperial college press, 200 pp., 2010

[2] Arora S., Barak B., Computational complexity - A modern Approach, Cambridge University Press, 573 pp., 2009

[3] Dattatreya G. R., Performance analysis of queuing and computer network, University of Texas, Dallas, USA, 472 pp., 2008

[4] Dubois M., Annavaram M., Stenstrom P., Parallel Computer Organization and Design, 560 pages, 2012

[5] Dubhash D.P., Panconesi A., Concentration of measure for the analysis of randomized algorithms, Cambridge University Press, UK, 2009

[6] Gelenbe E., Analysis and synthesis of computer systems, Imperial College Press, 324 pages, 2010

[7] Giambene G., Queuing theory and telecommunications, 585 pp., Springer, 2005

[8] Hager G., Wellein G., Introduction to High Performance Computing for Scientists and Engineers, 356 pages, July 2010

[9] Hanuliak M., Hanuliak P., Performance modeling of parallel computers NOW and Grid Vol. 2/5, Am. J. of Networks and Comm., Science GP, USA, 112-124 pp., 2013

[10] Hanuliak J., Hanuliak I., To performance evaluation of distributed parallel algorithms, Kybernetes, Volume 34, No. 9/10, UK, 1633-1650 pp., 2005

[11] Hanuliak P., Hanuliak I., Performance evaluation of iterative parallel algorithms, Kybernetes, Volume 39, No.1, UK, 107-126 pp., 2010

[12] Hanuliak P., Analytical method of performance prediction in parallel algorithms, The Open Cybernetics and Systemic Journal, Vol. 6, Bentham Open, UK, 38-47 pp., 2012

[13] Hanuliak P., Complex performance evaluation of parallel Laplace equation, AD ALTA – Vol. 2, issue 2, Magnanimitas, Czech republic, 104-107 pp.,2012

[14] Harchol-Balter Mor, Performance modeling and design of computer systems, Cambridge University Press, UK, 576 pp., 2013

[15] Hillston J., A Compositional Approach to Performance Modeling, University of Edinburg, Cambridge University Press, UK, 172 pages, 2005

[16] Hwang K. and coll., Distributed and Parallel Computing, Morgan Kaufmann, 472 pages, 2011

[17] John L. K., Eeckhout L., Performance evaluation and benchmarking, CRC Press, 2005

[18] Kshemkalyani A. D., Singhal M., Distributed Computing, University of Illinois, Cambridge University Press, UK, 756 pages, 2011

[19] Kirk D. B., Hwu W. W., Programming massively parallel processors, Morgan Kaufmann, 280 pages, 2010

[20] Kostin A., Ilushechkina L., Modeling and simulation of distributed systems, Imperial College Press, 440 pages, 2010

[21] Kumar A., Manjunath D., Kuri J., Communication Networking , Morgan Kaufmann, 750 pp., 2004

[22] Kushilevitz E., Nissan N., Communication Complexity, Cambridge University Press, UK, 208 pages, 2006

[23] Kumar A., Manjunath D., Kuri J., Communication Networking , Morgan Kaufmann, 750 pp., 2004

[24] Kwiatkowska M., Norman G., and Parker D., PRISM 4.0: Verification of Probabilistic Real-time Systems, In Proc. of 23rd CAV'11, Vol. 6806, Springer, 585-591 pp., 2011

[25] Le Boudec Jean-Yves, Performance evaluation of computer and communication systems, CRC Press, 300 pages, 2011

[26] McCabe J., D., Network analysis, architecture, and design (3rd edition), Elsevier/ Morgan Kaufmann, 496 pages, 2010

[27] Miller S., Probability and Random Processes, 2nd edition, Academic Press, Elsevier Science, 552 pages, 2012

[28] Misra Ch. S.,Woungang I., Selected topics in communication network and distributed systems, Imperial college press, 808 pages, 2010

[29] Natarajan Gautam, Analysis of Queues: Methods and Applications, CRC Press, 802 pages, 2012

[30] Peterson L. L., Davie B. C., Computer networks – a system approach, Morgan Kaufmann, 920 pages, 2011

[31] Resch M. M., Supercomputers in Grids, Int. J. of Grid and HPC, No.1, 1 - 9 pp., 2009

[32] Riano l., McGinity T.M., Quantifying the role of complexity in a system´s performance, Evolving Systems, Springer Verlag, 189 – 198 pp., 2011

[33] Ross S. M., Introduction to Probability Models, 10th edition, Academic Press, Elsevier Science, 800 pages, 2010

[34] Wang L., Jie Wei., Chen J., Grid Computing: Infrastructure, Service, and Application, CRC Press, 2009

[35] www pages

[36] www.top500.org

[37] www. intel.com

Simulation and evaluation of WIMAX handover over homogeneous and heterogeneous networks

Anmar Hamid Hameed[1, 2, *], Salama A. Mostafa[1, 2], Mazin Abed Mohammed[3, 4]

[1]College of Graduate Studies, Selangor, Malaysia
[2]Universiti Tenaga Nasional, Selangor, Malaysia
[3]Dept. of Planning and Follow up, Anbar, Iraq
[4]University of Anbar, Anbar, Iraq

Email address:

anmar_oio@yahoo.com(A. H. Hameed), semnah@yahoo.com(S. A. Mostafa), mazin_top_86@yahoo.com(M. A. Mohammed)

Abstract: Different aspects of wireless networks are changing as a result of the continuous needs in terms of speed, data rates and quality of service. Such aspects are required to be adaptable within the same network or among networks of different technologies and service providers. Consequently, Worldwide Interoperability for Microwave Access (WiMAX) is one of the future generation networks (4G) that needs further study. However, a major consideration for WiMAX to achieve mobility is the handover mechanism that concerns with the mobile station movement within the range of network coverage from one base station to another similar homogeneous or different homogeneous network. In this paper, an intensive analysis to handover mechanism in WiMAX followed by comparisons to WiMAX performance with UMTS and WiFi networks is carried out. QualNet 4.5.1 simulator is used to simulate the comparison process homogeneously and heterogeneously. Performance metrics of Throughput, End-to-End (E-2-E) Delay and Jitter in the comparison during handover process in/between the wireless networks are used. The simulation results are evaluated to identify the performance of the handover process over WiMAX-WiMAX, WiMAX-UMTS and WiMAX-WiFi with respect to the selected metrics. The environment of Wi-MAX-WiMAX has shown substantial enhancement of the system Throughput, reduction of E-2-E Delay and Jitter.

Keywords: Homogeneously and Heterogeneous Networks, WiMAX, WiFi, UMT, Handover Mechanism

1. Introduction

The world is moving to the age of velocity in every field especially the wireless networks field. To go along with it, it is needful to have faster facilities, more importantly in the wireless networks. The need is to provide mobile wireless with higher data rates, Quality of Service (QoS) and adaptability within the same network or among networks of different technologies and service providers. The users should be able to get its potential whatever they are using (PC, cell phone, electronic pad, etc. wise), and where ever they are sitting; at home, walking and even driving [1].

Worldwide Interoperability for Microwave Access (WiMAX) is one of the future generations (4G) promising networks to cover some of the consumers' needs. It is an emerging technology that is designed to deliver fixed, and more recently, mobile broadband connectivity [2]. The WiMAX trade name is used to group a number of wireless technologies that have emerged from the Institute of Electrical and Electronics Engineers (IEEE) 802.16 Wireless Metropolitan Area Network (MAN) standards [3]. The main standards introduce mobility and currently receive a great deal of interest in the telecoms world [2]. Figure 1 show environments that can be employed in WiMAX [1].

The IEEE community has defined the IEEE 802.16e amendment (i.e. mobile WiMAX) to support mobility. Mobile WiMAX of IEEE 802.16e defines wireless communication for mobiles, moving at speed of 125 KMPH in the range of 2-6 GHz (802.16e). IEEE 802.16e is implemented with Orthogonal Frequency-Division Multiple Access (OFDMA) as its physical layer scheme [4]. Full mobility technology is different from one network to another, and each technology has its own mobility characteristics [5]. To ensure high level of mobility, it is important for WiMAX to have an efficient handover mechanism [6].

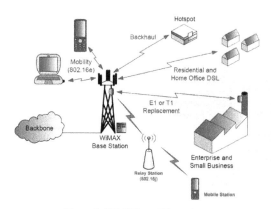

Figure 1. WiMAX possible environments

This paper investigates the WiMAX mobility capabilities and issues related to homogenous (between the same types of networks) and heterogeneous (between different types of networks) mobility. Specifically, the paper focuses on the handover performance of the WiMAX in comparison with similar WiMAX network or with other networks. For performance evaluation, Throughput, End-to-End (E-2-E) Delay and Jitter metrics are used as networks performance quality criteria to investigate the WiMAX mobility capabilities and efficiency of handover process in the comparison with other wireless technologies.

2. Objectives

Due to the importance and the complication of the mobile wireless networks, such as the WiMAX, UMTS and WiFi networks [7], an analysis study that aids in understanding the mobility problems, mobility insights from requirements, parameters and configurations in Mobile WiMAX network is presented in this research. The main objective focuses of the research are as follows:

1. To analysis mobile WiMAX network with regards to network topology acquisition, the base station and the main management messages that rules the mobility in WiMAX systems.
2. To evaluate mobile WiMAX network by performing an actual handover process to a new target base station.
3. To compare WiMAX performance with UMTS and WiFi wireless networks with respect to Throughput, E-2-E Delay and Jitter.

3. Problem Analysis

There are number of means to support mobility in communication devices such as roaming, portability, and covering mobility with a faster speed (vehicle speed) [9]. However, one of the important concerns of mobility concept for communication devices is how to motivate end users to use this technology, together with the emerging services for mobile devices presented by the service provider e.g. using packet data for streaming multimedia files [8].

As there are many wireless networks technologies, there

is a need to gather and analyze information about these technologies to be filtered and collected in a trendy presentation [10]. Communication devices such as smart phones and laptops capabilities are getting more complex and emergent services that are supplied by the network provider like streaming audio/video through packet data are more desirable [6]. More so, there are different methods behind the cutting edge of the overall system performance for instance portability, full mobility and roaming. These methods often differ among various network technologies as the mobility characteristics of each [10].

For the reasons above, it is vital to present a comprehensive study about mobile WiMAX and awareness of its mobility capabilities especially in the motion state and comparing it with other types of technologies. Consequently on one hand, knowing how WiMAX supports intra-network mobility, and is it better than the other contemporary technologies? And on the other hand, the handover mechanism as it is considered as an important process included within mobile WiMAX technology is not intensively reviewed [11]. As a matter of fact, researchers elaborated a handover in general or simulated the handover homogeneously within one network [7, 12].

4. Literature Review

Worldwide Interoperability for Microwave Access (WiMAX) is a new wireless technology that can be used to build a wide coverage area networks with high Throughput [13], and high security [14] as shown in Figure 1. It is intended for Wireless Metropolitan Area Network (WMAN). WiMAX used to be the buzzword of wireless communication industry for the last six years. It is able to provide Broadband Wireless Access (BWA) up to 30 miles (50 km) for fixed stations, and 3-10 miles (5-15 km) for mobile stations [14]. IEEE authorized in 1999 a new working group known as 802.16. The group published its first standard, IEEE 802.16a, in January 2003 [15]. The standards of WiMAX are identified as 802.16-2004 (October 2004), 802.16e (December 2005), 802.16j (Jun 2009) and 802.16m (still under development) [16]. Figure 2 shows the number of WiMAX deployments over time per frequency.

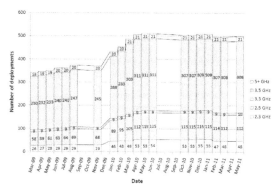

Figure 2. Cumulative number of WiMAX deployment per frequency [17]

Daan P. et al. [17], have elaborated the different activities that occurred within the three important organizations: The 802.16 working group of the IEEE for technology development and standardization, the WiMAX forum for product certification and the ITU (International Telcomunication Union) for international recognition. They presented a comprehensive overview about the evolution of WiMAX in terms of standardization and certification. Furthermore, they highlighted the steps that have been considered in cooperating with the ITU to improve the international esteem of the technology. Finally, WiMAX trend analysis has also been discussed.

Ray et al. [18], proposed a fast and simple scheme for Mobile Station (MS)-controlled handover mechanism for mobile WiMAX environment. They suggested Received Signal Strength (RSS) and path-loss formula in estimating MS distance from any neighboring Base Stations (BS). The MS monitors the received signal strength that serves specific BS periodically. The MS-controlled approach simulation of Mobile WiMAX network outcomes show an observable reduction to handover latency accompanied by increase in network scalability.

5. Methodology

Since the WiMAX is a new technology, it is difficult to get a realistic comparison regarding a specific area of concern under specific circumstances within its architecture. Because of the lack practical experience with WiMAX and most of the experiments or simulations already done were focused on one specific area of concern. This research contributes comprehend study on mobile WiMAX network handover process performance.

The simulation of the handover process in mobile WiMAX is implemented using QualNet 4.5.1 simulator. Three performance matrices selected to be used in the experiment are: Throughput E-2-E delay, and Jitter within WiMAX environment. The obtained results are compared with the results obtained from the handover simulation process of WiMAX to WiFi network and WiMAX to UMTS network. Finally, the Uplink Channel Descriptor (UCD) and the Down Link Channel Descriptor (DCD) are simulated within two different time intervals during the handover process. In this section the research methodology is presented to illustrate how the research is carried out to meet its objectives.

5.1. QualNet Simulator

QualNet [19] is a state-of-the-art simulator for large, heterogeneous networks and the distributed applications that execute on such networks. QualNet [20] is simulating software that can be run on several platforms like Windows, Linux, O SX, and Solaris and it is capable of simulating wireless networks such as WiMAX [21]. QualNet has been used to simulate high-fidelity models of wireless networks with as many as 50,000 mobile nodes [22]. It uses

architecture analogues of the TCP/IP network protocol stack which is a layered architecture.

The technology used in QualNet consists of 5 layers (top to bottom) as shown in Figure 3. The layers that are adjacent in the protocol stack communicate using well defined protocol. In general, the communication only occurs between adjacent layers (possible to be altered by the developers) [23].

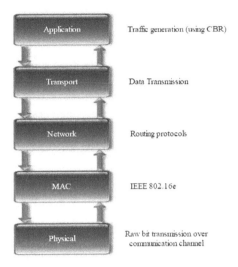

Figure 3. QualNet protocol stack

Figure 3 shows the QualNet layer model in which Constant Bit Rate (CBR) application and IEEE 802.16e MAC are used in the simulation. CBR application is generally used to simulate multimedia traffic (time critical traffic types). It can be configured to simulate a large number of real network applications by mimicking their traffic pattern and by filling traffic at a constant rate into the network. It can be accurately simulated by appropriately configuring the CBR application in QualNet. The simulated traffic sources CBR is generated using 4 mobile nodes performing handover process homogenously and heterogeneously with two time intervals, 5 and 10 seconds (UCD, DCD).

5.2. Performance Metrics

Performance metrics is used to measure and evaluate the handover processes homogeneously and heterogeneously within WiMAX environment. Three parameters have been selected for the evaluations which are Throughput, E-2-E and Jitter.

5.2.1. Throughput

The average Throughput represented by: the ratio of total amount of data that reaches its destination to the time taken for the data to transfer from the source to the destination. The data packets received at the physical layer are sent to the higher layers if they are destined for this station. Throughput is usually described by bytes or bits per second (Bps or bps).

$$Throughput = \frac{Total\ bytes\ received * 8[bit]}{End\ time[s] - Start\ time[s]} \quad (1)$$

5.2.2. End-to-End (E-2-E) Delay

The average End-to-End (E-2-E) Delay is the time taken for a packet to be transmitted across a network from source to destination. This metric describes the packet delivery time whereas the lower the E-2-E Delay, the better the application performance. The E-2-E Delay value is averaged over the number of packets. It is mathematically represented by the following equation.

$$Average\ E - 2 - E\ delay = \frac{Total\ E-2-E\ delay}{NO.of\ packets\ received} \quad (2)$$

5.2.3. Jitter

The Jitter is the deviation in the time among packets arriving at the destination side, caused by network congestion, timing drift, or route changes. It signifies the Packets from the source till they reach their destination with different delays. A packet's delays vary with its location in the queues of the routers along the path between source and destination. This location is not predictable due to network circumstances. Equation below describes the Jitter calculation:

$$Jitter = Receptiom\ Time\ of\ packet(I)\text{-}Reception\ Time\ of\ packet(I\text{-}1) \quad (3)$$

5.3. Simulation Parameters

Various parameters have been used in all simulation scenarios to analyze the handover behavior under specific circumstances both homogeneously and heterogeneously. The simulation parameters that are used in this research are listed in Table 1. The speed of mobile nodes during handover process is 80KMPH; the size of the field is 1500m x 1500m. During the simulation, all nodes start performing handover from 0 second until the time of simulation ends (25s).

5.4. Simulation Environments

Table 1. Simulation parameters

Simulation Parameters	
Parameter	**Value**
BS range radius *(m)*	1000
AP range radius *(m)*	500
Terrain-dimensions *(m)*	1500,1500
BS-AP distance *(m)*	1200
Frequency band *(GHZ)*	2.4
Handover RSS trigger	-78.0
Handover RSS margin	3.0
Channel bandwidth *(MHZ)*	20
Frame duration *(ms)*	20
FFT size	2048
MS velocity [*m.s⁻¹*]	20
BS transmit power $P_t = \frac{dBm}{height}(m)$	20/5
AP transmit power $P_t = \frac{dBm}{height}(m)$	20/1.5
MS transmit power $P_t = \frac{dBm}{height}(m)$	15/1.5
Somulation time (s)	25
Traffic	CBR

The QualNet simulator v.4.5.1 has been used to simulate the proposed scenarios for studying the handover process between WiMAX, WiFi and UMTS networks, as shown in Figure 4. UMTS and WiFi are chosen as they are common networks used nowadays and their technologies are almost similar. Therefore, it's possible to establish new connections between them heterogeneously.

The simulation of the handover process between the mentioned homogeneous and heterogeneous networks is conducted to evaluate the Throughput, E-2-E Delay and jitter. In each simulation scenario, the UCD and DCD messages are modified within two time intervals (5 and 10 seconds) during handover to find out which time interval performs better in each performance metric. The Constant Bit Rate (CBR) application is used to simulate the assumed scenarios in QualNet v.4.5.1. A snapshot for simulation scenario is shown in Figure 5.

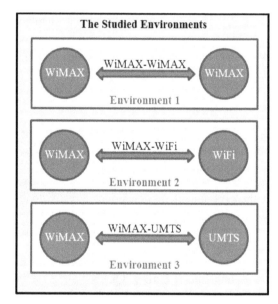

Figure 4. Simulated environments

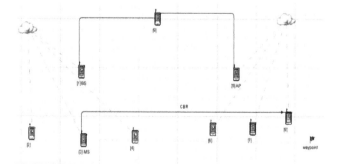

Figure 5. The QualNet simulation scenario of WiMAX and WiFi networks

The left side of Figure 5 illustrates the WiMAX network, the BS and three MSs associated to the BS. The other side, a WiFi network is configured with one access point (AP) node number 5, and three client nodes under the AP coverage.

The simulation setup corresponds to a connection of the WiMAX BS to another UMTS BS or WiFi AP (another network) by backhaul links T1 through a central gateway BS (node number 9). The MS (node number 3) is a mobile node that moves from its location on the waypoint and stops in the WiFi cell (for example of one moving node). However, the MS performs the handover process from WiMAX to WiFi coverage, while it is communicating or transmitting data to the destination node (node number 8) under WiFi access point (AP 5). In the meantime, the QualNet simulator collects the statistics for all performance metrics.

Four different experiments are conducted to evaluate the WiMAX handover mechanism. The first experiment is conducted to evaluate the average Throughput of the system during handover process. The second experiment compares the difference in the inter packet arrival times at the receiver (jitter) with respect to the total number of MS for every routing protocol. The purpose of the third experiment is to analyze the E-2-E Delay of the system during handover process. Finally, the fourth experiment is about comparing the time intervals of the UCD and DCD messages during handover process within two time intervals (5 and 10 seconds), regarding to the three performance metrics just mentioned above to find out which time interval performs the best during the handover process.

6. Results

Results obtained from the handover simulation process are discussed in this section. The Throughput, E-2-E Delay and Jitter of each simulated environment of the three (WiMAX-WiMAX, WiMAX-WiFi and WiMAX-UMTS) is discussed in the following sup-sections:

6.1. Throughput Result

Average Throughput is the ratio of total amount of data that reaches its destination to the time taken for the data to travel from the source to the destination. Figure 6 shows the average Throughput of the three handover environment with respect to the number of moving nodes.

The average Throughput in WiMAX-WiMAX handover is obviously the highest since it involves the same technology. In another words, the similarity in the BS type is high which saves any extra management signaling. WiMAX-UMTS comes in second, with slightly lower Throughput compared to WiMAX-WiMAX. WiMAX-WiFi comes in last, with much lower Throughput compared to WiMAX-WiMAX and WiMAX-UMTS. It is also observed that as the number of moving nodes increases, the average Throughput slightly decreases. Figure 6 depicts that the WiMAX-WiFi has much lower Throughput compared to the other two as a result of lower performance of the WiFi network compared with WiMAX and UMTS. It is also recorded that the Throughput drops dramatically with more moving nodes due to the high load on the network with more nodes.

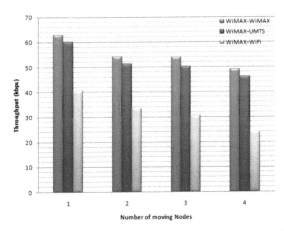

Figure 6. *Overall Throughput for various number of moving nodes*

Figure 7 below illustrates the system Throughput of WiMAX-WiMAX environment with variant values of UCD and DCD messages. Two time interval (5 and 10s) is configured in each simulation run. In first run, the time interval of UCD and DCD messages is 5seconds for each. In second run, the time interval is 10 seconds for each.

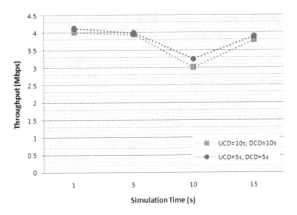

Figure 7. *System Throughput of WiMAX-WiMAX environment with variant values of UCD and DCD messages*

Consequently, the first case (with 5s) achieved higher Throughput than that of second case (with 10s). In fact, reducing acquisition time of such management messages (UCD and DCD) helps to enhance the network performance, thus, reducing the signaling time between BSs and MSs. Also, the MSs can be updated, by the BS, about the network information in shortest time (every 5s) which shows better performance, resulting here with higher Throughput. Figure 7 also shows that the performance of both cases is degraded in the simulation time 10s (X-axis); this is due to the occurrence of handover process at this time and conversely, the system Throughput is dramatically decreased.

6.2. End-to-End Delay Result

In general, the average of E-2-E Delay matrix for all simulation environments (WiMAX-WiMAX, WiMAX-UMTS and WiMAX-WiFi) increases as the number of MSs in the network increases (see Figure 8). This is expected in wireless environments due to queuing delays at every hop node (BS or AP). Neither the WiMAX-WiMAX nor the WiMAX-UMTS utilize their effects on average E-2-E Delay for traffic; hence it appears to be minimal. However, the average E-2-E Delay for the different numbers of MSs with the WiMAX-WiMAX and WiMAX-UMTS are lower than with the WiMAX-WiFi under all scenarios. Similar to the result in Section A, the WiMAX-WiMAX and WiMAX-UMTS handover give the best performance while the WiMAX-WiFi gives the worst performance due to WiFi network lower performance compared with WiMAX and UMTS in addition to the variety in the technology specifications quality.

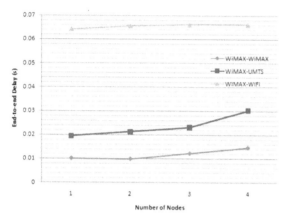

Figure 8. The average E-2-E Delay for the three simulated environments

Figure 9 below shows the average E-2-E Delay of WiMAX-WiMAX environment with variant time intervals of UCD and DCD messages. The first simulation run with 5s achieves lower E-2-E Delay compared to the second run with 10s. This is due to minimizing the required time for managing messages signaling.

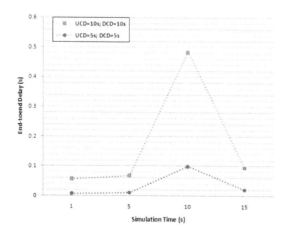

Figure 9. The average E-2-E Delay of WiMAX-WiMAX scenario with variant time intervals of UCD and DCD messages

In addition, the gap between both simulations run at the 10s (i.e. the moment of the handover process) is much wider compared to other simulation times (e.g. 1, 5, and 15s). This confirms that the 5s of UCD and DCD messages shows better performance compared to the 10s time interval.

6.3. Jitter Result

The simulation result of the overall mobility Jitter performance with various number of moving MSs for all simulated environments is shown in Figure 10. The WiMAX-WiMAX and WiMAX-UMTS environments perform with lower Jitter compared with WiMAX-WiFi environment due to the variance in the technology specifications quality.

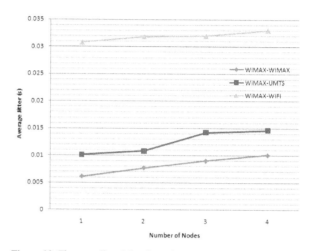

Figure 10. The overall mobility Jitter for versus number of moving MSs

The system with WiMAX-WiMAX environment in the first run of UCD and DCD messages with 5s time interval achieved lower Jitter compared to second run of 10s time interval as shown in Figure 11. This is due to the short time interval required for managing messages which reduces the time required for managing the messages signaling between nodes. More so, the first run with 5s, the simulation time at "10s" period shows better performance than the second run with 10s time interval.

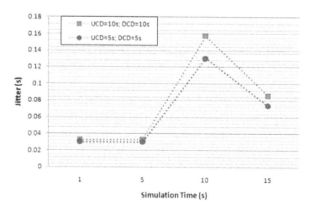

Figure 11. The system Jitter of the simulation scenario with WiMAX-WiMAX environment

7. Discussion

This paper presents an attempt to study and analyze the mobility in a collection of homogeneous and heterogeneous WiMAX, UMTS and WiFi environments regarding the handover process between such wireless networks. The main focus is to investigate the handover process in Mobile WiMAX in terms of its rules and address management and the main issues related to it. There is a limited related works that investigate the handover scenarios in networks' simulation analysis. They concentrate more on the transportation of subscribers (Mobile Stations) within WiMAX environment and with another environments (e.g. UMTS or WiFi), and suggest other fundamentals like scheduling and QoS as future works.

Two of the most important requirements for wireless communication technologies are to be applicable and universally desirable. Mobile WiMAX 802.16e handover mechanism analysis is the main target for this research. The focal point here is to introduce a complete understanding about handover process in WiMAX network, as it is considered as the most important process to achieve mobility within wireless networks. Work evaluation has been done by simulating handover process with different situations homogeneously and heterogeneously. The main task of the simulation process is to determine what parameters are affected during handover process in mobile networks homogeneously within WiMAX environment and heterogeneously to UMTS and WiFi networks.

The results show the ability to evaluate the performance of handover procedure within WiMAX and the other networks that have been chosen for this research. Based on the obtained results, a comparison among the three different networks has identified the best environment to establish a "handover".

Regarding to channel descriptor, the purpose of the DCD messages is to specify the characteristics of a given downlink physical channel. The BS transmits them at a given interval, which can have a maximum value of 10 seconds. It's the same for UCD, in that it specifies the characteristics of the uplink channel and is transmitted at a given interval, which cannot exceed 10 seconds.

The WiMAX-WiMAX case has been simulated using two time intervals (5 and 10) seconds under the three differential matrices (Throughput, E-2-E Delay and jitter) expecting that the time interval 5s will achieve the best results due to the short time it takes to update the current information compared to 10s.

Best results have been obtained from the simulation of WiMAX-WiMAX followed by WiMAX-UMTS while WiMAX-WiFi shows the worst. WiMAX-WiMAX has achieved the highest amount of Throughput and the lowest E-2-E Delay and jitter. Furthermore, a comparison has been done to find out which time interval achieves the best results within the same network WiMAX-WiMAX. The time interval 5s achieved the best outcomes comparing to the other time interval 10s regarding to the three parameter metrics.

8. Conclusion and Future Work

We discuss the simulation results of the handover process over WiMAX-WiMAX, WiMAX-UMTS and WiMAX-WiFi environments/networks. The simulation scenarios are simulated to compare the results of each environment's behavior with handover process using the QualNet simulator v4.5.1. The environment (WiMAX-WiMAX) has shown substantial enhancement of the system Throughput, reduction of E-2-E delay, and reduction of Jitter. This is mainly due to the similarity in the BS type which saves any extra management signaling. For example, between WiMAX BS and UMTS BS, the management messages signaling require longer time compared with WiMAX-WiMAX BSs.

In addition, the best environment (WiMAX-WiMAX) is simulated with various values of the time intervals of UCD and DCD management messages as a deep analysis. However, all related simulation results show that the shorter time interval of both UCD and DCD messages improves the overall network performance and the handover process.

Based on the analyses that have been done in this project, an optimization is required for the handover process of mobile WiMAX to improve its performance in terms of Delay and moving speed. Thus, the handover optimization in mobile WiMAX is the recommendation for future work, such as the proposed scheme that has been expressed in [24].

Acknowledgements

We would like to thank all the college of graduate studies staff of the Universiti Tenaga Nasional for their help and support during our study in the MIT program 2010/2012.

References

[1] IEEE 802.16 Working Group. "IEEE Standard for Local and Metropolitan Area Networks, Part 16: Air interface for fixed broadband wireless access systems," *IEEE* Std 802 (2004).

[2] L. M. Carlberg, and A Dammander, "WiMAX-A study of mobility and a MAC-layer implementation in GloMoSim," *Master's Thesis in Computing Science* (2006).

[3] D. H. Lee, K. Kyamakya and J. P. Umondi, "Fast handover algorithm for IEEE 802.16 e broadband wireless access system," Wireless pervasive computing, *2006 1st international Symposium on. IEEE*, 2006.

[4] A. F. Kabir, M. Khan, R. Hayat, A. A. M. Haque and M. S. I. Mamun, "WiMAX or Wi-Fi: The Best Suited Candidate Technology for Building Wireless Access Infrastructure," *arXiv preprint arXiv*:1208.3769 (2012).

[5] M. F. Finneran, "WiMAX versus WiFi a comparison of technologies markets and busniss plan," *dBrn associates Inc*, 2004.

[6] A. Ezzouhairi, Q. Alejandro and S. Pierre, "Towards cross layer mobility support in metropolitan networks," Computer Communications , *ELSEVIER* 33.2 (2010): 202-221.

[7] A. S. Rashid, A. A. Hassan, M. Amitava, F. Francisco and W. K. Daniel, "WiMAX, LTE, and WiFi interworking," *Journal of Computer Systems, Networks, and Communications*, 2010.

[8]   Q. B. Mussabbir, "Mobility management across converged IP-based heterogeneous access networks," PhD Thesis. *School of Engineering and Design*, Brunel Universit (2010).

[9]   International Telecommunication Union (ITU), "Broadband mobile communications towards a converged world," *Micworkshop on shaping the future mobile information society*, Seoul. 2004.

[10] I. F. Akyildiz, J. Xie and S. Mohanty. "A survey of mobility management in next-generation all-IP-based wireless systems," *Wireless Communications*, IEEE 11.4 (2004): 16-28.

[11] Z. Becvar, P. Mach and R. Bestak, "Initialization of handover procedure in WiMAX networks," *ICT-MobileSummit 2009 Conference Proceedings*, 2009.

[12] J. Liao, Q. Qi, X. Zhu, Y. Cao, and T. Li, "Enhanced IMS handoff mechanism for QoS support over heterogeneous network," *The Computer Journal* 53.10 (2010): 1719-1737.

[13] K. R. Santhi and G. S. Kumaran, "Migration to 4 G: Mobile IP based solutions," Telecommunications, 2006. AICT-ICIW'06. *International Conference on Internet and Web Applications and Services/Advanced International Conference on. IEEE*, 2006.

[14] WiMAX.com, "What is WiMAX," *WiMax.com Broadband Solutions, Inc.* 2011. < http://www.wimax.com/general/what-is-wimax>

[15] Chen, Kwang-Cheng, J. Roberto B. de Marca, and J. Roberto, eds. Mobile WiMAX. John Wiley, 2008.

[16] Z. Becvar and J. Zelenka, "Handovers in the mobile WiMAX," *Research in Telecommunication Technology* 1 (2006): 147-50.

[17] D. Pareit, B. Lannoo, I. Moerman and P. Demeester, "The history of WiMAX: A complete survey of the evolution in certification and standardization for IEEE 802.16 and WiMAX," *Communications Surveys & Tutorials, IEEE* 14.4 (2012) 1183-1211.

[18] K. R. Sayan, R. S. Kumar, K. Pawlikowski, A. McInnes, H. Sirisena "A fast and simple scheme for mobile station-controlled handover in mobile WiMAX," *Access Networks. Springer Berlin Heidelberg*, 2011. 32-44.

[19] A. Shami, M. Maier, and C. Assi. *Broadband Access Networks: Technologies and Deployments*. Springer Verlag, 2009.

[20] F. Muratore. *UMTS: Mobile communications for the future*. John Wiley & Sons, Inc., 2000.

[21] QualNet 4.5 Programmer's Guide, *Scalable Network Technologies, Inc.*, 2008. < http://www.eurecom.fr/-chenjlQualNet03.pdf.>

[22] QualNet Product Family, *Scalable Network Technologies, Inc.*, 2010. <http://www.scalable-networks.com/pdf/QualNet_Family.pdf>

[23] Y. F. Dong, and R. Liu, "Simulation studies of reliable data delivery protocols," *CSIRO ICT Centre*, Australia 2007.

[24] W. Jiao, P. Jiang and Y. Ma, "Fast handover scheme for real-time applications in mobile WiMAX," *Communications, 2007. ICC'07. IEEE International Conference on. IEEE*, (2007): 6038-6042.

4

Performance modeling of parallel computers NOW and Grid

Peter Hanuliak, Michal Hanuliak

Dubnica Technical Institute, Sladkovicova 533/20, Dubnica nad Vahom, 018 41, Slovakia

Email address:

phanuliak@gmail.com (P. Hanuliak), michal.hanuliak@gmail.com (M. Hanuliak)

Abstract: The paper describes development, realization and verification of more precise analytical models for the study of the basic performance parameters of parallel computers based on connected parallel computers (Cluster, NOW, Grid). At first the paper describes very shortly the developing steps of parallel computer architecture and then he summarized the basic concepts for performance modeling of mentioned parallel computers. To illustrate theoretical evaluation concepts the paper considers in its experimental part the achieved results on concrete analyzed examples and their comparison. The suggested model considers for every node of the NOW or Grid networks one part for the own workstation's activities and another one for node's communication channel modeling of performed data communications. In case of using multiprocessor system, as modern node's communication processor, the suggested model considers for own node's activities M/D/m queuing theory system and for every node's communication channel M/D/1 system. Based on these more realistic assumptions we have been developed improved analytical models to account the real no exponential nature of the inputs to the modeling queuing systems. The achieved results of the developed models were compared with the results of the common used analytical and simulation model to estimate the magnitude of their improvement. The developed analytical models could be used under various ranges of input analytical parameters, which influence the architecture of NOW or Grid computer networks and which are interested from the sight of practical using. These consequences are in relation to the developed analytical models and their verifications through simulation model.

Keywords: Parallel Computer, Network of Workstation (NOW), Cluster, Grid, Analytical Modeling, Queuing Theory, Performance Evaluation, Queuing Theory System

1. Developing Periods in Parallel Computers

In the first period of parallel computers between 1975 and 1995 dominated scientific supercomputers, which were specially designed for the high performance computing (HPC). These parallel computers have been mostly used computing models based on data parallelism. Those systems were way ahead of standard common computers in terms of their performance and price. General purpose processors on a single chip, which had been invented in the early 1970's, were only mature enough to hit the HPC market by the end of the1980s, and it was not until the end of the 1990's that connected standard workstation or even personal computers (PC) had become competitive at least in terms of theoretical peak performance. Increased processor performance was caused through massive using of various parallel principles in all forms of produced processors. Parallel principles were used so in single PC's and workstations (scalar or super scalar pipeline, symmetrical multiprocessor systems - SMP) [1] so as on POWER PC as in connected network of workstations (NOW). Gained experience with the implementation of parallel principles and intensive extensions of computer networks, leads to the use of connected computers for parallel solution. These trends are to be characterized through downsizing of supercomputers as Cray/SGI, T3E and from other massive parallel systems [16] (number of used processor >100) to cheaper and more universal parallel systems in the form of a network of workstations (NOW). This period we can name as the second developing period. Their large growth since 1980 have been stimulated by the simultaneous influence of three basic factors [10, 19]

- high performance processors and computers
- high speed interconnecting networks

• standardized tools for development of parallel algorithms (Shared memory, distributed memory).

Developing trends are actually going toward building of wide spread connected NOW networks with high computation and memory capacity (Grid). Conceptually Grid comes to the definition of metacomputer [31]. Metacomputer can be understood as the massive computer network of computing nodes built on the principle of the common use of existing processors, memories and other resources with the objective to create an illusion of one huge, powerful supercomputer. Such higher integrated forms of NOW's (Grid module) create various actually Grid systems or metacomputers we can define as the third period in developing trends of parallel computers.

2. Classification of Parallel Systems

It is very difficult to classify all existed parallel systems. But from the point of programmer-developer we can divide them [4, 10] to the two following different groups

• synchronous parallel architectures. These are used for performing the same or very similar computation on different sets of data. They are often used under central control, that means under the global clock synchronization (vector, array system etc.) or a distributed local control mechanism (systolic systems etc.). The typical examples of synchronous parallel computers illustrate Figure 1 on its left side. Some of used parallel principles in past time are step-by-step applied in actually modern personal computers (PC) for example in a form of SIMD (Single instruction multiple data) computer instructions within their computer set instruction (CSI)

• asynchronous parallel computers. They are composed of a number of fully independent computing nodes (processors, cores or computers. In programming parallel algorithms there are necessary to use inter process communications (IPC). To this group belong mainly various forms of computer networks (cluster), network of workstation (NOW) or more integrated Grid modules in the form as any networks of NOW networks (Grid). The typical examples of asynchronous parallel computers illustrate Figure 1 on its right side. According long-time trends asynchronous parallel computers based on PC computers (single, SMP) are dominant parallel computers [16, 27].

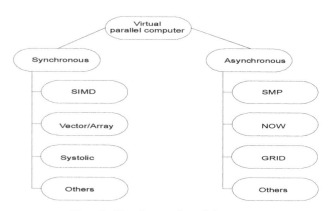

Figure 1. *Classification of parallel computers.*

3. Architectures of Parallel Computers

3.1. Symmetrical Multiprocessor System

Symmetrical multiprocessor system (SMP) is a multiple using of the same processors or cores which are implemented on motherboard in order to increase the whole performance of such system. Typical common characteristics are following

• each processor or core (computing node) of the multiprocessor system can access main memory (shared memory)

• I/O channels or I/O devices are allocated to individual computing nodes according their demands

• integrated operation system coordinates cooperation of whole multiprocessor resources (hardware, software etc.).

Concept of such multiprocessor system illustrates Figure 2.

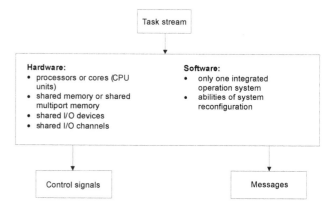

Figure 2. *Typical characteristics of multiprocessor systems.*

Typical practical architecture example of eight multiprocessor systems (Intel Xeon) illustrates Figure 3.

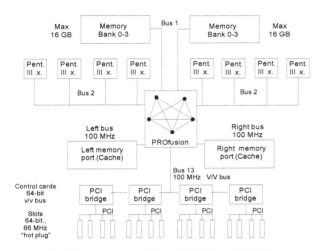

PROfusion - cross switch of 3 bus and 2 memory ports (parallel)

PCI cards - type Enthanced PCI (64 bit, 66 MHz, "Hot Plug" - on-line exchange)

Figure 3. Architecture of multiprocessor (8-Intel processor).

3.2. Network of Workstations

There has been an increasing interest in the use of networks of workstations (NOW) connected together by high speed networks for solving large computation intensive problems. This trend is mainly driven by the cost effectiveness of such systems as compared to massive multiprocessor systems with tightly coupled processors and memories (Supercomputers). Parallel computing on a cluster of workstations connected by high speed networks has given rise to a range of hardware and network related issues on any given platform [6]. With the availability of cheap personal computers, workstations and networking devises, the recent trend is to connect a number of such workstations to solve computation intensive tasks in parallel on such clusters. Network of workstations [13, 28] has become a widely accepted form of high performance computing (HPC). Each workstation in a NOW is treated similarly to a processing element in a multiprocessor system. However, workstations are far more powerful and flexible than processing elements in conventional multiprocessors (Supercomputers). To exploit the parallel processing capability of a NOW, an application algorithm must be paralleled. A way how to do it for an application problem builds its decomposition strategy. This step belongs to a most important step in developing effective parallel algorithm [13, 18].

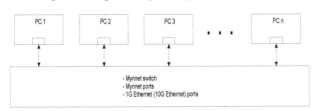

Figure 4. Architecture of NOW.

Principal example of networks of workstations is at Figure 4. The individual workstations are mainly powerful workstations based on multiprocessor or multicore platform.

3.3. Grid Systems

The In general Grids represent a new way of managing and organizing of computer networks and mainly of their deeper resource sharing (Figure 5.).

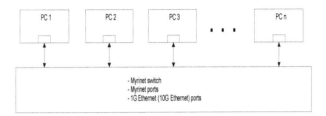

Figure 5. Architecture of Grid node.

Conceptually they go out from a structure of virtual parallel computer based on computer networks. In general Grids represent a new way of managing and organizing of resources like network of NOW networks. This term define massive computational Grid with following basic characteristics

- wide area network of integrated free computing resources. It is a massive number of interconnected networks, which are connected through high speed connected networks during which time whole massive system is controlled with network operation system, which makes an illusion of powerful computer system (Virtual supercomputer)
- grants a function of metacomputing that means computing environment, which enables to individual applications a functionality of all system resources
- system combines distributed parallel computation with remote computing from user workstations.

3.3.1. Conventional HPC Environment Versus Grid Environments

In Grids, the virtual pool of resources is dynamic and diverse, since the resources can be added and withdrawn at any time according to their owner's discretion, and their performance or load can change frequently over the time. The typical number of resources in the pool is of the order of several thousand or even more. An application in a conventional parallel environment (HPC computing) typically assumes a pool of computational nodes from (a subset of) which a virtual concurrent machine is formed [4, 24]. The pool consists of PC's, workstations, and possibly supercomputers, provided that the user has access (valid login name and password) to all of them. Such virtual pool of nodes for a typical user can be considered as static and this set varies in practice in the order of 10 – 100 nodes. At table 1 we summarize mine analyzed differences between conventional distributed and Grid systems. We can also generally say that

- HPC environments are optimized to provide maximal performance
- Grids are optimized to provide maximum of existed resource capacities.

Table 1. *Comparison of environments in HPC and Grid computing*

	Conventional HPC environments	Grid environments
1.	A virtual pool of computational nodes	A virtual pool of resources
2.	A user has access (credential) to all nodes in the pool	A user has access to the pool but not to individual nodes
3.	Access to a node means access to all resources on the node	Access to a resource may be restricted
4.	The user is aware of the applications and features of the nodes	User has little or no knowledge about each resource
5.	Nodes belong to a single trust domain	Resources span multiple trust domains
6.	Elements in the pool 10 – 100, more or less static	Elements in the pool >>100, dynamic

3.4. Integration of Parallel Computers

With the availability of cheap personal computers, workstations and networking devises, the recent trends are to connect a number of such workstations to solve computation intensive tasks in parallel on various integrated forms of clusters based on computer networks. We illustrated at Figure 6 typical integrated complex consisted of NOW networks modules. It is clear that any classical parallel computers (massive multiprocessor, supercomputers etc.) in the word could be a member of such NOW [29].

For the support of reaching connectivity to any of existed integrated parallel computers in Europe (supercomputers, NOW, Grid) we can use the European classical massive parallel systems by means of scientific visits of project participants in the HPC centers of EU. These HPC centers are EPCC Edinburgh (UK), BSC (Barcelona, Spain), CINECA (Bologna, Italy), GENCI (Paris, France), SARA (Amsterdam, Netherland), HLRS (Stuttgart, Germany), CSC (Helsinki, Finland).

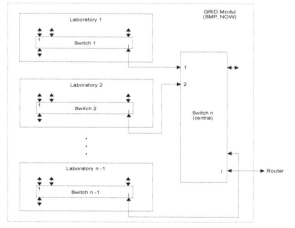

Figure 6. *Integration of NOW networks.*

4. The Role of Performance

Quantitative evaluation and modeling of hardware and software components of parallel systems are critical for the delivery of high performance. Performance studies apply to initial design phases as well as to procurement, tuning and capacity planning analysis. As performance cannot be expressed by quantities independent of the system workload, the quantitative characterization of resource demands of application and of their behavior is an important part of any performance evaluation study. Among the goals of parallel systems performance analysis are to assess the performance of a system or a system component or an application, to investigate the match between requirements and system architecture characteristics, to identify the features that have a significant impact on the application execution time, to predict the performance of a particular application on a given parallel system, to evaluate different structures of parallel applications.

The individual workstations are mainly powerful workstations based on multiprocessor or multicore platform.

4.1. Performance Evaluation Methods

The fundamental concepts have been developed for evaluating parallel computers. Trade-offs among these performance factors are often encountered in real-life applications. To the performance evaluation we can use following methods

- analytical methods
  - application of queuing theory [7, 8]
  - asymptotic (order) analysis [11, 13]
- simulation [20]
- experimental measurement
  - benchmarks [17, 23]
  - modeling tools [30, 22]
  - direct parameter measuring [11, 13]

When we solve a model we can obtain an estimate for a set of values of interest within the system being modeled, for a given set of conditions which we set for that execution. These conditions may be fixed permanently in the model or left as free variables or parameters of the model, and set at runtime. Each set of m input parameters constitutes a single point in m-dimensional input space. Each solution of the model produces one set of observations. Such a set of n values constitutes a single point in the corresponding n-dimensional observation space. By varying the input conditions we hope to explore how the outputs vary with changes to the inputs.

4.1.1. Analytic Techniques

There is a very well developed set of techniques which can provide exact solutions very quickly, but only for a very restricted class of models. For more general models it is often possible to obtain approximate results significantly more quickly than when using simulation, although the accuracy of these results may be difficult to determine. The techniques in question belong to an area of applied mathematics known as queuing theory, which is a branch of stochastic modeling [3, 25]. Like simulation, queuing theory depends on the use of powerful computers in order to solve its models quickly. We would like to prefer techniques which yield analytic solutions.

4.1.2. The Simulation Method

Simulation is the most general and versatile means of modeling systems for performance estimation. It has many uses, but its results are usually only approximations to the exact answer and the price of increased accuracy is much longer execution times. To reduce the cost of a simulation we may resort to simplification of the model which avoids explicit modeling of many features, but this increases the level of error in the results. If we need to resort to simplification of our models, it would be desirable to achieve exact results even though the model might not fully represent the system. At least then one source of inaccuracy would be removed. At the same time it would be useful if the method could produce its results more quickly than even the simplified simulation. Thus it is important to consider the use of analytic and numerical techniques before resorting to simulation. This method is based on the simulation of the basic characteristics that are the input data stream and their servicing according the measured and analyzed probability values simulate the behavior model of the analyzed parallel system. Its part is therefore the time registration of the wanted interested discrete values. The result values of simulation model have always their discrete character, which do not have the universal form of mathematical formulas to which we can set when we need the variables of the used distributions as in the case of analytical models. The accuracy of simulation model depends therefore on the accuracy measure of the used simulation model for the given task.

4.1.3. Asymptotic (Order) Analysis

In the analysis of algorithms, it is often cumbersome or impossible to derive exact expressions for parameters such as run time, speedup, efficiency, issoefficiency etc. In many cases, an approximation of the exact expression is adequate. The approximation may indeed be more illustrative of the behavior of the function because it focuses on the critical factors influencing the parameter. We have used an extension of this method to evaluate parallel computers and algorithms in [11, 13].

4.1.4. Experimental Measurement

Evaluating system performance via experimental measurements is a very useful alternative for parallel systems and parallel algorithms. Measurements can be gathered on existing systems by means of benchmark applications that aim at stressing specific aspects of the

parallel systems and algorithms. Even though benchmarks can be used in all types of performance studies, their main field of application is competitive procurement and performance assessment of existing systems and algorithms. Parallel benchmarks extend the traditional sequential ones by providing a wider a wider set of suites that exercise each system component targeted workload.

5. Little's Laws

One of the most important results in queuing theory is Little's law. This was a long standing rule of thumb in analyzing queuing systems, but gets its name from the author of the first paper which proves the relationship formally. It is applicable to the behavior of almost any system of queues, as long as they exhibit steady state behavior. It relates a system oriented measure - the mean number of customers in the system - to a customer oriented measure - the mean time spent in the system by each customer (the mean end-to-end time), for a given arrival rate. Little's law says

$E(q) = \lambda . E(t_q)$

or it's following alternatives

- $E(w) = \lambda . E(t_w)$
- $E(w) = E(q) - \rho$ (single service where m=1)
- $E(w) = E(q) - m . \rho$ (m – services).

We can use also following valid equation

$E(t_q) = E(t_w) + E(t_s)$.

where the named parameters are as

- λ - arrival rate at entrance to a queue
- m - number of identical servers in the queuing system
- ρ - traffic intensity (dimensionless coefficient of utilization)
- q - random variable for the number of customers in a system at steady state
- w - random variable for the number of customers in a queue at steady state
- $E(t_s)$ - the expected (mean) service time of a server
- $E(q)$ - the expected (mean) number of customers in a system at steady state
- $E(w)$ - the expected (mean) number of customers in a queue at steady state
- $E(t_q)$ - the expected (mean) time spent in system (queue + servicing) at steady state
- $E(t_w)$ - the expected (mean) time spent in the queue at steady state.

5.2. Queuing Networks

Continuing the examination of analytically tractable models, we look for useful results for networks of queues. These can be divided into two main groups, known as product form and non-product form. Product form networks have the property that they can be regarded as independently operating queues, where steady state can be expressed as both a set of global balance equations on customer flow in the whole network and a set of local balance equations on each queue. Local flow balance says that the mean number of customers entering any queue from all others must equal the number leaving it to go to all others, including customers which leave and rejoin the same queue immediately.

As an example of simplest queuing networks is serial connection of two queuing theory systems according Figure 7. (Tandem network), for which we can get following final solution, that the probability of k_1 demands at first node and k_2 demands at second node is

$$p(k_1,k_2) = p_1(k_1) \cdot p_2(k_2) = (1-\rho_1)\rho^{k_1} \cdot (1-\rho_2)\rho^{k_2}$$

where we assumed that

$$\rho_1 = \frac{\lambda}{\mu_1} < 1, \ \ \rho_2 = \frac{\lambda}{\mu_2} < 1$$

and

- p (k_1, k_2) is the probability of k_1 demands in the first queue and k2 demands in the second queue
- p (k_1) is the probability of k_1 demands in the first queue
- p (k_2) is the probability of k_2 demands in the second queue.

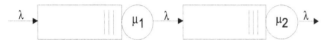

Figure 7. *Tandem network of two M/M/1 queuing systems.*

The final expression proves by evidence independence of both M/M/1 queuing theory systems. Generalization to the U queuing theory systems of M/M/1 or M/M/m made Jackson (Jackson theorem) [5, 9]. Several different ways of identifying this sort of behavior have been proposed, but the name product form comes from Jackson's theorem, which expresses the joint probability of the numbers of customers at each queue being a particular combination is the product of their individual probabilities of having that number. We begin by considering Jackson networks and then look at the extension to a more general queuing theory systems.

Based on Jackson result any network of queuing theory systems without feedback loop (they are not allowed to return any serviced demands again) with exponential service distribution, which are servicing independent Poisson input streams, generate also for a next node independent Poisson input stream. The whole demand probabilities for all nodes are given through multiplying of individual independent queuing theory systems and that in general as M/M/m systems. The outgoing stream will be exactly Poisson on the assumption of unlimited size of queue.

5.3. Jackson Theorem

Consider the case of a network of U queue/server nodes (Workstations). Customers enter the network at node j in a

Poisson stream with rate γ_j. Each node has a multiple servers m (workstations based on multiprocessor, for m = 1 workstations with single server) and service times are distributed exponentially, with mean $1/\mu_j$, (j = l,..., U). When a customer leaves node i it goes to node j with probability r_{ij}. Customers from i leave the network with probability

$$1 - \sum_{j=1}^{U} r_{ij}$$

Now let λ_i be the average total arrivals at node i, including those from outside (external input) and those from other nodes (Internal inputs). If the network is in steady state, λ_i is also the rate of customers leaving i node (including intern output). Overall we can formulate a set of „flow balance equations" which express these flows.

$$\lambda_i = \gamma_i + \sum_{i=1}^{U} \lambda_i \, r_{ij} \qquad j=1,2,\dots, U$$

As long as the network is open, i.e. at least one γ_i is non-zero, this represents a set of linear simultaneous equations with an obvious solution. Let traffic intensity at i be $\lambda_i / m_i \cdot \mu_j < 1$ for every i. The joint distribution of the number of customers p $(k_1, k_2, ... k_U)$ at each of the U nodes, $p_1(k_1), p_2(k_2), ... p_U(k_U)$, can be expressed as

$$p(k_1, k_2,..., k_U) = p_1(k_1) \cdot p_2(k_2) \cdot ... \cdot p_U(k_U) = \prod_{i=1}^{U} p_i . k_i$$

This is Jackson theorem for M/M/m system. The individual probabilities $p_i(k_i)$ are given as

$$p_i(k_i) = \begin{cases} p_0 \dfrac{(m\,\rho)^i}{i!}, & \text{pre } 1 \le i \le m \\ p_0 \dfrac{\rho^k m^m}{m!}, & \text{pre } i > m \end{cases}$$

, where

$$p_0 = \left[\sum_{i=0}^{m-1} \frac{(m\,\rho)^i}{i!} + \frac{(m\,\rho)^m}{m!(1-p)} \right]^{-1}.$$

Jackson's theorem describes each node as an independent single server system with Poisson arrivals and exponential service times. The total average number of customers in the whole NOW module is

$$E(q)_{now} = \sum_{i=1}^{U} E(q)_i$$

where $E(q)_i$ is given as

$$E(q)_i = \frac{(\rho\, m)^{m+1}}{(m-1)!\left[\sum_{i=0}^{m} \frac{(m\,\rho)^i}{i!}\left[(m-i)^2 - i\right]\right]}$$

Then from Little's law, total time spent by customers in the network $E(t)_q$ is

$$E[t_q]_{now} = \sum_{i=1}^{U} \frac{E(q)_i}{\lambda_i}$$

Jackson theorem assumes for its applying verification of assumed independence of individual network nodes. Every element on its right side is a solution of independent M/M/m geeing system with their average input value λ_i. We can get the intensities of this individual inputs λ_i with solving a system of linear differential equations for concrete values of extern inputs λ_i and for given transition matrix r_{ij}.

6. Modeling of the NOW and Grid

NOW is a basic module of any Grid system (Network of NOW networks as for example Internet). Structure of essential parts in any workstation (i-th node) of NOW based on single processor (m=1) or multiprocessor system (m - processors or cores) is illustrated at Figure 8. Inter process communication (IPC) represents all needed communication in NOW as

- communication among parallel processes
- control communication.

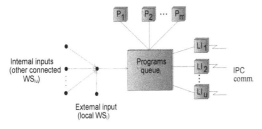

Figure 8. Structure of i – th computing node (WSi).

In principle we are assumed any constraints on structure of communication system architecture. Then we are modeling one workstation as a system with two dominant overheads

- computation overheads (processor's latency)
- communication latency [21, 26].

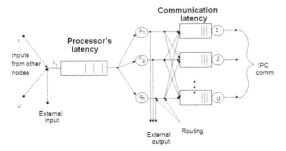

Figure 9. Mathematical model of i – th node of NOW.

To model these overheads through applying queuing theory we created mathematical model of one i-th computing node according Figure 9, which models

- computation overheads (processor's latency) [2] as queuing theory system
- every communication channel of i-th node LI_i i=1,2,…U (Link interface) as next queuing theory systems (communication system).

Such communication network in NOW module we can represent by a weighted graph where their nodes are individual workstations. IPC data arrive at random at a source node and follow a specific route in the networks towards their destination node. Data lengths of communicated parallel processes in data units (for example in words) are considered to be random variables following distributions according Jackson theorem. Those data units are then sent independently through the communication network nodes towards the destination node. At each node a queue of incoming data units is served according to a first-come first-served (FCFS) discipline. The defined communication network generally creates oriented graph (communication network) with U-nodes according to the Figure 10, where

- $\gamma_1, \gamma_2, …, \gamma_U$ represent total extern intensities of input data stream to the given WS_i
- r_{ij} is a relation probabilities from node i to the neighbouring connected nodes j (WS_{ij}) for (i= 1,…,U, j = 1,…U)
- u_i – the number of communication channels at i-th node
- U – number of computing nodes (workstations)
- $\beta_1, \beta_2, …, \beta_U$ are the individual total external output streams of data units from WS_i.

Such a model corresponds in queuing theory to the model of open servicing network. Adjective "open" characterize the extern input and output data stream to the servicing transport network [14, 26]. In common they are the open Markov servicing networks, in which the demand are mixed together at their output from one queuing theory system to another connected queuing theory system in a random way to that time as they are leaving the network. To the given i-th node the demand stream enter extern (from the network side), with the independent Poisson arrival distribution and the total intensity γ_i demands in seconds. After servicing at i-th node the demand goes to the next j-th node with the probability r_{ij} in such a way that the demand walks to the j-th node intern (from the sight network). At this time the demand departures from i-th node to the other nodes are defined with probability

$$1 - \sum_{j=1}^{U} r_{ij}$$

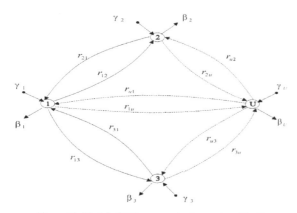

Figure 10. Model of IPC communication system (U=4).

6.1. Analytical Model of Workstations as M/M/m System

Let U be a node number of the whole transport system. For every node of NOW (i-th node according Figure 11.) we define the following parameters

- λ_i - the whole number of incoming demands to the i-th node, that is the sum both of external and internal inputs to the i-th node

- $\gamma = \sum_{i=1}^{U} \gamma_i$ represent the sum of individual total extern intensities in the NOW

- λ_{ij} - the whole input flow to the j-th communication channel at i-th node
- $E(t_q)_i$ - the average servicing time in the program queue (the waiting in a queue and servicing time) in the i-th node
- $E(t_q)_{ij}$ - the average servicing time of the j-th queue of the communication channel (the queue waiting time and servicing time) at i-th node.

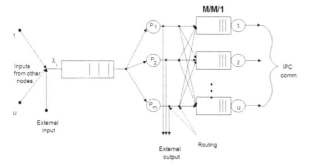

Figure 11. Mathematical model of i-th node.

Then the whole extern input flow to the transport network is given as

$$\gamma = \sum_{i=1}^{U} \gamma_i \quad \text{and} \quad \lambda_i = \sum_{i=1}^{u} \lambda_{ij} + \beta_i$$

where β_i represents the intern output from i-th node (finished parallel programs in this node) which is not further transmitted and is therefore not entering to the $(LO)_i$. Then the whole delay we can modeled as

$$E(t_q)_{now} = \frac{1}{\gamma} \left[\sum_{i=1}^{U} \left(\lambda_i \cdot E(t_q)_i + \sum_{j=1}^{u_i} \lambda_{ij} \cdot E(t_q)_{ij} \right) \right]$$

, where $\dfrac{\lambda_i \cdot E(t_q)_i}{\gamma}$ and $\dfrac{\lambda_{ij} \cdot E(t_q)_{ij}}{\gamma}$

define individual contribution of computation queue delay (M/M/m) and communication channel delay (M/M/1) of every node to the whole delay. For establishing $E(t_q)_i$ for computation queue delay it is necessary to know λ_i as the whole intensity of the input flow to the message queue where $\lambda_i = \gamma_i +$ all intern inputs flow to i-th node.

The intern input flow to i-th node is defined as the input from other connected nodes. We can express it in two ways

- through solving a system of linear equations in matrix form as $\lambda = \gamma + \lambda \cdot R$

- using of two data structures in form of tables and that is the routing table (RT) and destination probability tables (DPT).

In related model the routing table creates deterministic logical way from source i to the destination j. Concretely RT(i,j) has index (1,...,N) of the next node on the route from i to j. This assumption of the fixed routing is not rare. We have proved also experimental, that the fix routing produces good analytical results in comparison to the alternate adaptive routing in a concrete communication network. The destination probability table destiny for each i,j pair the probability, that the message which outstands in node i is destined for node j. This table with n x n dimension and elements DPT (i,j) terminates which fraction of the whole extern input γ_i has the destination j, that is $\gamma_i \cdot DTP(i,j)$. A path through the transport network we can define as the sequence $(x_1, x_2, ... , x_m)$ in which

1) exist physical communication channel, which connects x_k a $x_{k1}, k = 1,2,...,m-1$
2) x_j a $x_k, \forall j,k \ j \neq k$ (they do not exist loops).

We can define path with record "path (j→k,i)" as expression of the ordered sequence nodes, which are on the route from node j to the node k and they pass step by step through nodes i. That is $x_i=j$, $x_m=k$, $x_p=i$ and $1 < p \leq m$.

We define then $\displaystyle\sum_{k \in path(j \rightarrow k,i)}^{U}$

as the summation over the set of all destination nodes k so, that node i lies on the route from the source node j. Then we get the following relation

The intern input flow to i-th node $=$ $\displaystyle\sum_{j=1}^{U} \sum_{k=1}^{U} \gamma_j \cdot DTP\ (j,k)$, for $j \neq i, k \in$ path $(j \rightarrow k, i)$

and whole input flow to node i as

$$\lambda_i = \gamma_i + \sum_{j=1}^{U} \sum_{k=1}^{U} \gamma_i \cdot DTP\ (j,k),$$

for $j \neq i$, $k \in$ path $(j \rightarrow k, i)$

We supposed also that the incoming demands are exponential distributed and that queue servicing algorithm is FIFO (First In – First Out). The program queue PQ_i is servicing through one or more the same computation processors, which performed incoming demands (parallel processes). In demand servicing in a given node could be two possibilities

- demand will be routed to another node of the transport networks by their placing to the one of the used communication channel (IPC communication)
- demand is in the addressed node and she will leave communication network (finishing of parallel algorithms in this node) of the given node).

To every communication channel is set the queue of the given communication lines (LQ), which stores the demands (their pointers) who are awaiting the communication through this communication channel. Also in this case we supposed its unlimited capacity, exponential interarrival time distribution of input messages and the servicing algorithm FIFO. Every communication line queue has its communication capacity S_{ij} (in data units per second). Because we supposed the exponential demand length distribution the servicing time is exponential distributed too with average servicing time $1/\mu S_{ij}$, where μ is the average message length and S_{ij} is the communication capacity of node i and of communication channel j. For simplicity we will assume, as it is obvious, that S_{ij} is a part of μ. To find the average waiting time in the queue of the communication system we consider the model of one communication queue part node as M/M/1 queuing theory system according Figure 12.

Figure 12. *Model of one M/M/1 communication channel of the i-th node.*

The total incoming flow to the communication channel j at node i which is given through the value λ_{ij} and we can determine it with using of routing table and destination probability table in the same way as for the value λ_i. Then ρ_{ij} as the utilization of the communication channel j at the node i is given as

$$\rho_{ij} = \frac{\lambda_{ij}}{\mu \, S_{ij}}$$

The total average delay time in the queue $E(t_q)_i$ is

$$E(t_q)_{ij} = \frac{1}{\mu_{ij} - \lambda_{ij}}$$

If we now substitute the values for T_i and T_{ij} to the relation for T we can get finally the relation for the total average delay time of whole transport system as

$$E(t_q)_{now} = \frac{1}{\gamma}\left[\sum_{i=1}^{U}\left(\lambda_i \cdot \frac{1}{\mu_i - \lambda_i} + \sum_{j=1}^{u_i}\lambda_{ij}\cdot\frac{1}{\mu_{ij} - \lambda_{ij}}\right)\right]$$

6.2. Suggestion and Derivation of more Precise Models

6.2.1. Model with M/D/m and M/D/1 Systems

The used model were build on assumptions of modeling incoming demands to program queue as Poisson input stream and of the exponential interarrival time between communication inputs to the communication channels. The idea of the previous models were the presumption of decomposition to the individual independent channels together with the independence presumption of the demand length, that is demand lengths are derived on the basis of the probability density function $p_i = \mu \, e^{-\mu t}$ for $t > 0$ and $f(t)=0$ for $t \leq 0$ always at its input to the node. On this basis it was possible to model every used communication channel as the queuing theory system M/M/1 and to derive the average value of delay individually for every channel too. The whole end-to-end delay was then simply the sum of the individual delays of the every used communication channel.

These conditions are not fulfilled for every input load, for all architectures of node and for the real character of processor service time distributions. These changes could cause imprecise results. To improve the mentioned problems we suggested the behavior analysis of the modeled NOW module improved analytical model (Figure 13), which will be extend the used analytical model to more precise analytical model supposing that

- we consider to model computation activities in every node of NOW network as M/D/m system
- we consider an individual communication channels in i-th node as M/D/1 systems. In this way we can take into account also the influence of real non exponential nature of the interarrival time of inputs to the communication channels.

These corrections may to contribute to precise behavior analysis of the NOW network for the typical communication activities and for the variable input loads. According defined assumption to modeling of the computation processors we use the M/D/m queuing theory systems according Figure 13. To find the average program queue delay we used the approximation formula for M/D/m queuing theory system [14, 15] according as

$$E(t_w)(M/D/m_i)=$$
$$\left[1+(1-\rho_i)\cdot(m_i-1)\cdot\frac{\sqrt{45m_i}-2}{16\rho_i m_i}\cdot\frac{E(t_w)\,(M/D/1)}{E(t_w)\,(M/M/1)}\cdot E(t_w)\,(M/M/m_i)\right]$$

, in which

- ρ_i - is the processor utilization at i-th node for all used processors
- m_i - is the number of used processors at i-th node
- $E(t_w)(M/D/1)$, $E(t_w)\,(M/M/1)$ and $E(t_w)\,(M/M/m)$ are the average queue delay values for the queuing theory

systems M/D/1, M/M/1 and M/M/m respectively

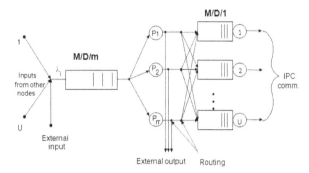

Figure 13. Precise mathematical model of i-th node.

The chosen approximation formulae we selected from two following points
- for his simply calculation
- if the number of used processors equals one the used relation gives the exact solution, that is W(M/D/1) system. Such number of processors is often used in praxis
- if the number of processors greater than one ($m_i>1$) the used relation generate a relative error, which is not greater as 1%. This fact we verified and confirmed through simulation experiments.

Let $\overline{x_i}$ define the fixed processing time of the i-th node processors and $E(t_w)_i$ (PQ) the average program queue delay in the i-th node. Then ρ_i, as the utilization of the i-th node, is given as

$$\rho_i = \frac{\lambda_i \cdot \overline{x_i}}{m_i}$$

Then the average waiting time in PQ queue $E(t_w)_i(M/D/m_i)$ is given trough the following relations

$$E(t_w)_i(M/D/1) = \frac{\rho_i \cdot \overline{x_i}}{2(1-\rho_i)}$$

$$E(t_w)_i(M/M/1) = \frac{\rho_i \cdot \overline{x_i}}{1-\rho_i}$$

$$E(t_w)_i(M/M/m_i) = \cfrac{\cfrac{(m_i \cdot \rho_i)^{m_i}}{m_i!(1-\rho_i)}}{\sum_{j=0}^{m_i-1}\left[\cfrac{(m_i \cdot \rho_i)^j}{j!} + \cfrac{(m_i \cdot \rho_i)^{m_i}}{m_i!(''1-\rho_i)}\right] \cdot \cfrac{\overline{x_i}}{m_i}}{(1-\rho_i)}$$

By substituting relations for ρ_i, $E(t_w)_i(M/D/1)$, $E(t_w)_i(M/M/1)$ and $E(t_w)_i(M/M/m_i)$ in the relation for $E(t_w)_i(M/D/m_i)$ we can determine $E(t_w)_i(PQ)$. Then the total average delay for the communication activities in i-th node is simply the sum of average message queue delay (MQ) plus the fixed processing time

$$E(t_w)_i = E(t_w)_i(PQ) + \overline{x_i}$$

To find the average waiting time in the queue of the

communication system we consider the model of one communication queue part node as M/M/1 queuing theory system according Figure 12. Let $\overline{x_{ij}}$ determine the average servicing time for channel j at the node i. Then ρ_{ij} as the utilization of the communication channel j at the node i is given as

$$\rho_{ij} = \frac{\lambda_{ij} \overline{x_{ij}}}{S_{ij}}$$

where S_{ij} is the communication channel speed of j-th node. For simplicity we will assume that $S_{ij} = 1$. The total incoming flow to the communication channel j at node i which is given through the value λ_{ij} and we can determine it with using of routing table and destination probability table in the same way as for a value λ_i. Let $E(t_w)_{ij}(LQ)$ be the average waiting queue time for communication channel j at the node i. Then

$$E(t_w)_{ij}(LQ) = \frac{\rho_{ij} \cdot \overline{x_{ij}}}{(1-\rho_{ij})}$$

The total average delay value is the queue $E(t_w)_{ij}$ is given then as

$$E(t_w)_{ij} = E(t_w)_{ij}(PQ) + \overline{x_{ij}} = \frac{\rho_{ij} \cdot \overline{x_{ij}}}{(1-\rho_{ij})} + \overline{x_{ij}}$$

There If we now substitute the values for $E(t_q)_i$ and $E(t_q)_{ij}$ to the relation for $E(t_q)_{now}$ we can get finally the relation for the total average delay time of whole NOW model is given as

$$E(t_q)_{now} = \frac{1}{\gamma}\left[\sum_{i=1}^{U}\left(E(t_w)_i(PQ) + \overline{x_i}\right) + \sum_{j=1}^{u_i}\left(E(t_w)_{ij}(LQ) + \overline{x_{ij}}\right)\right]$$

6.2.2. Other Real Analytical Models

6.2.2.1. Analytical Model with M/M/m and M/D/1 Systems

This model is mixture of analyzed model. The first part of final total average time $E(t_q)_i$ we get from chapter 6.1 and second part from 6.2.1 one. Then for $E(t_q)_{now}$ we can get finally

$$E(t_q)_{now} = \frac{1}{\gamma}\left[\sum_{i=1}^{U}\left(\lambda_i \cdot \frac{1}{\mu_i - \lambda_i} + \sum_{j=1}^{u_i}\left(E(t_w)_{ij}(LQ) + \overline{x_{ij}}\right)\right)\right]$$

6.2.2.2. Model with M/D/m and M/M/1 Systems

In this model the first part of final total average time $E(t_q)_i$ we can also get from chapter 6.2.1 and second part from 6.1 respectively. Then for $E(t_q)_{now}$ we get for this model finally

$$E(t_q)_{now} = \frac{1}{\gamma}\left[\sum_{i=1}^{U}\left(\left(E(t_w)_i(PQ) + \overline{x_i}\right) + \sum_{j=1}^{u_i}\lambda_{ij} \cdot \frac{1}{\mu_{ij} - \lambda_{ij}}\right)\right]$$

7. Analytical Model of Real Grid Systems

We have defined Grid system as network of NOW network modules. Let N is the number of individual NOW networks or similar clusters. Then final total average time $E(t_q)_{grid}$

$$E(t_q)_{grid} = \frac{1}{\alpha}\left[\sum_{i=1}^{N} E(t_q)_{i\,now}\right]$$

where

- $\alpha = \sum_{i=1}^{N} \gamma_i$ represent the sum of individual total extern intensities to the i-th NOW module in the Grid

- $E(t_q)_{i\,now}$ correspondent to individual average times in i- th NOW module (i=1, 2, ... N).

8. Results

Figure 14 and Figure 15 represent results and relative errors for the average value of the total message delay in the 5-noded communication network so for classical analytical model (M/M/m + M/M/1) as for developed more precise analytical model (M/D/m + M/D/1) in which for multiprocessor's node activities we consider very real fixed latency. The same fixed delay was included to the average communication delay at each node and in simulation model too. These assumptions correspondence to the same communication speeds in each node's communication channel. If used communication channels do not have the same communication speeds then communication latencies are different constants. In both considered analytical models (M/M/m + M/M/1, M/D/m + M/D/1) performed experiments have proved that decreasing of processor utilization ρ cause decreasing of total average delay in NOW module $E(t_q)_{now}$. Therefore parallel processes are waiting in parallel processes queues shorter time. In contrary decreasing of node's communication channel speed increase communication channel utilization and then data of parallel processes have to wait longer in communication channel queues and increase the total node's latency. Tested results have also proved the influence of real non exponential nature of the input inter-arrival time to node's communication channels. In relation to it the analytical model M/D/m + M/D/1 provides best results and the analytical model M/M/m + M/M/1 the worst ones. The results for other possible mixed analytical models (M/M/m + M/D/1, M/D/m + M/M/1) provide results between the best and worst solutions. For simplicity deterministic time to perform parallel processes at node's multiprocessor activities (the servicing time of PQ queue) was settled to 8μs and the extern input flow for each node was the same constant too.

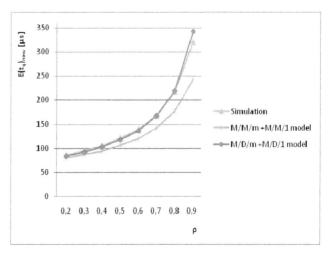

Figure 14. Comparison of analyzed models.

To vary node's processor utilization we modified the extern input flow in the same manner for each node of NOW module. For both analytical models (the best and the worst cases) are at Figure 15 the relative errors in relation to simulation results. The best analytical model (M/D/m + M/D/1) provides very precision results in the whole range of input workload of multiprocessors and every communication channel's utilization with relative error, which does not exceed 6.2% and in most cases are in the range up to 5%. This is very important to project heavily loaded NOW network module (from about 80 to 90%), where the accurate results are to be in bad need of to avoid any bottleneck congestions or some other system instabilities. The performed comparison of this best analytical model to analytical model (M/M/m + M/M/1) according Figure 12 show improvements in all range of input node's multiprocessor loads (from 20 to 90%).

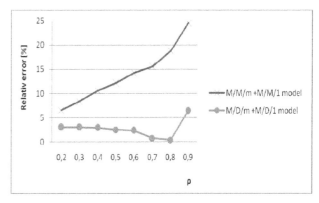

Figure 15. Relative errors of analyzed models.

The relative errors of worst analytical model are from 7 to 25%. This is due influences of processes queues delays, the nature of interarrival input to the communication channel in the case of high processor utilization. All developed analytical models could be applied also for large NOW networks practically without any increasing of the computation time in comparison to simulation method because of their explained module's structure based on NOW module. Simulation models require oft three orders of magnitude more computation time for testing such a massive metacomputer. Therefore limiting factor of the developed analytical models will not be computation time, but space complexity of memories for needed RT and DPT tables. These needed RT and DPT tables require $O(n^2)$ memory cells, thus limiting the network analysis to the number of N nodes about 100-200 for the common SMP multiprocessor. In case of possible solving system of linear equations (SLE) to find in analytical way node's λ_i and λ_{ij} respectively input intensities, most parallel algorithms use to its solution Gauss elimination method (GEM). These GEM parallel algorithms have computation complexity given as $O(n^3)$ floating point multiplications and a similar number of additions [2, 9]. These values are however adequate to handle most existing communication network of based NOW module. In addition to it also for any future massive metacomputers we would be always used hierarchically modular architecture, which consist on such simpler NOW modules.

9. Conclusion and Perspectives

Performance evaluation of computers generally used to be a very hard problem from birthday of computers. It was very hard to apply any analytical methods (for example known results of queuing theory) to performance evaluation of sequential computers because of their high number of not predictable parameters. Secondly endless user demands to increase computer performance were to be done more quickly through continues technology improvements and computer architecture changes. Application incorporation of various forms of parallel principles for a long time create more stabile conditions to apply performance evaluation methods mainly for parallel computers (actually dominant using of SMP multiprocessors and multicores, NOW and Grid systems) open more possibilities to apply mainly a queuing theory results to analyze performance of parallel structured computers. This implies one of known queuing theory results that many inputs to queuing theory system, which create shared stream and which are generating at various independent resources by chance, could be a very good approximation of Poisson distribution as a basic assumption to solve such systems in analytical way. Therefore we are able to model parallel computing nodes (multiprocessor, multicores, workstations etc.) of any actually dominant or perspective parallel computer (SMP, NOW, Grid, metacomputer) as M/D/m queuing theory systems and computing node's communication channels as M/D/1 queuing theory systems respectively.

Then such very flexible modeling tool (queuing theory), based on preferred analytical solutions show real paths to a very effective and practical performance analysis tool including massive NOW networks or another types of massive computer networks (metacomputer, Grid).

In summary developed more precise analytical models could be applied to performance modeling of dominant parallel computers and that in following typical cases
- single computing node based on SMP parallel computer (multiprocessors or multicores)
- NOW based on workstations (multiprocessors or multicores)
- Grid (Network of NOW network modules)
- mixed parallel computers (SMP, NOW, Grid)
- metacomputers (massive Grid).

For our further research work in relation to dominant trends in parallel computers (SMP, NOW, Grid), based of powerful workstations, we will be looking for preferred analytical models in which could be to study load balancing, inter-process communication (IPC) in both parallel and distributed computing, effective transport protocols, influence of various parallel computer architectures, performance prediction etc. We would be also like to analyze
- role of adaptive routing in considered analytical models [5]
- to prove, or to indicate experimentally, the role of the independence assumption, if we are looking for higher moments of overhead latencies (IPX communication – parallel and distributed computing, synchronization, parallelization, architecture etc.)
- to verify analytical models also for node's limited resources capacities – buffers, communication channels etc., and for other existing queue servicing algorithms than standard assumed FIFO (First in First out) [12, 15].

Acknowledgements

This work was done within the project "Modeling, optimization and prediction of parallel computers and algorithms" at University of Zilina. The authors gratefully acknowledge crucial help of project supervisor Prof. Ing. Ivan Hanuliak, PhD.

References

[1] Abderazek A. B., Multicore systems on-chip – Practical Software/Hardware design, Imperial college press, 200 pp., August 2010

[2] Arora S., Barak B., Computational complexity - A modern Approach, Cambridge University Press, 573 pp., 2009

[3] Cepciansky G., Schwartz L., Stochastic processes with discrete states, LAP Lambert, Germany, 109 pp., 2013

[4] Coulouris G., Dollimore J., Kindberg T., Distributed Systems – Concepts and Design (5 - th Edition), Addison Wesley, 800 pp., 2011

[5] Dattatreya G. R., Performance analysis of queuing and computer network, University of Texas, Dallas, USA, 472 pp., 2008

[6] Dubois M., Annavaram M., Stenstrom P., Parallel Computer Organisation and Design, 560 pages, 2012

[7] Gautam Natarajan, Analysis of Queues: Methods and Applications, CRC Press, 802 pages, 2012

[8] Gelenbe E., Analysis and synthesis of computer systems, Imperial College Press, 324 pages, April 2010

[9] Giambene G., Queueing theory and telecommunications, Springer, 585 pages, 2005

[10] Hager G., Wellein G., Introduction to High Performance Computing for Scientists and Engineers, CRC Press, 356 pages, 2010

[11] Hanuliak P., Analytical method of performance prediction in parallel algorithms, The Open Cybernetics and Systemics Journal, United Kingdom, pp. 38-47, 2012

[12] Hanuliak P., Hanuliak M., Performance modeling of SMP parallel computers, pp. 1-18, Int. Journal of Science, Commerce and Humanities (IJSCH), , United Kingdom, Vol. 1, No 5, pp. 243/261, July 2013

[13] Hanuliak J., Hanuliak I., To performance evaluation of distributed parallel algorithms, Kybernetes, Volume 34, No. 9/10, United Kingdom, pp. 1633-1650, 2005

[14] Hanuliak M., Hanuliak I., To the correction of analytical models for computer based communication systems, Kybernetes, Volume 35, No. 9, 1492-1504, United Kingdom, 2006

[15] Harchol-Balter Mor, Performance modeling and design of computer systems, Cambridge University Press, 576 pages, 2013

[16] Hwang K. and coll., Distributed and Parallel Computing, Morgan Kaufmann, 472 pages, 2011

[17] John L. K., Eeckhout L., Performance evaluation and benchmarking, CRC Press, 2005

[18] Kshemkalyani A. D., Singhal M., Distributed Computing, University of Illinois, Cambridge University Press, United Kingdom, 756 pages, 2011

[19] Kirk D. B., Hwu W. W., Programming massively parallel processors, Morgan Kaufmann, 280 pages, 2010

[20] Kostin A., Ilushechkina L., Modelling and simulation of distributed systems, Imperial College Press, 440 pages, Jun 2010.

[21] Kushilevitz E., Nissan N., Communication Complexity, Cambridge University Press, United Kingdom, 208 pages, 2006

[22] Kwiatkowska M., Norman G., and Parker D., PRISM 4.0: Verification of Probabilistic Real-time Systems, In Proc. 23rd Int. Conf. on CAV'11, Vol. 6806 of LNCS, Springer, pp. 585-591, 2011

[23] Le Boudec Jean-Yves, Performance evaluation of computer and communication systems, CRC Press, 300 pages, 2011

[24] McCabe J., D., Network analysis, architecture, and design (3rd edition), Elsevier/ Morgan Kaufmann, 496 pages, 2010

[25] Miller S., Probability and Random Processes, 2nd edition, Academic Press, Elsevier Science, 552 pages, 2012

[26] Misra Ch. S., Woungang I., Selected topics in communication network and distributed systems, Imperial college press, 808 pages, April 2010,

[27] Patterson D. A., Hennessy J. L., Computer Organization and Design (4th edition), Morgan Kaufmann, 914 pages, 2011

[28] Peterson L. L., Davie B. C., Computer networks – a system approach, Morgan Kaufmann, 920 pages, 2011

[29] Resch M. M., Supercomputers in Grids, Int. Journal of Grid and HPC, No.1, p 1 - 9, January- March 2009

[30] Riano l., McGinity T.M., Quantifying the role of complexity in a system's performance, Evolving Systems, Springer Verlag, pp. 189 – 198, 2011

[31] Wang L., Jie Wei., Chen J., Grid Computing: Infrastructure, Service, and Application, CRC Press, 2009.

5

Authority ring periodically load collection for load balancing of cluster system

Sharada Santosh Patil[1, 2, *], Arpita N. Gopal[2]

[1]MCA, Deptt. SIBAR Kondhwa, Pune, Maharshtra, INDIA
[2]Director MCA, SIBAR, Kondhwa,Pune, Maharashtra INDIA

Email address:

sharada_jadhao@yahoo.com(S. S. Patil), arpita.gopal.@gmail.com(A. N. Gopal)

Abstract: Now a day clusters are more popular because of their parallel processing and super computing capabilities. Generally complicated processes needs more amount of time to run on single processor but same can give quick result on clusters because of their parallel execution capabilities due to their compute nodes. Typically this performance depends on which load balancing algorithm is running on the clustered system. The parallel programming on the cluster can achieve through massage passing interface or application programming interface (API). Though the load balancing algorithm distributing load among the compute nodes of the clusters, it needs parallel programming, hence MPI library plays very important role to build new load balancing algorithm. The workload on a cluster system can be highly variable, increasing the difficulty of balancing the load across its compute nodes. This paper proposes new dynamic load balancing algorithm, which is implemented on Rock cluster and maximum time it gives the better performance as compares with previous dynamic load balancing algorithm.

Keywords: MPI, Parallel Programming, HPC Clusters, DLBA ARPLCLB, ARPLC

1. Introduction

Scheduling of processes among the compute nodes of the cluster system has always been an important and challenging area of research. The research is more challenging because of the various factors involved in implementing a load balancing algorithm in clustered system. Some of these influencing factors are the parallel workload, presence of any sequential and/or interactive jobs, native operating system, node hardware, network interface, network, and communication software. The main objective of load balancing algorithm is to speed up the system and enhance super computing power within the clustered system. There are two main types of performing load balancing – static policy and dynamic policy. [9].

1.1. Static Load Balancing Policies

Static load balancing policies judge system and application status statically and apply this information in decision making. The two renowned static policies are mentioned below.

(1). Load-dependent static policy
(2). Speed–weighted random splitting policy[8].

1.2. Dynamic Load Balancing Policies

In dynamic policies, workload is distributed among the processors at run time. New processes are assigned to the processors based on the runtime information that is collected from each node. If each node in the system becomes overloaded, the task that causes this overloading should be transferred to an under loaded node and run there. Although the dynamic policies have many benefits and can adopt with the current state of the system, sometimes they will incur extra overhead on the system because of the process migration and reservation resources of the system for collecting the information of the current status of the system. The information exchange policy can obey one of the following policies [8].

(1). Periodic policies
(2). Demand-driven policies
(3). State-change-driven policies

Load imbalance has three main sources; application imbalance, workload imbalance and heterogeneity of hardware resources. Application imbalance occurs when different parallel threads of computation take varying times to complete the super step. [20] [7]

Load balancing in the application level concentrates on minimizing the completion time of an application while load balancing in the system level is known as a distributed scheduling that is used for maximizing the throughput or utilization rate of the nodes. [1]

1.3. Centralized and Decentralized Scheme

In Centralized schemes, a central manager collects the information of each node and performs the decision making based on overall knowledge of the system. In a distributed load balancing scheme, each node makes decision based on its local knowledge. A neighbour based scheme is a distributed scheme that makes efficient load decisions without having any overall knowledge about the system.

In Decentralized scheme each node can participate in load balancing decision, but distribution of load is depend upon complete knowledge i.e. load of each node. Here, each node uses same load distribution policy. Hence accordingly it distributes load using process transfer policy. (Parimah Mohammadpour, Mohsen Sharifi, Ali Paikan-2008) (Janhavi B,Sunil Surve ,Sapna Prabhu-2010) [8][5]

1.4. Pre-Emptive and Non Pre-Emptive Load Balancing

In non pre-emptive load balancing approach, only those processes of heavily loaded node are migrated to lightly loaded node whose execution has not started yet. In other words only new born processes can be migrated, running processes cannot be migrated in this system. In pre-emptive load balancing approach processes in any state of heavily loaded nodes are migrated to lightly loaded node where the process state could be new born or running or waiting. [17][19][20].

2. The Message Passing Interface (MPI)

Message Passing Interface (MPI) is a standardized and portable message-passing system designed by a group of researchers from academia and industry to function on a wide variety of parallel computers. The standard defines the syntax and semantics of a core of library routines useful to a maximum size of users writing portable message-passing programs in Fortran 77 or the C programming language. According to R. Butler and E. Lusk P4 [2] is a third-generation parallel programming library, including both message passing and shared-memory components, portable to a great many parallel computing environments, including heterogeneous networks. Although p4 contributed much of the code for TCP/IP networks and shared-memory multiprocessors for the early versions of MPICH, most of that has been rewritten. P4 remains one of the "devices" on which MPICH can be built (see Section 41, but in most cases more customized alternatives are available.

Chameleon Written by W.D. Gropp and B. Smith [3][4][18] is a high-performance portability package for message passing on parallel supercomputers. It is implemented as a thin layer (mostly C macros) over vendor

message-passing systems (Intel's NX, TMC's CMMD, IBM's MPL) for performance and over publicly available systems (~4 and PVM) for portability. A substantial amount of Chameleon technology is incorporated into MPICH.

3. The Idea of New Research

As it has been said before, the distribution rule can be based on remote execution or in process migration. Many authors have discussed about the benefits and drawbacks of process migration for the algorithm distribution rule.

Some authors have concluded that migration cannot provide better performance than other alternatives like remote execution due to its costs, while others have argued that even when the costs of migration is high, it can significantly improve systems' performance, especially when the systems are highly variable.[5]

The dynamic load balancing algorithm mainly removes the bottle necks presented by the static co-scheduling approach thus making the cluster co-scheduling scalable. But it presents a larger communication overhead as compared to static load balancing algorithm because dynamic load balancing algorithm performs process migration. In a centralized load balancing algorithm, load is distributed uniformly among the processors. The only disadvantage is maximum time of the central processor is wasted in load balancing rather than process execution. Hence performance of the server decreases. A decentralized load balancing algorithm decision of load distribution is taken by all the node hence, each node has load of other nodes and communication overhead increases tremendously.

In order to balance the load uniformly over a cluster system, one has to choose a mix of centralized and decentralized approach.

This communication overhead and load balancing time depends upon the approach selected in the algorithm. Those approaches are given below;

The load balancing algorithm for the clusters can be made more robust by scheduling all jobs irrespective of any constraints so as to balance the load perfectly. This work is extended to develop a new algorithm that modify dynamic decentralized approach so as to reduce the communication overhead as well as reduce migration time and also make it scalable.

4. Authority Ring Periodically Load Collection for Load Balancing Algorithm (ARPLCLB)

As said in above discussion, this algorithms mix two approaches - centralized and decentralized. The authority packet is circulated among the compute nodes. Whenever system is completely imbalanced, any lowly loaded processor can pick up this authority packet and get authority to become master node. Master node is responsible to balance the system. Every compute node can broadcast load to others at certain period such that they can evaluate their current state to find whether the system is balanced or imbalanced. Hence the name of this algorithm is Authority Ring Periodically Load Collection for Load Balancing Algorithm in short it is (ARPLCLB)

This section explains overall procedure, different policies used in the algorithm, Data structure used to build algorithm, and parallel algorithm

4.1. Overall Procedure

The Overall Procedure of this algorithm is given below;

Step 1: After completion of every ring period, every processor passes or broadcasts information packet to all processor which consists of;

1. Current status of the node
2. Current load of the node with load factor

Step 2: Every processor can store the current information as well as past information of all the processors.

Step 3: Every processor collects authority packet from previous processor and circulate it to next processor.

Step 4: When any Idle node or Low load node get authority packet then immediately it take the charge of master node and performs following activities;

1. Create workload Distribution table using following process criteria;
   a. It chooses newly arrived processes. (That means new born Processes)
   b. It chooses processes which needs 80 % time for execution.
2. Create Order packets according to Workload Distribution table
3. Send order packet of all node to that appropriate node.
4. After load distribution send authority packet to next node.

Step 5: As soon as any node gets order packet they should follow the order of order packet to perform process migration.

Step 6: Master node again starts authority ring means authority packet is circulated to each node of the LAN one by one again.

Step 7: Repeats Steps 1 to 7 till cluster is not shut down

5. Policies Used in ARPLCLB Algorithm

Following Policies used in proposed dynamic load balancing algorithm.

5.1. Load Information Policy

Load information serves as one of the most fundamental elements in the load balancing process. Every dynamic load balancing algorithm is based on some type of load information. According to this algorithm, each load of cluster system has some current state. These CPU states can be idle, lowly loaded, normal or heavily loaded CPU.

1. The CPU state can be idle state, if ready queue is empty

and it is not executing any process and hence 100 % memory is available.

$$\sum_{i=0}^{Qtotal} Pi \simeq 0 \ \& \ MEMfree \simeq 100\%$$

2. The CPU state can be low state, if total number of processes < (less than) LOW_LOAD_THRESHOLD_VALUE (L) * size of queue (Q_s) and more than 75 % memory (Memory$_{free}$) is available.

$$\sum_{i=0}^{Qtotal} Pi \leq L * Qs \ \& \ 75\% \leq MEMfree$$

3. The CPU state can be normal state if total number of processes < (less than) NORMAL_LOAD_THRESHOLD_VALUE (N) * size of queue and 25 % to 75% memory is available.

$$\sum_{i=0}^{Qtotal} Pi \leq N * Qs \ \& \ 25\% \leq MEMfree < 75$$

4. The CPU state can be heavy state if total number of processes > (greater than) NORMAL_LOAD_THRESHOLD_VALUE * Size of Queue and less than 25% memory is available.

$$\sum_{i=0}^{Qtotal} Pi \leq N * Qs \ \& \ MEMfree < 25$$

5. The system balance depends on following criteria;
* When all nodes are heavily loaded then system can be called as heavily balanced system.
* When heavily loaded nodes are 1% to 85% and remaining are lowly loaded or idle or normal processors then system is imbalanced and need process migration
* When no node is heavily loaded and may be idle or lowly loaded or normal loaded then system is called slightly balanced or slightly imbalanced system, and it does not require any process migration.

5.2. Information Exchange Policies of ARPLCLB

This information exchange policy depends on how node can be exchange load information with others. This algorithm uses periodic policy means it exchanges this load information after every certain number of time slot. This information policy is executed by all nodes.

5.3. Process Transfer Policies of ARPLCLB

Process can be transfer from heavily loaded processor to idle or lowly loaded or normally loaded processor. This policy is executed on master node. This policy can be mentioned as follows;

1. System is heavily balanced if all CPU in the clusters are heavily loaded then it execute delay of 1000 such that all CPU can execute their load to get relief from authority token ring.

2. When system is slightly balanced or slightly imbalanced, then system is in normal condition

3. When system is completely imbalanced then this algorithm decides decision of process migration.

4. For process transfer activity , it calculates the ideal load of each processor as follows;

$$Ideal\_load = \frac{Total\_System\_load}{Total\_Computenode\_cluster}$$

5. It transfers total ideal load processes from heavily loaded processor to lightly loaded processor

5.4. Selection Policies of ARPLCLB

A selection policy decides which process is selected for transfer that means process migration. The chosen process could be a new process which has not started, that means, new born process or an old process which is already starting its execution. If it is old process then it should satisfy following condition.

1. If ((rbt/bt*100>=80) Process is selected for migration.

$$\frac{Re\,maining\_Burst\_Time}{Burst\_Time}*100 >= 90?: Pr\,ocess\_Migration$$

This selection policy is executed by master node.

5.5. Location Policies of ARPLC

Location policy decides selected processes for migration is migrated to which CPU For this activity , it select ideal CPU to migrate processes from heavily loaded CPU to lightly loaded CPU using following steps

1. Select heavy loaded CPU
2. Select first idle CPU
   a. If found go to 3 else b
   b. Select first low loaded CPU
      i. If found go to step 3 else ii
      ii. Select first normal loaded CPU
3. Select process for migration to selected CPU
4. Update load of that CPU
5. Repeat 3 and 4 till Load of CPU < ideal load of the system

This location policy is executed by master node.

6. Parallel Sub Algorithms of ARPLCLB)

This algorithm ARPLCLB is divided in to three parallel sub algorithms, that are;

1. Authority token ring with Periodically Load collection algorithm(ARPLC)
2. Load Distribution With Order Packet Algorithm(LDOP)
3. Process Migration Algorithm(PM)

6.1. Authority Token Ring with Periodically Load Collection Algorithm (ARPLC)

This algorithm is similar to musical ball passing game. Similar to musical ball authority packet is moved around the logical ring of the compute nodes of the cluster. This algorithm circulates authority tokens between the processors such that they can choose their new master node when cluster system is imbalanced. This algorithm collects load information periodically. The master node CPU can be responsible to start authority ring, in this it passes authority packet to next nearby neighbor. When system is imbalanced then any lowly loaded or idle node picks up this authority packet to become new master node. This node conveys this information to all using new master indication packet. This is explained in following figure 1

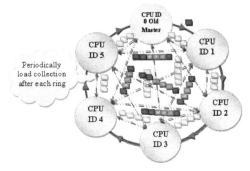

Figure 1: *Authority token Ring with Periodically Load collection Algorithm (ARPLC)*

6.2. Load Distribution with Order Packet Algorithm (LDOP)

When cluster system is imbalanced and new master node is decided by ARPLC parallel sub algorithm, then this new master node has load information of each node. Hence it decides load distribution using load information and distribution policy, process selection and location policy, accordingly creates order packets, and distributes or broadcasts this order packet to each node, and calls process migration algorithm which follows order of master node blindly. Maximum part of this LDOP algorithm is executed on master node and very minimum part of algorithm is executed on other compute node. This is explained in following figure 2.

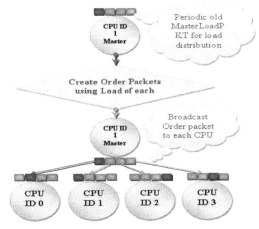

Figure 2: *Load Distribution with Order Packet Algorithm (LDOP)*

6.3. Process Migration of Algorithm (PM)

Once order packet is distributed by the new master node among all other compute nodes of the cluster then other compute nodes are automatically divided in to heavy load CPU group and idle or low load CPU group. And heavy loaded CPU transfers their own load to lightly loaded CPU. This is explained in figure 3. The load is transferred using two types of the packets , that means process control block packet whose size is fixed and then it transfers actual process to destination CPU such that it can be easily start its remaining execution on new processor.

When process migration is over it again execute the ARPLC parallel sub algorithm to monitor system imbalance and distribute authority packet among the compute nodes

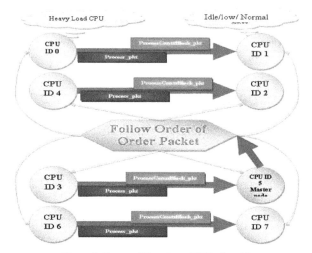

Figure 3: *Process Migration of (PM) Algorithm*

As it has been said before, the distribution rule can be based on remote execution or in process migration

7. Performance of the Authority Ring with Periodically Load Collection Algorithm

Table 6.3: *Performance of Algorithm1 (ARPLCLB)*

Iteration of Outer Loop	Total Process Migration	Total Number of Rings	New Master	Old master
1	9	2	2	0
2	9	1	2	2
3	8	6	2	2
4	8	2	0	2
5	8	1	0	0
6	8	4	3	0
7	9	4	0	3
8	8	4	3	0
9	8	1	0	3
10	8	4	3	0

For evaluating the performance of the above algorithm we implement algorithm run it on Rock cluster on centos operating system then it produces some output in data file. These data files are too lengthy hence only some result are given, following figure shows screen shot of the same during the execution.

Figure 4: *Screen shots of ARPLCLB(a)*

Figure 4: *Screen shots of ARPLCLB(b)*

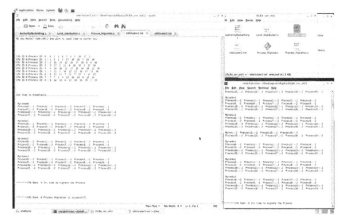

Figure 4: *Screen shots of ARPLCLB(c)*

The authority rings continues, means system is currently in balanced state. When system is in imbalanced stage, then it performs process migration. The graph of outer loop iteration against process migration is given in figure 5;

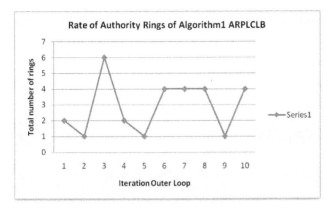

Figure 5: Performance of ARPLCLB using Process migration

As result shows, this algorithm migrates maximum number of processes in each number of iterations of the loop. Hence Maximum time of the CPU is wasting to perform process migration rather than process execution. Hence it is degradation of the system.

The graph shows total authority rings verses iteration of outer loop. When total authority rings are more means system is currently in balanced state.

The graph shows total authority rings verses iteration of outer loop. When total authority rings are more means system is currently in balanced state.

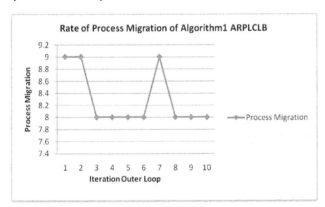

Figure 5: Performance of Algorithm 1 Using Total Rings (ARPLCLB)

The result of above graph states that every iteration of the ring migrates to too much processes. And this algorithm uses periodically load collection policy, hence, it uses too much communication overhead. The advantages and disadvantages of the algorithm are given below;

8. Conclusion

In order to balance the load uniformly over a cluster system, our proposed algorithm has used a mix of centralized, decentralized, approach. The performance of this algorithm gives better result many a time but due to heavy communication overhead and heavy process migration, affect the performance. All previous algorithms may either use centralized approach or decentralized approach. But this proposed algorithm uses both approaches. Hence in future work is extended to improve the result of

this algorithm. The load balancing algorithm for the clusters can be made more robust by scheduling all jobs irrespective of any constraints so as to balance the load perfectly

9. Future Enhancement

This work is extended to remove all disadvantages of this proposed algorithm. As well as policies used in this algorithm is also improved. In future, this work can be extended to develop a two new dynamic load balancing algorithm to modify dynamic decentralized approach so as to reduce the communication overhead as well as to reduce migration time and also make it scalable

Acknowledgment

We sincerely express our thanks to laboratory research cell of SIBAR-MCA and SKNCOE for their full support. We are very grateful to all friends who have directly and indirectly supported to this research work

References

[1] Bernd F reisleben Dieter Hartmann Thilo Kielmann "Parallel Raytracing A Case Study on Partitioning and Scheduling on Workstation Clusters" 1997 Thirtieth Annual Hawwaii International Conference on System Sciences.

[2] Blaise Barney, (1994) Livermore Computing, MPI Web pages at Argonne National Laboratory http://www-unix.mcs.anl.gov/mpi "Using MPI", Gropp, Lusk and Skjellum. MIT Press

[3] Erik D. Demaine, Ian Foster,Carl Kesselman, and Marc Snir. "Generalized Communicators in the Message Passing Interface" 2001 IEEE transactions on parallel and distributed systems pages from 610 to 616.

[4] Hau Yee Sit Kei Shiu Ho Hong Va Leong Robert W. P.Luk Lai Kuen Ho" An Adaptive Clustering Approach to Dynamic Load balancing" 2004 IEEE 7th International Symposium on Parallel Architectures, Algorithms and Networks (ISPAN'04)

[5] Janhavi B,Sunil Surve ,Sapna Prabhu "Comparison of load balancing algorithms in a Grid" 2010 International Conference on Data Storage and Data Engineering Pages from 20 to 23.

[6] M. Snir, SW. Otto, S. Huss-Lederman, D.W. Walker and J. Dongarra,(1996) MPI: The Complete Reference (MIT Press, Cambridge, MA, 1995). 828 W. Gropp et al./Parallel Computing 22 (1996) 789-828.

[7] Marta Beltr´an and Antonio Guzm´an "Designing load balancing algorithms capable of dealing with workload variability" 2008 International Symposium on Parallel and Distributed Computing Pages from 107 to 114.

[8] Parimah Mohammadpour, Mohsen Sharifi, Ali Paikan," A Self-Training Algorithm for Load Balancing in Cluster Computing", 2008 IEEE Fourth International Conference on Networked Computing and Advanced Information Management , Pages from 104 to 110.

[9]   Paul Werstein, Hailing Situ and Zhiyi Huang „Load Balancing in a Cluster Computer" 2006 Proceedings of the Seventh International Conference on Parallel and Distributed Computing, Applications and Technologies.

[10] Sharada Patil, Dr Arpita Gopal,[2012], Ms Pratibha Mandave "Parallel programming through Message Passing Interface to improving performance of clusters " – International Docteral Conference (ISSN 0974-0597) SIOM, Wadgoan Budruk in Feb 2013.

[11] Sharada Patil,Arpita Gopal "Comparison of Cluster Scheduling Mechanism using Workload and System Parameters" 2011 ISSN 0974-0767 International journal of Computer Science and Application.

[12] Sharada Patil,Arpita Gopal "STUDY OF DYNAMIC LOAD BALANCING ALGORITHMS FOR LINUX CLUSTERED SYSTEM USING SIMULATOR" 2011 ISSN 0974-3588 International journal of Computer Applications in Engineering Technology and Sciences.

[13] Sharada Patil,Dr Arpita Gopal, [2011] "Study of Load Balancing ALgorithms" – National Conference on biztech 2011, Selected as a best paper in the conference, got first rank to the paper,DICER, Narhe ,Pune in year March 2011.

[14] Sharada Patil,Dr Arpita Gopal, [2013] "Cluster Performance Evaluation using Load Balancing Algorithm" – INTERNATIONAL CONFERENCE ON INFORMATION COMMUNICATION AND EMBEDDED SYSTEMS ICICES   2013,978-1-4673-5788-3/13/$31.00©2013IEEE (ISBN 978-1-4673-5786-9) Chennai, India in year Feb 2013.

[15] Sharada Patil,Dr Arpita Gopal,[2012] "Need Of New Load Balancing Algorithms For Linux Clustered System" – International Conference on Computational techniques And Artificial   intelligence   (ICCTAI'2012)   (ISBN 978-81-922428-5-9) Penang Maleshia in year Jan 2012.

[16] Sharada   Patil,Dr   Arpita   Gopal,[2013]   "Enhancing Performance   of   Business   By   Using   Exctracted Supercomputing   Power   From   Linux   Cluster's   " – International Conference on FDI 2013 (ISSN 0974-0597) SIOM, Wadgoan Budruk in Jan 2013

[17] Sun Nian1, Liang Guangmin2 "Dynamic Load Balancing Algorithm for MPI Parallel Computing" 2010 Pages 95 to 99

[18] William Gropp, Rusty Lusk, Rob Ross, and Rajiv Thakur (2005)   "MPI   Tutorials   "   Retrieved   from www.mcs.anl.gov/research/projects/mpi/tutorial Livermore Computing specific information:

[19] Yanyong Zhang, Anand Sivasubramaniam, JoseÂ Moreira, and Hubertus Franke" Impact of Workload and System Parameters on Next Generation Cluster Scheduling Mechanisms" 2001 IEEE transactions on parallel and distributed systems Pages from 967 to 985.

[20] Yongzhi Zhu Jing Guo Yanling Wang "Study on Dynamic Load Balancing Algorithm Based on MPICH" 2009 MPI_COMM_RANK: World Congress on Software Engineering. Pages from 103 to 107.

6

Cryptography: salvaging exploitations against data integrity

A. A. Ojugo[1], A. O. Eboka[2], M. O. Yerokun[2], I. J. B. Iyawa[2], R. E. Yoro[3]

[1]Department of Mathematics/Computer, Federal University of Petroleum Resources Effurun, Delta State
[2]Department of Computer Sci. Education, Federal College of Education (Technical) Asaba, Delta State
[3]Department of Computer Science, Delta State Polytechnic Ogwashi-Uku, Delta State

Email address:

ojugo_arnold@yahoo.com(A. A. Ojugo), maryarnoldojugo@gmail.com(A. A. Ojugo), an_drey2k@yahoo.com(A. O. Eboka), agapenexus@hotmail.co.uk(M. O. Yerokun), iyawaben@hotmail.com(I. J. B. Iyawa), rumerisky@yahoo.com(R. E. Yoro)

Abstract: Cryptography is the science and art of codes that makes it possible for two people to exchange data in such a way that other people cannot understand the message. In this study – we are concerned with methods of altering data such that its recipient can undo the alteration and discover the original text. The original text is called plaintext (PT) while altered text is ciphertext (CT). Conversion from PT to CT is called encoding/enciphering as codes that result from this process are called ciphers. The reverse operation is called decoding/deciphering. If a user tries to reverse the cipher by making meaning of it without prior knowledge of what method is used for encoding as the data was originally, not intended for the user, the process is called cracking, while such a user is a called a cryptanalyst. Cryptography is about communicating in the presence of an adversary (cryptanalyst) – and it embodies problems such as (encryption, authentication, key distribution to name a few). The field of cryptography and informatics provides a theoretical foundation based on which we may understand what exactly these problems are, how to evaluate protocols that purport to solve them, and how to build protocols in whose security we can have confidence. Thus, cryptography is the only practical means of sending and receiving data over an insecure channel from source to destination in such a way that other users cannot understand the message unless it was intended for them. Data sent over public network is not safe and the more ciphertext a cryptanalyst has, the easier it is to crack the ciphers. Thus, it is good to change the coding mechanism regularly – because, every coding scheme has a key set.

Keywords: Ciphertext, Plaintext, Encryption, Decryption, Pseudo-Random Numbers, Ciphers

1. Introduction

Information as a veritable tool for decision-making has been an integral part of our society and its transfer has led to advancements in data processing activities with advent of information and communication technology (ICT) devices. Conceptually, how data is recorded has not changed over-time (from paper to electronic). The dramatic change has been in the way it is copied and altered. Originally, paper data can be copied but its original can be distinguished from duplicates. With electronic data, it is impossible to distinguish original from copies [5]. Advent of network to ease data transfer has raised security alerts – though Internet was first used in military applications, but later employed in other facets to allow the sharing of expensive hardware and software resources. Now, data integrity and security has become a threat due to e-initiatives (such as e-Commerce, e-banking etc). These changes have upgraded security alert as well as its consciousness in users – as there are always intruders whose job, is to steal data. The use of Internet for data transfer has demerits that have led to the field of *Data Security* and *Cryptography* [5, 6].

Cryptography is the science and art of cipher/codes that allows two users to exchange data in such a way that other users cannot understand – through the use of data altering schemes such that only an intended recipient can undo and discover the original text sent by the sender. The original text is a plaintext (PT) while altered text is called ciphertext (CT). The conversion from PT to CT is encoding; while its reconversion is decoding. An unintended user that tries to undo the change without prior knowledge of encoding/decoding method used is a cryptanalyst – and the process of undoing alterations is *cracking* [1].

Cryptography is the only computationally secure and

practical means of transferring data over an insecure channel in such a way that other users cannot understand the message unless it was intended for them. Data sent over public network is not safe and the more ciphertext a cryptanalyst has, the easier it is to crack. Thus, it is a good idea to change the coding scheme regularly – as every coding scheme has a key set. If a different key-set is used daily, there may never be enough ciphertext to decode sent data. Though effective, its demerits is that the user also has to device a means of generating new keywords and making sure such keywords are sent over secure network to the intended recipient [1]. Cryptography is as old as writing but until the advent of computers – a major constraint in cryptography (as used during the war) has been the ability of the code clerk to perform needed transformations and switch from one cryptographic method to another (on battlefield) with little equipment, as it entails retraining a large number of persons. There was danger of the code clerk being captured, making it essential to change the cryptographic method instantly as the need arose [2].

System and elements used in these processes that make it impossible for a cryptanalyst to deduce meaningful data is called a *cryptosystem* – such that in encrypting, the transmitted data and encryption key is fed into the encryption algorithm before being sent so that on arrival of the data at its destination, the recipient passes the data via a decryption algorithm in order to have access to the transmitted data [5].

The goals of cryptography includes to aid data confidentiality and privacy, integrity, authentication and non-repudiation – all services that prevents an entity from denying previous actions taken. All these are resolved via the use of digital signatures and some tools as in fig. 2 below, which are evaluated based on: (a) Security Level – is upper bound on the amount of work necessary to defeat the objective, (b) Functionality – primitives can be combined to meet various security goals, (c) Operation – tools can be applied in various ways with various input to exhibit different characteristics and different primitives help provides different functionality depending on its mode of operation, (d) Performance – is the efficiency of the primitive's mode of operation and the number of bits per seconds that it can encrypt, (e) Key Space Size – a set of keys generated and sent over secure line for encryption and decryption and (f) Implementation – difficulty of implanting its complexity in software and/or hardware environment [4].

The importance of these criteria is much dependent on the application and resources available that in some cases, there are tradeoffs of high-level security for better performance/memory. Fig. 2 are primitives and types of ciphers available, generally classified into *symmetric* and *public* key. Symmetric key ciphers include *block* (substitution, transposition and product) and *stream* ciphers.

1.2. Substitution Ciphers

In a substitution cipher, one character is substituted for another. Here is a simple example:

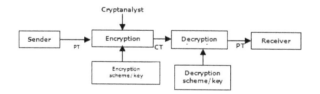

Figure 3 shows the encoding key set E

1.1. Cryptographic Goals and Tools

Figure 1: A Cryptographic System

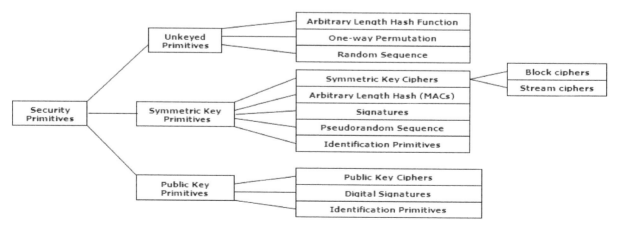

Fig. 2: Primitive tools for an encryption model

To encode data, we use E as in fig. 3 by selecting a character from upper line, and replacing it with those characters oppositely below – so that we encode the word "FORE-CAST" to be "KNOQBRXD". To decode, the user reverse the process by taking "K" in the lower line, and finding its matching letter above to get "F", which is same as first letter

– and so on. If the intended user has a lot of data to decode, it is easier to invert fig. 3 to get fig. 4, making it easier to decode.

```
    A B C D E F G H I J K L M N O P Q R S T U V W X Y Z
E = ----------------------------------------------------------------
    W C H T K G P X V U F M Z O R I E A N Y D L Q S J B
```

Fig. 4: *Inverted key E rearranged with their corresponding opposite keys*

Substitution ciphers are easy to crack since English (or any other language) have certain letters that appear more frequently. A list of English letters approximately in percentage order of usage are: E: 12.7%, T: 9.1%, A: 8.2%, O: 7.5% and others are J: 0.2%, Q: 0.1%, Z: 0.1%. Thus, E T A O I N S H R D L U C M W F G Y P B V K X J Q Z.

Take the short passage that follows:

NOT NUA JMPETZ UTST ZENNELV EL NOT JE-GELV SAAW, UMENELV KAS NOTES OAZNTZZ, UOA UMZ ZJEVONJX PTJMXTP. NOT PMCVONTS AK NOT KMWEJX UMZ UENO NOTW, AL NOT NO-TASX NOMN ZOT UACJP QTTY NOT GEZENASZ ADDCYETP PCSELV NOT UMEN. NOT DOEJP UMZ YTSOMYZ ZEI XTMSZ AJP, ZLCR LAZTP, RCDQ NAANOTP MLP RTZYTDNMDJTP. ZOT WMELN-MELTP M PTTY ZEJTLDT MLP NOT NUA JMPETZ YTTSTP PACRNKCJJX MN OTS. KELMJJX, ALT AK NOTW WCNNTSTP NA NOT ANOTS, "LAN GTSX Y-S-T-N-N-X, E KTMS," DMSTKCJJX ZYTJJELV NOT QTX UASP. UOTSTCYAL NOT DOEJP YEYTP CY, "RCN MUKCJ Z-W-M-S-N!"

The character frequencies for the passage above are shown in figure 5 below:

A	B	C	D	E	F	G	H	I	J	K	L	M	N	O	P	Q	R	S	T	U	V	W	X	Y	Z
25	0	14	10	28	0	3	0	1	23	10	18	25	44	31	24	4	5	22	60	14	7	7	11	13	23

To cryptanalyse the sample text, we start by guessing that: (1) Letter E and T is represented by either "T" or "N" – as they are two most frequently occurring letters and (2) O is either A or H. With the substitutions, we have the text below:

```
EAT E      T    TT  EE     EAT            E        EAT
THE T      E    EE  TT     THE            T        THE
NOT NUA JMPETZ UTST ZENNELV EL NOT JEGELV SAAW, UMENELV KAS NOTES

A  ET  A          E  T    T    EAT      ET     EAT      EA
H  TE  H          T  E    E    THE      TE     THE      TH
OAZNTZZ, UOA UMZ ZJEVONJX PTJMXTP. NOT PMCVONTS AK NOT KMWEJX UMZ UENO

EAT     EAT EAT   EA    AT       TT EAT   E            T
THE     THE THE   TH    HE       EE THE   T            E
NOTW, AL NOT NOTASX NOMN ZOT UACJP QTTY NOT GEZENASZ ADDCYETP PCSELV

EAT    E. EAT       T      T          T        E   EA
THE    T. THE       E      E          E        T   TH
NOT UMEN. NOT DOEJP UMZ YTSOMYZ ZEI XTMSZ AJP, ZLCR LAZTP, RCDQ NAANOTP MLP

T  TE   T    AT E   T     TT   T     EAT E    T  TT T
E  ET   E    HE T   E     EE   E     THE T    E  EE E
RTZYTDNMDJTP. ZOT WMELNMELTP   M   PTTY ZEJTLDT MLP NOT NUA JMPETZ YTTSTP

E      E AT       T    EAT    EET T  E  EAT   EAT
T      T HE       E    THE    TTE E  T  THE   THE
PACRNKCJJX MN OTS. KELMJJX, ALT AK NOTW WCNNTSTP NA NOT ANOTS,

"  E   T    -T-E-E-   T      T      T      EAT  T     AT T    EAT
"  T   E    -E-T-T-   E      E      E      THE  E     HE E    THE
"LAN GTSX Y-S-T-N-N-X, E KTMS," DMSTKCJJX ZYTJJELV NOT QTX UASP. UOTSTCYAL NOT

A     T    E        -E!"
H     E    T        -T!"
DOEJP YEYTP CY, "RCN MUKCJ Z-W-M-S-N!"
```

"THE" appears more in English than "EAT" – only a guess. Now: (1) Line 5, only A or I stands alone in English and A appears more than I. So "M" is A, (2) Line 6, two-word that starts E "EX/EN" or starts with T is "TO". TO appears more, and (3) Line 6, word OTHE_ is missing R – so "S" is letter R. Thus, having:

```
THE T  O   A   E ERE  TT    THE           ROO      A  T  R THE R
NOT NUA JMPETZ U TST ZENNELV EL NOT JEGELV SAAW, UMENELV KAS   NOTES

HO TE    HO   A   T    E AE. THE  A  TER   THE   A   A     TH
OAZNTZZ, UOA UMZ ZJEVONJX PTJMXTP. NOT PMCVONTS AK NOT KMWEJX UMZ UENO
```

```
THE    O THE THE R   THAT    HE        EE THE      T R    E     R
NOTW, AL NOT NOTASX   NOMN   ZOT   UACJP QTTY   NOT   GEZENASZ ADDCYETP PCSELV

THE    A T. THE         A    ERHA          EAR E      T  TH    A
NOT  UMEN. NOT DOEJP UMZ YTSOMYZ ZEI XTMSZ AJP, ZLCR LAZTP, RCDQ NAANOTP MLP

E    E TA    E HE A TA  E  A   EE     E      A  THE T    A  E  EERE
RTZYTDNMDJTP. ZOT WMELNMELTP    M  PTTY ZEJTLDT MLP   NOT NUA JMPETZ YTTSTP

T    AT H    A    E  THE       TTERE   TO THE   OTHER,
PACRNKCJJX MN OTS. KELMJJX, ALT AK NOTW WCNNTSTP NA   NOT   ANOTS,

"  T  E  -R-E-T-T-    EAR   ARE        E    THE    E   R.  HERE     THE
"LAN GTSX Y-S-T-N-N-X, E KTMS," DMSTKCJJX ZYTJJELV NOT   QTX UASP. UOTSTCYAL NOT

H    E     T A        -A-R-T!"
DOEJP YEYTP CY, "RCN   MUKCJ Z-W-M-S-N!"
```

Also: (1) Line 1, "_ERE" is missing W. Replace "U" with W, (2) Line 1, "ROO_" is missing M, so replace "W" with M, and (3) Line 3, a two-word starting with O are OF, ON. "L" is either F or N. Replace L as N on line 4 in the triple word "AN_" to give AND. Thus, "L" is N and "P" is letter D.

```
THE TWO  AD E   WERE  TT N  N  THE      N   ROOM, WA  T N  OR   THE R
NOT NUA JMPETZ  UTST ZENNELV EL   NOT JEGELV   SAAW, UMENELV KAS    NOTES

HO TE  , WHO WA       HT   E A ED. THE DA    HTER O THE    AM   WA  W TH
OAZNTZZ, UOA   UMZ ZJEVONJX PTJMXTP.  NOT PMCVONTS AK NOT   KMWEJX UMZ   UENO

THEM, ON  THE THEOR  THAT  HE  WO  D EE THE      TOR  O    ED  D R N
NOTW, AL   NOT NOTASX NOMN ZOT   UACJP QTTY NOT GEZENASZ ADDCYETP   PCSELV

THE  WA T. THE    D WA   E A       EAR O D N   O ED,    T  THED  AND
NOT UMEN. NOT DOEJP  UMZ YTSOMYZ ZEI XTMSZ AJP, ZLCR LAZTP, RCDQ NAANOTP   MLP

E  E TA   ED HE MA  NTA NED A DEE    EN   AND THE TWO   A   E EERED
RTZYTDNMDJTP. ZOT WMELNMELTP   M  PTTY ZEJTLDT MLP   NOT NUA  JMPETZ YTTSTP

T      AT HER    NA    NE O  THEM M   TTERED TO THE  OTHER,
PACRNKCJJX MN  OTS. KELMJJX, ALT AK NOTW WCNNTSTP  NA  NOT ANOTS,

"N  T  E  -R-E-T-T-    EAR," A E     E    THE    E   WORD. WHERE  ON THE
"LAN GTSX Y-S-T-N-N-X, E KTMS," DMSTKCJJX ZYTJJELV NOT   QTX  UASP.  UOTSTCYAL NOT

H    E     T A       -M-A-R-T!"
DOEJP YEYTP CY, "RCN   MUKCJ  Z-W-M-S-N!"
```

Again we can see that in: (1) Line 2, the last two words (triple and quad) (UMZ/UENO) are WA_ and W_TH. Replace "Z" with S and "E" with I. (2) Line 5, replace "Y" with P and (3) Line 7, replace "A" with O. Thus:

```
THE TWO  ADIES  WERE  SITTIN IN  THE     I IN   ROOM, WAITIN  OR    THEIR
NOT NUA JMPETZ  UTST ZENNELV EL   NOT JEGELV   SAAW, UMENELV KAS    NOTES

HOSTESS  WHO WAS S I  HT   DE A ED. THE DA   HTER O THE    AMI    WAS WITH
OAZNTZZ, UOA   UMZ ZJEVONJX PTJMXTP.  NOT PMCVONTS  AK  NOT   KMWEJX UMZ   UENO

THEM, ON THE THEOR  THAT SHE  WO  D EEP THE    ISITORS  O    PIED D  RIN
NOTW, AL NOT NOTASX NOMN ZOT   UACJP QTTY  NOT GEZENASZ ADDCYETP   PCSELV

THE WAIT. THE    I D  WAS  PERHAPS  I  EARS  O D, SN   NOSED,       TOOTHED  AND
NOT UMEN. NOT DOEJP  UMZ  YTSOMYZ ZEI XTMSZ  AJP,  ZLCR LAZTP,  RCDQ NAANOTP   MLP

ESPE TA   ED. SHE MAI NTAINED   A  DEEP SI  EN E AND THE  TWO   ADIES PEERED
RTZYTDNMDJTP. ZOT WMELNMELTP   M  PTTY ZEJTLDT MLP   NOT NUA   JMPETZ YTTSTP

O  T    AT HER   INA   ONE O  THEM M   TTERED TO  THE  OTHER,
PACRNKCJJX MN  OTS. KELMJJX, ALT  AK NOTW WCNNTSTP  NA  NOT ANOTS,
"NOT  ER  P-R-E-T-T- I  EAR,"   ARE     PE  IN THE    E  WORD. WHERE PON  THE
"LAN GTSX Y-S-T-N-N-X, E KTMS," DMSTKCJJX ZYTJJELV NOT   QTX  UASP.  UOTSTCYAL    NOT

HI D PIPED P   T A        S-M-A-R-T!"
DOEJP YEYTP CY, "RCN   MUKCJ  Z-W-M-S-N!"
```

We see also in: (1) Line 1, replace "V" with G and "J" with L, (2) Line 2, replace "X" with Y, "C" letter U and "K" with F, (3) Line 3, replace "G" with V and "Q" with K, and (4) Line 4, replace "I" with X and "D" with C to give us the test below.

```
THE TWO LADIES  WERE SITTING IN   THE   LIVING   ROOM, WAITING FOR   THEIR
NOT NUA JMPETZ  UTST ZENNELV EL   NOT JEGELV   SAAW, UMENELV KAS   NOTES

HOSTESS  WHO WAS SLIGHTLY DELAYED.  THE DAUGHTER  OF  THE   FAMILY  WAS  WITH
OAZNTZZ, UOA   UMZ ZJEVONJX PTJMXTP.  NOT PMCVONTS  AK   NOT   KMWEJX UMZ  UENO

THEM, ON THE THEORY THAT  SHE  WOULD KEEP  THE  VISITORS  OCCUPIED  DURING
NOTW, AL NOT NOTASX NOMN ZOT   UACJP  QTTY  NOT GEZENASZ ADDCYETP  PCSELV

THE WAIT.  THE  CHILD  WAS  PERHAPS SIX  YEARS  OLD, SNU  NOSED,   UCK  TOOTHED AND
NOT UMEN. NOT DOEJP  UMZ  YTSOMYZ ZEI  XTMSZ AJP,  ZLCR  LAZTP, RCDQ NAANOTP  MLP

ESPECTACLED. SHE MAINTAINED   A  DEEP SILENCE AND  THE  TWO  LADIES  PEERED
RTZYTDNMDJTP. ZOT WMELNMELTP  M   PTTY ZEJTLDT MLP  NOT  NUA  JMPETZ  YTTSTP

DOU TFULLY AT  HER.  FINALLY, ONE OF THEM  MUTTERED  TO  THE   OTHER,
PACRNKCJJX MN  OTS. KELMJJX, ALT  AK NOTW WCNNTSTP  NA  NOT  ANOTS,

"NOT VERY P-R-E-T-T-Y, I  FEAR,"  CAREFULLY SPELLING THE   KEY WORD. WHEREUPON THE
"LAN GTSX Y-S-T-N-N-X, E KTMS," DMSTKCJJX ZYTJJELV  NOT   QTX  UASP.  UOTSTCYAL  NOT

CHILD PIPED UP, "  UT   AWFUL   S-M-A-R-T!"
DOEJP YEYTP CY, "RCN   MUKCJ   Z-W-M-S-N!"
```

Having come thus far, we make one last substitution by: replacing "R" with letter B.
```
THE TWO LADIES  WERE SITTING IN  THE  LIVING  ROOM, WAITING FOR   THEIR
NOT NUA JMPETZ  UTST ZENNELV EL  NOT JEGELV  SAAW, UMENELV KAS    NOTES

HOSTESS  WHO WAS SLIGHTLY DELAYED.  THE DAUGHTER  OF  THE   FAMILY  WAS  WITH
OAZNTZZ, UOA   UMZ ZJEVONJX PTJMXTP.  NOT PMCVONTS  AK  NOT   KMWEJX UMZ  UENO


THEM,  ON THE THEORY  THAT  SHE  WOULD KEEP  THE  VISITORS  OCCUPIED  DURING
NOTW, AL  NOT NOTASX NOMN ZOT   UACJP  QTTY  NOT GEZENASZ ADDCYETP  PCSELV

THE WAIT.  THE  CHILD  WAS  PERHAPS SIX  YEARS  OLD, SNUB NOSED, BUCK  TOOTHED  AND
NOT UMEN. NOT DOEJP  UMZ  YTSOMYZ ZEI  XTMSZ AJP,  ZLCR  LAZTP,  RCDQ NAANOTP  MLP

BESPECTACLED.  SHE  MAINTAINED A  DEEP SILENCE AND  THE  TWO  LADIES  PEERED
RTZYTDNMDJTP. ZOT WMELNMELTP  M   PTTY ZEJTLDT MLP  NOT  NUA  JMPETZ  YTTSTP

DOUBTFULLY AT HER.  FINALLY, ONE OF THEM  MUTTERED  TO  THE  OTHER,
PACRNKCJJX MN  OTS. KELMJJX, ALT  AK NOTW WCNNTSTP  NA  NOT  ANOTS,

"NOT VERY P-R-E-T-T-Y, I  FEAR,"  CAREFULLY SPELLING THE KEY WORD. WHEREUPON THE
"LAN GTSX Y-S-T-N-N-X, E KTMS," DMSTKCJJX ZYTJJELV NOT QTX UASP.  UOTSTCYAL  NOT

CHILD PIPED UP, "BUT AWFUL  S-M-A-R-T!"
DOEJP YEYTP CY, "RCN MUKCJ Z-W-M-S-N!"
```

From the above sample data, the key E is as seen in figure 6, though letters Z, J and Q are used as thus:

```
   A B C D E F G H I J K L M N O P Q R S T U V W X Y Z
E = -----------------------------------------------------
   M R D P T K V O E F Q J W L A Y B S Z N C G U I X H
```

Fig. 6: *Key E for example above*

This cipher is simple. [3] notes to secure such ciphers are:

a. Punctuations and separators can be encoded with alternative symbols.

b. Permutation simply breaks up data into equal letter chunk(s), adding junk letters where necessary, to extend and make equal the last chunk – so that each chunks encodes itself. For example, 21-letter chunk "THISCIPHERIS-NOTSECURE" is broken into 25-letter, extended with

AAAA in 5X5 matrix. Thus, a user reads text by column instead of row as: TIISEHPSEAIHNCASEOUACRTRA.

Table 1: 5 x 5 matrix of letters for permutation

T	H	I	S	C
I	P	H	E	R
I	S	N	O	T
S	E	C	U	R
E	A	A	A	A

a. Using 127 different symbols for "E", 91 for "T" etc. A cryptanalyst notices equally common symbols – making it subject to frequency attack if he/she has access to lots of text and letter pairs.

b. A user can also have codes for letter pairs and common-words that help make the task still more difficult. Its demerit is that with more combinations, the cipher becomes more difficult to decode. Addition of "nulls" to the cipher (garbage encoding) that must be ignored and tossed during the decoding can help balance frequencies as well as the use of special items that mean things like "ignore the next item", or "delete the previous item". In spite of all these suggestions, if a cryptanalyst has sufficient text, it is only a matter of time before someone could break it.

1.3. The Vigen`ere Cipher

This is a mixture of 26 different alphabets. Thus, this cipher technique has ciphers called "A", "B", "C" etc, which are read off for encoding in rows and column format. It can be used in two ways: either as a set using a particular keyword (to encode "ARNOLD" using cipher "G", we look up each of letter under the row "G" to derive the cipher "GXTURJ"), or we use individual ciphers for each letter in the plaintext through a keyword as in table 2 below [3]:

Table 2: *The Vigenere Cipher*

	A	B	C	D	E	F	G	H	I	J	K	L	M	N	O	P	Q	R	S	T	U	V	W	X	Y	Z
A	A	B	C	D	E	F	G	H	I	J	K	L	M	N	O	P	Q	R	S	T	U	V	W	X	Y	Z
B	B	C	D	E	F	G	H	I	J	K	L	M	N	O	P	Q	R	S	T	U	V	W	X	Y	Z	A
C	C	D	E	F	G	H	I	J	K	L	M	N	O	P	Q	R	S	T	U	V	W	X	Y	Z	A	B
D	D	E	F	G	H	I	J	K	L	M	N	O	P	Q	R	S	T	U	V	W	X	Y	Z	A	B	C
E	E	F	G	H	I	J	K	L	M	N	O	P	Q	R	S	T	U	V	W	X	Y	Z	A	B	C	D
F	F	G	H	I	J	K	L	M	N	O	P	Q	R	S	T	U	V	W	X	Y	Z	A	B	C	D	E
G	G	H	I	J	K	L	M	N	O	P	Q	R	S	T	U	V	W	X	Y	Z	A	B	C	D	E	F
H	H	I	J	K	L	M	N	O	P	Q	R	S	T	U	V	W	X	Y	Z	A	B	C	D	E	F	G
I	I	J	K	L	M	N	O	P	Q	R	S	T	U	V	W	X	Y	Z	A	B	C	D	E	F	G	H
J	J	K	L	M	N	O	P	Q	R	S	T	U	V	W	X	Y	Z	A	B	C	D	E	F	G	H	I
K	K	L	M	N	O	P	Q	R	S	T	U	V	W	X	Y	Z	A	B	C	D	E	F	G	H	I	J
L	L	M	N	O	P	Q	R	S	T	U	V	W	X	Y	Z	A	B	C	D	E	F	G	H	I	J	K
M	M	N	O	P	Q	R	S	T	U	V	W	X	Y	Z	A	B	C	D	E	F	G	H	I	J	K	L
N	N	O	P	Q	R	S	T	U	V	W	X	Y	Z	A	B	C	D	E	F	G	H	I	J	K	L	M
O	O	P	Q	R	S	T	U	V	W	X	Y	Z	A	B	C	D	E	F	G	H	I	J	K	L	M	N
P	P	Q	R	S	T	U	V	W	X	Y	Z	A	B	C	D	E	F	G	H	I	J	K	L	M	N	O
Q	Q	R	S	T	U	V	W	X	Y	Z	A	B	C	D	E	F	G	H	I	J	K	L	M	N	O	P
R	R	S	T	U	V	W	X	Y	Z	A	B	C	D	E	F	G	H	I	J	K	L	M	N	O	P	Q
S	S	T	U	V	W	X	Y	Z	A	B	C	D	E	F	G	H	I	J	K	L	M	N	O	P	Q	R
T	T	U	V	W	X	Y	Z	A	B	C	D	E	F	G	H	I	J	K	L	M	N	O	P	Q	R	S
U	U	V	W	X	Y	Z	A	B	C	D	E	F	G	H	I	J	K	L	M	N	O	P	Q	R	S	T
V	V	W	X	Y	Z	A	B	C	D	E	F	G	H	I	J	K	L	M	N	O	P	Q	R	S	T	U
W	W	X	Y	Z	A	B	C	D	E	F	G	H	I	J	K	L	M	N	O	P	Q	R	S	T	U	V
X	X	Y	Z	A	B	C	D	E	F	G	H	I	J	K	L	M	N	O	P	Q	R	S	T	U	V	W
Y	Y	Z	A	B	C	D	E	F	G	H	I	J	K	L	M	N	O	P	Q	R	S	T	U	V	W	X
Z	Z	A	B	C	D	E	F	G	H	I	J	K	L	M	N	O	P	Q	R	S	T	U	V	W	X	Y

Thus, if we wish to encode "SEND THE ONE MILLION THROUGH UNION BANK" with keyword "PASSWORD", it is as below. But first, note that the keyword "password" is not as long as the phrase – thus, the keyword "password" is repeated so that it becomes as long as the phrase. To encode,

"P" cipher on "S" of "SEND", "A" cipher on "E", the "S" code on the "N", and so on to yield the figure 9 below – where the first column represents the text to be encoded, the second column is the keyword password repeated; while the third column is the ciphertext as given in table 3.

Table 3: *Encoded Plaintext using the Vigenere cipher to yield the Ciphertext*

S	E	N	D	T	H	E	O	N	E	M	I	L	L	I	O	N	T	H	R	O	U	G	H	U	N	I	O	N	B	A	N	K
P	A	S	S	W	O	R	D	P	A	S	S	W	O	R	D	P	A	S	S	W	O	R	D	P	A	S	S	W	O	R	D	P
H	E	F	V	P	V	V	R	C	E	E	A	H	Z	Z	R	C	T	Z	J	K	I	X	K	J	N	A	G	J	P	R	Q	Z

Vigenere cipher is more difficult to cryptanalyse and the longer the keyword, the more difficult to crack. If its keyword were too long and random, it becomes impossible to decode, as any decoding is as likely as any others. To use such cipher, keyword is usually an English word/phrase, not hard to remember. A long keyword of random letters is unbreakable – because any encoding can represent any text with some keyword. Thus, the word "WEASEL", with an appropriate "keyword" – can represent any 6 characters word. Clearly, if the key is allowed to be arbitrarily long and composed of arbitrary letters, then anything can stand for anything, and thus, code is quite secure.

2. Converting Text to Numbers

Most mathematical encryption methods are integer transformations – no matter how they are achieved. Some, have slight merits over others, and the most commonly used is ASCII (assigns a number to each character with 128 different characters, including upper/lower case alphabets, digits, punctuation amongst other characters as octal implementation). Thus, H in row 11 and column 0 has octal value 110. To convert to decimal, 110 (octal) = $1 \cdot 8^2 + 1 \cdot 8^1 + 0 \cdot 8^0 = 64 + 8 + 0 = 72$ (decimal) [3].

Table 4: ASCII Codes

	0	1	2	3	4	5	6	7
00	^@	^A	^B	^C	^D	^E	^F	^G
01	^H	^I	^J	^K	^L	^M	^N	^O
02	^P	^Q	^R	^S	^T	^U	^V	^W
03	^X	^Y	^Z	^[	^\	^]	^^	^_
04		!	"	#	$	%	&	'
05	(	)	*	+	,	-	.	/
06	0	1	2	3	4	5	6	7
07	8	9	:	;	<	=	>	?
10	@	A	B	C	D	E	F	G
11	H	I	J	K	L	M	N	O
12	P	Q	R	S	T	U	V	W
13	X	Y	Z	[	\	]	^	_
14	`	a	b	c	D	e	F	g
15	h	i	j	K	l	m	N	o
16	p	q	r	s	t	u	V	W
17	x	Y	Z	{			}	~	DEL

ASCII encodes 7-bits data, whereas computers manipulate data as 8-bit. When ASCII is used, a character is put in one byte effectively but losses 1-bit; and with concatenation, the 7-bit or 8-bit is put side-by-side to represent group of letters – so that if a user wishes to encode two ASCII characters, he simply place their binary values next to each other to get a 16-bit number. Users are at home with base 10 (decimal) giving the encoding of figure 10. Note: XX entry is not used, while "SP" is Space character. U is 31; 7 is 97, etc. To encode the word: "CAR3", it is represented by the 8-digit number 13-11-28-93 (i.e. 13112893).

Table 5: Decimal Alphanumeric Encoding

	0	1	2	3	4	5	6	7	8	9
0	XX	XX	XX	XX	XX	XX	XX	XX	XX	XX
1	SP	A	B	C	D	E	F	G	H	I
2	J	K	L	M	N	O	P	Q	R	S
3	T	U	V	W	X	Y	Z	a	B	C
4	D	e	f	g	h	i	j	k	L	M
5	N	o	p	q	r	s	t	u	"	W
6	X	y	z	,	;	:	'	:	"	'
7	!	@	#	$	%	^	&	*	-	+
8	(	)	[	]	{	}	?	/	<	>
9	0	1	2	3	4	5	6	7	8	9

2.1. Quasi/Psuedo-Random Numbers

Generating a very long key to use in a Vigenere has its problem, in that the key has to be transmitted in a secure way to the decoding user. If key is random, the more secure it is. A truly, random key can only be generated in radioactivity. Other methods are via quasi or Psuedo random mode (Knuth, 2000). This is achieved as thus: if p is a prime number, n is a number in range $0 < n < p$. Then, the sequence of numbers nK (mod p) cycles all numbers between 1 and p−1, in a somewhat random order as K goes through 1, 2, .., p − 1. For example, if p = 17 and n = 3:

$3 \cdot 1 \ (\text{mod } 17) = 3 \ \text{mod } 17 = 3$
$3 \cdot 2 \ (\text{mod } 17) = 6 \ \text{mod } 17 = 6$
$3 \cdot 3 \ (\text{mod } 17) = 9 \ \text{mod } 17 = 9$
$3 \cdot 4 \ (\text{mod } 17) = 12 \ \text{mod } 17 = 12$
$3 \cdot 5 \ (\text{mod } 17) = 15 \ \text{mod } 17 = 15$
$3 \cdot 6 \ (\text{mod } 17) = 18 \ \text{mod } 17 = 1$
$3 \cdot 7 \ (\text{mod } 17) = 21 \ \text{mod } 17 = 4$
$3 \cdot 8 \ (\text{mod } 17) = 24 \ \text{mod } 17 = 7$

$3 \cdot 9 \ (\text{mod } 17) = 27 \ \text{mod } 17 = 10$
$3 \cdot 10 \ (\text{mod } 17) = 30 \ \text{mod } 17 = 13$
$3 \cdot 11 \ (\text{mod } 17) = 33 \ \text{mod } 17 = 16$
$3 \cdot 12 \ (\text{mod } 17) = 36 \ \text{mod } 17 = 2$
$3 \cdot 13 \ (\text{mod } 17) = 39 \ \text{mod } 17 = 5$
$3 \cdot 14 \ (\text{mod } 17) = 42 \ \text{mod } 17 = 8$
$3 \cdot 15 \ (\text{mod } 17) = 45 \ \text{mod } 17 = 11$
$3 \cdot 16 \ (\text{mod } 17) = 48 \ \text{mod } 17 = 14$
$3 \cdot 17 \ (\text{mod } 17) = 51 \ \text{mod } 17 = 0$

The numbers cycles: 0, 3, 6, 9, 12, 15, 1, 4, 7, 10, 13, 16, 2, 5, 8, 11 and 14. 0 is not a number between 1 and p-1. This sequence appears to be random and the larger the prime, the longer the sequence. Using this to encrypt data: "Password: Ojugo" plus the space will have the numeric codes: 26, 37, 55, 55, 59, 51, 54, 40, 65, 10, 25, 46, 57, 43, 51 (as from figure 11). We use a combination of this method, to generate a sequence of pseudo-random numbers to use as the key. Let p = 16139 and n = 4352. As k goes 1 to 15, 4352*k(mod 16139) becomes the 15-numbers:

$4352 \cdot 1 \ (\text{mod } 16139) = 4352 \ \text{mod } 16139 = 4352$
$4352 \cdot 2 \ (\text{mod } 16139) = 8704 \ \text{mod } 16139 = 8704$
$4352 \cdot 3 \ (\text{mod } 16139) = 13056 \ \text{mod } 16139 = 13056$
$4352 \cdot 4 \ (\text{mod } 16139) = 17408 \ 16139 = 1269$
$4352 \cdot 5 \ (\text{mod } 16139) = 21706 \ \text{mod } 16139 = 5621$
$4352 \cdot 6 \ (\text{mod } 16139) = 26112 \ \text{mod } 16139 = 9973$
$4352 \cdot 7 \ (\text{mod } 16139) = 30464 \ \text{mod } 16139 = 14325$
$4352 \cdot 8 \ (\text{mod } 16139) = 34816 \ \text{mod } 16139 = 2538$
$4352 \cdot 9 \ (\text{mod } 16139) = 39168 \ \text{mod } 16139 = 6890$
$4352 \cdot 10 \ (\text{mod } 16139) = 43520 \ \text{mod } 16139 = 11242$
$4352 \cdot 11 \ (\text{mod } 16139) = 47872 \ \text{mod } 16139 = 15594$
$4352 \cdot 12 \ (\text{mod } 16139) = 52224 \ \text{mod } 16139 = 3807$
$4352 \cdot 13 \ (\text{mod } 16139) = 56576 \ \text{mod } 16139 = 8159$
$4352 \cdot 14 \ (\text{mod } 16139) = 60928 \ \text{mod } 16139 = 12511$
$4352 \cdot 15 \ (\text{mod } 16139) = 65280 \ \text{mod } 16139 = 724$

Thus: 4352, 8704, 13056, 1269, 5621, 9973, 14325, 2538, 6890, 11242, 15594, 3807, 8159, 12511, 724.

To encode, we add these to the numbers in our sequence and take its result of modulo 100 as thus:

$26 + 4352 \ (\text{mod } 100) = 4378 \ \text{mod } 100 = 78$
$37 + 4352 \ (\text{mod } 100) = 4387 \ \text{mod } 100 = 87$
$55 + 4352 \ (\text{mod } 100) = 4407 \ \text{mod } 100 = 7$
$55 + 4352 \ (\text{mod } 100) = 4407 \ \text{mod } 100 = 7$
$59 + 4352 \ (\text{mod } 100) = 4411 \ \text{mod } 100 = 11$
$51 + 4352 \ (\text{mod } 100) = 4403 \ \text{mod } 100 = 3$
$54 + 4352 \ (\text{mod } 100) = 4406 \ \text{mod } 100 = 6$
$40 + 4352 \ (\text{mod } 100) = 4392 \ \text{mod } 100 = 92$
$65 + 4352 \ (\text{mod } 100) = 4412 \ \text{mod } 100 = 12$
$10 + 4352 \ (\text{mod } 100) = 4362 \ \text{mod } 100 = 62$
$25 + 4352 \ (\text{mod } 100) = 4377 \ \text{mod } 100 = 77$
$46 + 4352 \ (\text{mod } 100) = 4398 \ \text{mod } 100 = 98$
$57 + 4352 \ (\text{mod } 100) = 4509 \ \text{mod } 100 = 9$
$43 + 4352 \ (\text{mod } 100) = 4395 \ \text{mod } 100 = 95$
$51 + 4352 \ (\text{mod } 100) = 4403 \ \text{mod } 100 = 3$

The encoded sequence becomes 78, 87, 7, 7, 11, 3, 6, 92, 12, 62, 77, 98, 9, 95 and 3.

The user only decodes data if he knows the value of p and n, and thus – generates same exact sequence of keys to undo the encryption. To generate first number: subtract 78 from 4352 (= -4274), take modulo 100 (= -72). Since it is negative, we subtract from 100 to get 26 etc. Note, the only data passed to the recipient as key is the pair of p and n. No need to transmit long sequence of keys. The generator is modified by choosing a large prime p, two integers (m, n) and seed "starting number" X_0 – so that $Xi = (m*X_{i-1} + n)(\bmod\ p)$, if i > 0. So, if p = 161, m = 91, n = 541, and $X_0 = 0$, we generate the sequence (no matter how large p, m and n – the sequence cycles p thus) as:

$X_0 = 11$ and $X_1 = (91 \cdot 11 + 49)(\bmod\ 161) = 1050 \bmod 161 = 84$

$X_2 = (91 \cdot 84 + 49)(\bmod\ 161) = 7693 \bmod 161 = 77$ X_3 = $(91 \cdot 77 + 49)(\bmod\ 161) = 7056 \bmod 161 = 133$

To decipher (p = 16139, m = 91, n = 541, $X_0 = 0$) the data 23, 52, 85, 91, 15, 6, 53, 61, 30, 72, 23:

With $X_0 = 0$. Thus, our first number is 23 = M.

$X_1 = (91 \cdot 0 + 541)(\bmod\ 16139) = 541 \bmod 16139 = 541$

$(52 - 541) \bmod 100 = -89$ and $100 - 89 = 11 = A$

$X_2 = (91 \cdot 541 + 541)(\bmod\ 16139) = 49772 \bmod 16139 = 1355$

$85 - 1355 \bmod 100 = -70$ but $100 - 70 = 30 = T$

$X_3 = (91 \cdot 1355 + 541)(\bmod\ 16139) = 10873 = 1355$

$91 - 10873 \bmod 100 = -82$ but $100 - 82 = 18 = H$

$X_4 = (91 \cdot 10873 + 541)(\bmod\ 16139) = 5505$

$15 - 5505 \bmod 100 = -90\ (100 - 90) = 10 = SPACE$

$X_5 = (91 \cdot 5505 + 541)(\bmod\ 16139) = 1187$

$6 - 1187 \bmod 100 = -81$ but $100 - 81 = 19 = I$

$X_6 = (91 \cdot 1187 + 541)(\bmod\ 16139) = 11724$

$53 - 11724 \bmod 100 = -71$ but $100 - 71 = 29 = S$

$X_7 = (91 \cdot 11724 + 541)(\bmod\ 16139) = 2251$

$61 - 2251 \bmod 100 = -90\ (100 - 90) = 10 = SPACE$

$X_8 = (91 \cdot 2251 + 541)(\bmod\ 16139) = 11714$

$30 - 11714 \bmod 100 = -84$ but $100 - 84 = 16 = F$

$X_9 = (91 \cdot 11714 + 541)(\bmod\ 16139) = 1341$

$72 - 1341 \bmod 100 = -69$ but $100 - 69 = 31 = U$

$X_{10} = (91 \cdot 1341 + 541)(\bmod\ 16139) = 9599$

$23 - 9599 \bmod 100 = -76$ but $100 - 76 = 24 = N$

3. Public-Key Cryptography

A major problem in cryptography is keys distribution as same keys cannot be used over and over – else, it becomes possible for cryptanalysts to crack the cipher. It is painstaking to transfer the keys – as incidents can occur, if the keys are mailed, cryptographically sent or hand-delivered. Examples include RSA, AES, Triple DES amongst others, are based on *trap-door* scheme that allows each cipher to have set of encoding and different decoding keys – so that if user A knows the decoding key, it is easy to make the encoding key. But user B is unable to make the decoding key if he/she only has the encoding key. To communicate, user A uses his trap-door to generate a decoding D_a and corresponding encoding key E_a. User B does same, but tells user A the encoding key E_b (but not D_b). User does same by telling B encoding key E_a (but not D_a). Thus, user A can send messages by encoding using E_b (only user B can decode) and vice-versa – because user B is the only one with access to decoding key D_b just as A is also the only user with access to

D_a. To change to a new key, users make up new pairs and exchange encoding keys. If the encoding keys are stolen, the eavesdropper can only encode but not decode. Thus, encoding keys are public keys – since they are published, while the decoding keys are made private or secret [4].

3.1. RSA Scheme

Commonly used trap-door is RSA (Rivest, Shamir, and Adleman), based on number factoring. Number multiplication with computers is quite simple but it can be difficult to factor numbers. A computer can factor numbers – knowing the possible combinations by checking the order of the size of the square root of number to be factored. For example, it takes seconds for the computer to try out 38000 possibilities, but the larger the number to be factored the longer it takes to factor them. It is also not hard to check if a number is prime or to see if it cannot be factored. If not prime, it is difficult to factor. Thus, the RSA scheme finds two huge prime numbers (m and n) with up to 200 digits each, which the user keeps s secret (private key) – and multiplies them to get N (public key). It is easy for a cryptanalyst to get N by multiplying m and n; but impossible to find m and n [4, 8]. RSA works as thus (though the prime numbers are very large):

User A makes a public key, which B uses to send A messages. User A sends B just a number, which is assumed that A and B have agreed on a method to encode text as numbers using these steps below:

1. User A selects two primes p and q (say 23 and 41). Note: User A will use *much* larger in real life scenario and gets N = pq ((23)(41) = 943), known as public key that B (and rest of the world, if he wishes) knows.

He then also chooses another number relatively prime to (p−1)(q−1) – in this case (22)(40) = 880. Thus, e = 7 and also part of the public key, so B also is told the value of e.

2. B has enough data to encode data to User A. If the data is the number W = 33, User B calculates the value of C = W^e (mod N) = 33^7(mod 943). $W^e = 33^7 = 42618442977$ and 42618442977 (mod 943) = 244. The number 244 is the encoding that User B sends to User A.

3. User A then decodes 244 by finding a number d in that ed = 1(mod (p−1)(q −1)).

7d = 1(mod 880) and d = 503, since in reverse – we have 7*503 = 3521 = 4(880)+1 = 1(mod 880).

4. User A finds the decoding by calculating C^d (mod N) = 244^{503} (mod 943) – just the binary expansion

$503 = 256 + 128 + 64 + 32 + 16 + 4 + 2 + 1$.

$244^{503} = 244^{256+128+64+32+16+4+2+1} \rightarrow 244^{256} \cdot 244^{128} \cdots 244^1$

We are only interested in the result (mod 943), calculate all partial results via repeated squaring of 244 and get all exponents in powers of 2 as:

244^2 (mod 943) $\rightarrow$ 244*244 = 59536 (mod 943) = 127 – with the workings below:

244^1(mod 943) = 244

244^2(mod 943) = 127

244^4(mod 943) = 98

244^{16}(mod 943) = 100

$244^{32}(\mod 943) = 570$
$244^{64}(\mod 943) = 508$
$244^{128}(\mod 943) = 625$
$244^{256}(\mod 943) = 223$
This becomes:
$244^{503} (\mod 943)$
$= 223 \cdot 625 \cdot 508 \cdot 570 \cdot 100 \cdot 98 \cdot 127 \cdot 224 (\mod 943)$
$= 33$.
User A decodes B's data to obtain the original data N = 35.

4. Discussions

Studies have shown that the Internet as a medium of data transfer is insecure and unsafe. Being aware that data sent over such public media lacks integrity – this has led to the field of cryptography. Our study thus, defines cryptography, its methods (symmetric and public key) and notes problem with each scheme.

Alternatively, public-key ciphers are safer but their major problem is that with the keys being public, a user can forge a message to authentic users pretending to be one of them. Thus, *digital* signature and certificate makes it possible to send messages so that a user is certain it is from a friend. This can be accomplished as thus: User A takes his name, pretends it is an encoded message, which can only be decoded using D_a (accessible only to User A) – and includes it in the real message encoded using E_b, which only user B can decode. User B receives it, decodes using D_b but discovers additional junk data (gotten by User A by decoding his name). User B simply encodes junk using A's public key E_a and makes certain that it is user A's name. User A alone, can make the text to encode to his name, such that when he receives the data – he is sure it is User B.

5. Summary / Conclusion

The proliferation of communication systems brought with it, demands from the private sector, for means of protecting digital data as well as data security services. Beginning with the work of Feistel at IBM in early 1970s, culminated in 1977 when U.S Federal information processing standard adopted Data Encryption Standard (DES) for encrypting unclassified data – making DES the most well-known cryptographic method and standard in history. Diffie and Hellman's work "New Directions in cryptography" saw the birth of public-key cryptography that today provides data security based on the intractability of the discrete logarithm problem. The search for public-key schemes and its im-

provements, continue with proofs of security at rapid pace. Security products, schemes and algorithms are developed to address the security needs of an information society.

6. Recommendation

Because keys in a public key scheme are made public, a user can forge message to users, pretending to be another users, which often can mislead authentic users. Thus, digital signatures/certificates are used to make it possible to send messages so that a user is certain that the sent message is from a friend (and not foe). An easy way to do this is to take user A's name and pretend it is an encoded message, which will be decoded using D_a. User A is the only one with D_a, so he includes this in the real message and encodes using E_b, which only user B can decode. When B receives it, he decodes using D_b but discovers additional junk characters (derived when user A decodes his name). User B simply encodes the junk using user A's public key E_a and makes certain that it is user A's name. Since user A is the only one who can make meaning of the cipher, he knows the message is from user B. Additional information can be encoded for certification to assure privacy.

References

[1] Adewumi, S.E and Garba, E.J., Data security: cryptosystems algorithm using data compression and systems of non-linear equations, 2002, Nigerian Computer Society, Vol. 13

[2] Zeng, K., Psuedorandom bit generators in stream cipher cryptography, 1991, IEEE J. Computers, 24(2), pp 45 – 54.

[3] Davis, K.L., The art of ciphers, 2000, J. Comp and Info. Syst., ISSN: 2032-3765, pp 32

[4] Knuth, D., The art of computer programming — seminumerical algorithms, 2000, New Jersey, Prentice Hall publications

[5] Mendez, A., Van Oorschot, P and Vanstone, S Handbook of applied cryptography, 1997, New Jersey, CRC Press.

[6] Stallings, W and Van Slyke, R., Introduction to business data communication, 2008, McGraw hill publications.

[7] Tanenbaum, A.S., Computer networks, 1996 New Jersey, Prentice Hall publications.

[8] Aghware, F.O., A modified RSA encryption as implemented in the Nigerian banking software, 2005, Unpublished Masters thesis, Nnamdi Azi kiwe University: Awka-Nigeria

Social interactions and query analysis in an online forum

Oladosu Olakunle Abimbola, Okikiola Folasade Mercy, Alakiri Harrison Osarenren, Oladiboye Olasunkanmi Esther

Department of Computer Technology, Yaba College of Technology, Yaba, Lagos

Email address:

kunledosu@gmail.com(Oladosu O. A.), olakunlerejoicemysoul@yahoo.com(Oladosu O. A.), sade.mercy@yahoo.com(Okikiola F. M.), halakiri@yahoo.com(Alakiri H. O.), estherworld2001@yahoo.com(Oladiboye O. E.)

Abstract: Online social networks (forum) have become extremely popular; numerous sites allow users to interact and share content using social links observing that there are some limitations on most of the social interaction networks. A recent study has demonstrated that the "strength of ties" varies widely, ranging from pairs of users who are best friends to pairs of users who even wished they weren't friends. In order to distinguish between these strong and weak links, researchers have suggested examining the activity networks, the network that is formed by users who actually interact using one or many of the methods provided by the social networking site. Also for effective study, we collect data on both friendship relationship and interactions for a large subset of the Facebook social network and forum, Twitter, 2go, m.eskimi and other social interactive networks. From a technological standpoint, forums or boards are web applications managing user-generated content. Early Internet forums could be described as a web version of an electronic mailing list or newsgroup (such as exist on Usenet); allowing people to post messages and comment on other messages. It has been discovered that Chat forum is one of the newest way in which organization and the business world use now to reach out to their numerous customers in order to get feedback and responses concerning its entire product lines to share of their opinion.

Keywords: Content, Blog, Bookmark, Video Conference, Forum, Social Network, Chat

1. Introduction

A discussion forum is hierarchical or tree-like in structure: a forum can contain a number of sub forums, each of which may have several topics. Within a forum's topic, each new discussion started is called a thread, and can be replied to by as many people as so wish. Depending on the forum's settings, users can be anonymous or have to register with the forum and then consequently log in (i.e. in order to post messages). On most forums, users do not have to log in to read existing messages. Social networks have become a popular way for users to connect, express themselves, and share content. Popular social networks have hundreds of millions of registered users and are growing at a rapid pace. As these networks grow and mature, users have been observed to form hundreds to even thousands of friendship links. For example, a user in Orkut has on average of over 100 friends [23] and the average number of friends in Facebook is over 120. Certain individuals have much higher degrees than the average; in fact, in Flickr we found multiple users who have more than 25,000 friends! While most social networking sites allow only a binary state of friendship, it has been unsurprisingly observed that not all links are created equal. A recent study [14] has demonstrated that the "strength of ties" varies widely, ranging from pairs of users who are best friends to pairs of users who even wished they weren't friends. In order to distinguish between these strong and weak links, researchers have suggested examining the activity networks, the network that is formed by users who actually interact using one or many of the methods provided by the social networking site. While the initial studies on activity networks [24] have brought great insights into how an activity network is structurally different from the social network and how system designers can utilize such information, little attention has been paid to a rather natural and important aspect of user interaction: the fact that the level of interaction between two individuals can vary over time. Also for effective study, we collect data on both friendship relationship and interactions for a large subset of the Facebook social network and forum, Twitter, 2go, m.eskimi and other social interactive networks. . In total, we examine over 60,000 users and over 800,000 logged interactions between those users over a period of months. This data set has been made available to the research community in an anonymized form. We make a number of interesting observations.

2. Online Forum and Social Network Interaction

The modern forums originated from bulletin boards, and are a technological evolution of the dialup bulletin board system [25]. From a technological standpoint, forums or boards are web applications managing user-generated content [16]. Early Internet forums could be described as a web version of an electronic mailing list or newsgroup (such as exist on Usenet); allowing people to post messages and comment on other messages. Later developments emulated the different newsgroups or individual lists, providing more than one forum, dedicated to a particular topic [14]. Internet forums are prevalent in several developed countries. Japan posts the most with over two million per day on their largest forum called "2channel". China also has many millions of posts on forums such as "Tianya Club". Forums perform a function similar to that of dial-up bulletin board systems and Usenet networks that were first created starting in the late 1970s [13].

Early web-based forums date back as far as 1994, with the WIT [26] project from W3 Consortium and starting from this time, many alternatives were created [9]. A sense of virtual community often develops around forums that have regular users. Technology, video games, sports, music, fashion, religion, and politics are popular areas for forum themes, but there are forums for a huge number of topics. Internet slang and image macros popular across the Internet are abundant and widely used in Internet forums.

3. Forum Structure

A forum consists of a tree like directory structure. The top end is "Categories". A forum can be divided into categories for the relevant discussions. Under the categories are sub forums and these sub forums can further have more sub forums. The topics (commonly called threads) come under the lowest level of sub-forums and these are the places under which members can start their discussions or posts. Logically forums are organized into a finite set of generic topics (usually with one main topic) driven and updated by a group known as members, and governed by a group known as moderators. It can also have a graph structure [3]. All message boards will use one of three possible display formats. Each of the three basic message board display formats is;

    a.   Non-Threaded

    b.   Semi-Threaded

    c.   Fully Threaded

These formats or threads have their own advantages and disadvantages. If messages are not related to one another at all a Non-Threaded format is best. If you have a message topic and multiple replies to that message topic a semi-threaded format is best. If you have a message topic and replies to that message topic, and replies to replies, then a fully threaded format is best [18].

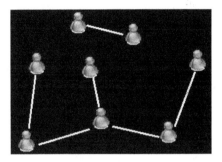

Fig 3.1. *Social network*

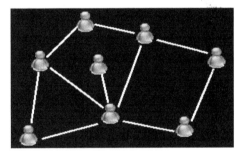

Fig 3.2. *Interaction network*

4. Email, IM, Discussion Forums, and Chat

    a.   Chat: The term 'chat' refers to an interactive discussion in which all participants are online at the same time, similar to a telephone conversation. Instead of talking into a receiver, however, the chat participants communicate using their keyboards to type comments. Chat is completely interactive, i.e., "live". You can ask questions and make comments and other participants in the chat "room" will be able to see them and reply immediately. Our Women's History Month chats will take place at predetermined times and once they are over, you will not be able to add additional comments or questions to them.

    b.   Discussion: The term 'discussion' refers to the ongoing forums, like this one, in which you can post comments and questions which will be saved for others to read and comment on in turn. The discussion forums will be available for as long as the WHM website remains online.

    c.   Email: Is the acronym for "electronic mail" (email is an official English word that requires no hyphen).

Email is like an old fashioned letter but in electronic format sent from one computer to another. There is no going to the metal mailbox down the road, no envelopes to address and stamps to lick, yet email very much resembles the classic post office mail process. Most importantly: the email recipient does not have to be at their computer for an email to successfully send. Recipients retrieve their email on their own time. Because of this lag between sending and receiving, email is called "non-real time" or "asynchronous time" messaging.

d. Instant Messaging: Unlike email, instant messaging is a real-time messaging format. IM is really a specialized form of 'chat' between people who know each other. Both IM users must be online at the same time for IM to fully work. IM is not as popular as email, but it is popular amongst teenagers and people in office places that allow instant messaging.

e. Discussion Forum: Discussion forums are really a slow-motion form of chat. Forums are designed to build online communities of people with similar interests. Also known as a "discussion group", "board" or "newsgroup", a forum is an asynchronous service where you can trade non-instant messages with other members. The other members reply on their own schedule, and do not need to be present while you are sending. Every forum is also dedicated to some specific community or subject, such as travel, gardening, motorcycles, vintage cars, cooking, social issues, music artists, and more. Forums are very popular, and are renowned for being quite addictive because they put you in touch with many similar-minded people.

5. Results and Analysis

Table 5.1. List of queries posted and responses

Forum Topic	Number of Posts	Number of Responses
Education	10	30
Technology	8	24
Fashion	12	36
Electronic Gadget	7	21
Automobile	5	15
Shares	6	18
Marriage	11	33

The graph below shows the responses of posts against queries in the chart above. Number of posts is at the Y-axis while number of queries is at the X-axis. They are drawn in ascending order of posts and queries. The graph simply shows that there is tremendous rate at which people respond to topics like Fashion, Marriage, Education, e.t.c. against topic such as Automobile, Shares, e.t.c. based on

professionalism, interest and usability of the available topics.

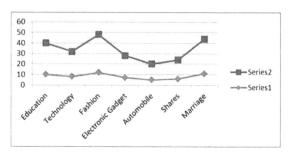

Fig 5.1. Line graph of Queries and Responses

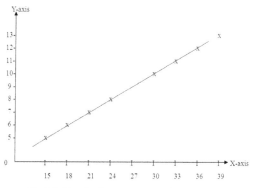

Fig 5.2. Graph of query posts against responses

6. Interpretation

Both graphs actually shows that there is a relationship between the number of response received and the query posted i.e. the relative response time to any query post depends on the how interesting is the subject of discourse. Hence, the rate at which users of social network and forum respond to query post is dependent on user's interest in the subject.

7. Summary and Conclusion

The study has been all about finding an easier, safer and educative ways of sharing opinions and interact with other people which is commonly refers to as social network (forum). In addition to chatting, this is a new and dynamics means of talking to different people all over the world within a slip of time. It has been discovered that Chat forum is one of the newest way in which organization and the business world use now to reach out to their numerous customers in order to get feedback and responses concerning the entire product lines to share of their opinion.

References

[1] UITS - Instant Messaging Chatiquette, 2012 - University of Arkansas". Uits.uark.edu. Retrieved 2012-01-19.

[2] IRC Chatiquette – Chat Etiquette, 2012. Livinginternet.com. 1995-11-28. Retrieved 2012-01-19

[3] LessWrong, 2011 Debate tools: http://wiki.lesswrong.com/wiki/Debate_tools#Explore-Ideas.com. Retrieved 2011-03-31

[4] http://www.chelseafcforums.com, Retrieved 2011.

[5] Hurrchan/Courtney Wade, 2010 Tripcode Explorer - Hurr Network". http://www.hurrchan.net/wiki/Tripcodes. Retrieved 2010-08-22.

[6] Michael Herman, 2010 "Chat room user guilty of web rage". The Times (London). Retrieved 2010-05-20.

[7] Administrator seeking moderators to control spam abuse, 2010. ddforums.com. http://www.ddforums.com/showpost.php?p=12566&postcount=9. Retrieved 2010-04-20.

[8] Non-Threaded vs Semi-Threaded vs Threaded Formats, 2010 - BulletinBoards.com". BulletinBoards.com. http://www.bulletinboards.com/ThreadHelp.cfm. Retrieved 2010-05-27.

[9] Forum Software Timeline, 2010. Forum Software Reviews. http://www.forum-software.org/forum-software-timeline-from-1994-to-today. Retrieved 2010-12-24.

[10] TechnoFyed.com, 2009: Postcount Information: http://www.technofyed.com/showthread.php?tid=219. Retrieved 2009-11-11.

[11] Message Board Features - Website Toolbox, 2009. Website Toolbox. http://www.websitetoolbox.com/message_board/features.html. Retrieved 2009-07-12

[12] Categories of moderators, 2009: moderators can be global moderators who have rights for the complete forum and only specific to a sub-forum.

[13] Bulletin Community Forum, 2008 - FAQ: What is a bulletin board?". vBulletin.com. http://www.vbulletin.com/forum/faq.php?faq=vb3_board_usage#faq_vb3_forums_threads_posts Retrieved 2008-09-01. "A bulletin board is an online discussion site. It is sometimes also called a 'board' or 'forum'. It may contain several categories, consisting of sub-forums, threads and individual posts."

[14] Gilbert E., Karahalios, Ethan Feerst and Dylan Stewart group, 2008. What is an "Internet forum and video entry?" http://www.videojug.com/expertanswer/internet-communities-and-forums-2/what-is-an-internet-forum. Retrieved 2008-11-04.

[15] L. Backstrom, R. Kumar, C. Marlow, J. Novak, and A. Tomkins, 2008. Preferential behavior in online groups. In WSDM,

[16] http://bugclub.org/glossary.html. Retrieved 2008

[17] D. Crandall, D. Cosley, D. Huttenlocher, J. Kleinberg, and S. Suri, 2008 Feedback e_ects between similarity and social inuence in online communities. In SIGKDD.

[18] Bulletin FAQ: Moderators and Administrators, 2008. vBulletin.com. http://www.vbulletin.com/forum/faq.php?faq=vb3_reading_posting#faq_vb3_mods_admins. Retrieved 2008-09-14.

[19] phpBB FAQ: What is COPPA?". phpBB.com. http://www.phpbb.com/community/faq.php#f07. Retrieved 2008-09-14

[20] L. Backstrom, D. Huttenlocher, J. Kleinberg, and X. Lan.2006 Group formation in large social networks: membership, growth, and evolution. In SIGKDD, 2006.

[21] Merchant, Guy, 2001. "Teenagers in cyberspace: an investigation of language use and language change in internet chatrooms." Journal of Research in Reading. 2001 Volume 24, Issue 3. ISSN 0141-0423

[22] WIT - WWW Interactive Talk, 1995. W3 Consortium. http://www.w3.org/WIT/. Retrieved 1995-05. http://www.fanforum.com/faq.php?faq=vb_user_maintain#faq_vb_why_register

[23] Mislove A. et al, 2007.

[24] Chun H. et al, 2008 and Wilson C. et al 2009.

[25] http://www.videojug.com/expertanswer/internet-communities-and-forums, Retrieved

[26] http://www.w3.org/WIT/, Retrieved 1995.

8

Robust cellular network for Rural Telephony in Southern Nigeria

A. A. Ojugo.[1], R. E. Yoro.[2], D. Oyemade.[1], M. O. Yerokun.[3], A. O. Eboka[3], E. Ugboh[3], F. O. Aghware[3]

[1]Department of Mathematics/Computer, Federal University of Petroleum Resources Effurun, Delta State
[2]Department of Computer Science, Delta State Polytechnic Ogwashi-Uku
[3]Department of Computer Science, Federal College of Education Technical Asaba

Email address:

ojugo_arnold@yahoo.com(A. A. Ojugo.), aojugo@mail.fupre,edu.ng(A. A. Ojugo.), rumerisky@yahoo.com(R. E. Yoro),
davidoyemade@yahoo.com(D. Oyemade.), an_drey2k@yahoo.com(A. O. Eboka), agapenexus@hotmail.co.uk(M. O. Yerokun),
ugbohh@gmail.com(E. Ugboh), aghwarefo@yahoo.com(F. O. Aghware)

Abstract: The paper identifies the state of telecommunication's services and its growth cum economic relevance in Nigeria, while proffering a cost-effective and less problematic telephone network that is readily affordable and available to the poverty-stricken populace predominant in rural areas. The design and implementation of an efficient, robust network based on existing Global Satellite for Mobile communication) cum Code Division Multiple Access, as means to cater and provide for telecommunications in rural Delta State in Nigeria. A major hinderance to the provision of such services/facilities in rural area for decades, has been the high operational cost and little profit margins. Thus, the robust network aims to create an effective, affordable option for such rural and semi-urban settler; while on the part of operators, a system with minimal cost to implement, deploy and maintain for Delta State Senatorial District in Delta State, Nigeria, is used as the case study to test-run the analysis and design of such a novel cellular network.

Keywords: Topography, Robust, Telephony, Tele-Penetration, Tele-Density

1. Introduction

Telephony in rural Nigeria and those of developing nations have been a recurring subject – as a country's economic growth/development as measured via its gross national product, is strongly correlated to communication (namely telephone density and her tele-penetration – a function of the number of lines per 100 persons). It is common knowledge that rural telecommunication users do not generate same amount of traffic and revenue as their urban counterpart, which tends to lower the incentives to invest by many operators. The cost of wiring a vast area for low data traffic causes providers to delay or ignore service provision to such areas except with government's strict intervention and regulation[1].

The increased interest worldwide in the growth of rural areas has led to feasibility studies towards the provision of telecommunication services in rural areas for national, economic and social interactions. The International Telecommunications Union (ITU) has defined a rural area as "that which has one or more of these feats: (a) area's primary source of power is scarce, unavailable, uncoordinated and scattered, (b) Local scarcity of qualified technical personnel, (c) Topographical conditions exists (like mountains etc) hindering construction of switching cum transmission system, and (d) Economic constraint on investment that renders service non-profitable due to high-cost construction and maintenance, especially if the cost is borne by the rural zone settlers alone [1].

For the purpose of providing telecommunication services, ITU and CCITT (International Telephone and Telegraph Consultative Committtee), an arm of ITU – has set these rules for the classification of an area as rural [1]: (a) Average subscriber density of 50km maximum, (b) subscribers range from about 5km to 50km and (c) to be served communities are isolated with maximum of 1000subscribers. Hence we note, *rural* telephony is defined as provision of telecomm services in sparsely populated areas with economic and geographical disadvantages.

Telepenetration measures the user percentage covered by

telecomm services in a country's entire population; while teledensity measures percentage of such users with telephone lines. The importance of rural communication in Nigeria is continually demonstrated via *NCC* (Nigerian Communications Commission) stakeholders' forum that identifies the existence of rural dwellers, rural-urban dichotomy, need to bridge such divide via service provision extended and enforced to underserved and unserved areas. Africa's telecomms huge and steady growth (with Nigeria in focus) this last few years, has been dramatic with highlights for African market that saw a growth rate of 66% in 2005 (the highest ever experienced in any world region) and 150million subscribers milestone exceeded in 2006, 12years after its first launch in Africa [7].

In Nigerian, Zain came in 2001 as Buddie (and later rebranded as Vodafone, Celtel and Airtel) and now has a customer-base of over 12million. Next, came MTN in 2003, now with over 16million users; Globacom in 2005 now with over 10million and lastly, Etisalat, now with over 8million users. Thus, GSM penetration in Nigeria is presently over 40million in the past 10years. Despite this progress, the extent of the geographical telepresence of GSM and the technology itself, is still among the lowest.

Lastly, most territories in the North, South West, South East and the swampy coastal areas of Niger Delta are yet to be properly served and covered by such telecomm operators. This is not unconnected with the population density in such regions, the rural/urban dichotomy as well as in some cases, the unfriendly terrain that makes the huge investment of terrestrial infrastructure grossly unprofitable.

2. Literature Review

There are many issues to be defined when designing rural telecommunications network – all of them of varying importance, and falls into three categories: (a) Geopolitical – deals with the areas development level based on its social, educational, political and economic history, (b) Economic – defines the inital investment required to deploy such communication services via installation of facilities and equipment in such a remote region, its operational cost and the user's revenue generating capacity – so that such a service can be provided for, supported financially and keeping its subsidies to a negligible minimum, and (c) Technical – deals with availability of technology to implement the design, legal and regulatory framework, rural communities'size and the expected traffic, type of services and availability of technical workforce [8]. However, study centers on technical issues involved in analysis, design and deployment of a robust network for rural telephony.

2.1. Wireless Technology

With application of electronic data systems in every facet of our daily endeavor, it becomes increasingly bothersome to be tethered by wires. Thus, wireless technology comes to the rescue with these merits: (a) low-cost of devices, (b) broadcast same data to many locations at same time, (c) easily deploy in hostile and difficult environment, and (d) mobile communication. Its demerits: (a) lesser data rates, (b) operates in less controlled environ, (c) more susceptible to interference, eavesdropping and/or noise, and (d) lesser reusable frequencies than in the case of guided wired media.

Wireless Technologies are grouped as thus:

1. 1st Generation – such as AT and T's advanced mobile phone services, commonly used in South America, China and Australia. Its system comprise of three basic devices: a mobile unit, base transceiver and mobile telephone switching office (MTSO). The mobile unit interacts with the trasceiver; which then interacts with MTSO. The MTSO controls all trasceivers and connects calls to public telephone network. Its demerit is its limitation to reuse same frequency in various communication, as its signals if not constrained, interferes with one another (even when separated geographically). It supports a very large number of communications simultaneously and needs spectrum conservation. It has two 25-MHz bands, one transmits from station to mobile unit and the other transmits from the mobile terminal to the base station. Each is split in two, to ease and encourage competition. Its channels are also spaced 30kHz apart, with total of 416channels for an operator. 21channels for call control, and the other 395 for calls.

2. 2nd Generation – GSM and CDM/FDM access (code/frequency division multiplexing) falls into this generation. *GSM* was developed to solve the incompatibility of cellular technologies so that a number of subscriber terminals can be used. Its basic feats includes: Subscriber Identity Module (SIM) a portable smart-card like, plug-in device that stores a subscriber's identification number, the networks the subscriber is authorized to use, encryption keys, and other information specific to the subscriber. Each terminal is generic until a SIM is inserted. Thus, the SIMs roam but not the terminal or subscriber device. Its transmissions are encrypted with A5 cipher from the user to the base trasceiver and another A3 cipher is used to authenticate calls so that it is private. GSMs support data and image services based on ISDN (Integrated Service Digital Network) model with user data rates upto 9.6kbps.

Multiple Access like GSM – uses digital method to enhance advanced call processing feats. 1st generation systems were successful but its use was constrained by spectrum. Thus, there is a premium on its efficiency use of spectrum. Thus, multiple access details how the spectrum can be divided among active users in its planned systems via four methods: *frequency* division multplexing FDM, *space* division multiplexing SDM, *time* division multiplexing TDM and *code* division multiplexing CDM.

Besides pure form channel splitting in multiple access,

there can also be hybrids. For example, GSM uses FDM to divide the allotted spectrum into 124 carriers and each carrier is then split into eight parts using TDM so that the number of potential users in any one cell is potentially enormous and any subscriber in any area can enter the cell. In addition, lots of roamers can show up and fortunately, the system can identify the number of customers who are in a given cell at any time and have their terminals registered with the system in a modest form. Its only problem is how to determine active users in a cell and assigning them to vacant subchannels. It is resolved with MTSO as mobile terminal, entering a cell via handover is quickly allotted a channel directly via MTSO. The access method to use is depedent on the system to be designed, factors or constraints in the natural environment among others. Many providers now sign up for CDM as it offers more/additional feats.

3. 3rd-Generation offers mobile unit and personal communication services incorporating into a set of standard, services that a handheld terminal can support (voice, data, image, video). Example is Motorola's Iridium project capable of paging, data and facsimile in hand-held phones/pagers worldwide.

2.2. Satellites

Satellites are characterized by the orbit they keep into three (3) basic types namely [5, 6]:

1. GeoSynchronous (GEOS) circularly orbits at 22,300miles above, rotating at earth's equatorial planes, exactly same angular speed as the Earth. So, it rotates at same spot and speed above the equator. Its merit as a stationery satellite relative to earth are: (a) averts frequency change due to relative motion of satellite on earth, (b) simplified tracking by its earth station, (c) its height allows it communicate with roughly a fourth of earth and (d) its assigned frequency covers very large area. Its demerits also includes: (a) weak signal strenght due to height, (b) polar areas and hemispeheres are poorly served, (c) with signal traveling at light speed, delays exist, computed as $(2*22300)/186000 = 0.24sec$. Thus, stations that receive/resend, delay is longer, (d) for TV broadcast, application to coverage area is wasted and special spots and steer beam antenna is used, to restrict coverage area and control footprint.

2. Highly Elliptic Orbit Satellite (HEOS): At low angles, satellite signals are easily obstructed by hills, mountains and buildings. HEOS is used to orbit elliptically inclined with respect to equator. At farthest apogee of 24856miles over Northern hemisphere and 500km at its perigee in Southern, to provide good coverage. Thus, it spends 12hrs over Northern sphere and with its orbit elliptic, multiple staellites are needed for 24-hr coverage. The extent its apogee is greater than geosynchronous altitude

allows signal attenuation and/or loss to expand.

3. Low-Earth Orbit Sats (LEOS) is constellation of inexpensive satellites orbiting 200 – 700miles above earth – creating *lightsats* of stronger cum better signals, better localized coverage area and conserve spectrum. We have two LEOS namely: *Small* LEOS (works at frequency below 1GHz using 5MHz to support data upto 10kbps used in paging, tracking and low-rate messages) and *Big* LEOS (with frequency above 1Ghz to support data a few mbps. It offers same services as small LEOS alongside voice and positioning services. Many of it is needed to provide 24-hrs coverage.

3. Materials / Method

3.1. Study Purpose/Objective

Study aims to design a robust telephone network to aid rural telephony; and via this, improve Nigeria's socio-economic and gross national product with increased tele-penetration density – for Delta North Senatorial District in Nigeria – estimated as 3984.34km2 in fig 2 using a scale of 1:100.

3.2. Proposed Satellite System

Satellite technologies employed in rural telephony needs a different analysis of its technical parameters – becasue the chosen satellite network technology heavily influences choice of a user's terminal (small earth stations) feats. its overall costs and lifetime. A remote user can access a private switch telephone network (PSTN) via satellite in any of two methods: (a) *Direct* access allow users to transmit from mobile terminal directly to the satellite, and (b) *Indirect* access allow users transmit from wired/wireless land line (WLL) user terminal via VSAT terminal, private branch exchange (PBX) and a GEOS as in fig. 1.

This study adopts an *indirect* user access approach that is seamlessly integrated with TIW FlexiDAMA II and VSAT network into a CDMA-2000 wireless system (CDMA will serve as WLL segment patform for our robust network and the reason for our choice as in sub-section

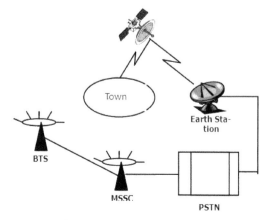

Fig. 1: *Direct user access to Satellite*

Other reasons for CDMA as suitable choice are:

a. Voice compression on the satellite link at rates as low as 9.6kbps (with fax, image, video and data)

b. Uses a robust-forward error correction coding on a very-narrow band, which also has full-digital operation capabilities on the satellite access.

c. GPS-derived frequency and timing references

d. Demand Assigned Multiple Access for all voice calls that allows for roaming and allows just active users to be authenticated on handover, registering of subscriber terminals as they roam and random access channel allocation directly by MTSO via a control message that assigns the user terminal a dedicated channel when a conversation or data transfer is necessary.

e. True voice activation so that satellite energy is only used during voice peaks.

f. Full-mesh with distributed call-routing operation so that calls originating from any station is routed directly to the destination station without having to be routed via a hub, gateway or another station – and maximize efficiency of the satellite links as well as minimize propagation delay.

3. Planning / Implementation

A cost-effective solution will result in proper radio network planning with a focus on service delivery, quality, coverage and capacity [9], discussed under:

3.1. Pre Planning Coverage and Capacity

Involves a theoretical coverage and capacity plans. In theory, as the cells are split into smaller sizes – interference received from cell frequency reuse will remain; and irregular propagation and non-circular cell shapes will lead to greater interference being received from the surrounding cells that all use the same channel set [10]. A way to reduce the level of interference is to use a *directional* antenna at base station – each antenna illuminates a sector of the cell, and a separate channel set allocated to each sector.

Study adopts 3-sectored, triangulation cell design of 7-cell repeated pattern (N=7 trunking efficiency) – resulting in the overall requirement of 21channels sets. The base station to be used are 3-sectored and each sector (cell) will cover a range of 4.0km. Thus, area covered by site is as thus:

$$Area\ Covered = \frac{3\phi}{360}\pi R^2 = \pi R^2$$

ϕ = sector angle $120°$, R = One antenna coverage range = (4km). Area $\pi R^2 = 3.142 * 4^2 = 50.28km$

Total number of sites to be setup is given by:

$$V = \frac{Coverage\ Area}{Cell\ sites\ coverage} = \frac{3984.34}{51} = 78.124$$

Thus, the total average number of cell sites to be constructed are approximately 78sites.

3.2. Bandwidth Limit to Subscribers

Cell layout in fig 4 helps formulate number of radios to be used in the plan via the simple equation below using the parameters in a 3-sector, GOS = 0.1:

Delta North Senatorial District area = 3984.34km.
Population projection for 2012 = 10,820,123
Average radius of cells (km) = 4
Number of radio channels in each n_c = 93
Number of Subscribers per Channel = 28
Number of Subscriber / Cell = 2604

N_c depends on what cell repeat pattern is used and spectrum allocated to the service [10]. We use a wireless CDMA on a cellular network with dynamic assignment cum trunking – interpreted as channel gain [10] allowing 28-subscribers individual access to one channel. Number of subscribers supported by each channel in the cells is multiplied by factor (30). With channel gain, number of subscribers on each cell is given by:

$$Subscribers = \frac{Base\ station\ GOS = 0.01}{subscriber\ Erlang\ 0.03} = \frac{0.1}{0.03} = 3.333$$

n_c = 30 * Subscribers = 30 * 3.333 ➔ 99.9 ≈ 100
Total Subscribers supported simultaneously:
= $10\pi\ n_c$ * Subscriber/Cell
= 10 * 3.142 * 100 * 2604 = 8,181,768subscribers

Thus, the proposed network can approximately cater for 8,181,768subscribers at same time in with effective service quality. These result shows that sparsely populated areas are easier to serve. If each subscriber has an average call and holding time of 6-calls/daily at an average of 3mins (180secs) duration, as it is unlikely for a user or any subscriber to talks for 24hrs, generated traffic tends to bunch up – at busy (peak) hours and will fall drastically as more users try to access the network [10]:

$$Erlang = \frac{Calls/Day * Duration}{Mins * Hrs *} = \frac{6 * 180}{60 * 60 * 4} = 0.125$$

Erlangs/Subscriber = 0.125Erlangs/subscriber.

3.3. Site Surevey and Choice

Radio site selection criteria consists of:

a. Location chosen by propagation analysis
b. Availability of capacity existing sites
c. Site compatibility
d. Environmental and planning factors

Sites in the nooks of a Delta North Senatorial District are determined considering these criteria [11]: (a) buildings and other structures can be used as suitable antennas system erection sites, (b) accounting for effects on coverage by physical obstruction so as to ensure site stability, (c) good radiation pattern via sufficient height for antenna installation and avoid obstruction of the transmitted signals, (d) provision of service to required area, and (e) centrally located sites to ensure omin-direction of radiation pattern for minimum coverage area.

The use of the highest point or available site is not usually considered the best option as it degrade the services, due to co-channel users. The choice of site is an interactive process that is repeatedly computed with several constraints and parameters checked to validate as well as yield an optimal result or choices of selected sites [12].

3.4. Propagation Analysis of Site Location

For radio site choices, we need precise locations of construction, bearing in mind feats of Delta North Senatorial District, which are as follows:
- ✓ Rural area in focus, is not connected to national power grid – though work is currently in progress
- ✓ Buildings within semi-urban and urbanized areas of the district are bungalows and storey buildings that are quite high (between 5 – 8m).
- ✓ Its land mass is covered by valleys and trees of various heights that can block out signals.
- ✓ Settlers are predominantly farmers, with clusters and handful of civil and public servants.

Via propagation analysis, an average of 78-sites is selected for effectively coverage of the said District. Each selected cell site coordinate and elevation were taken using the Global Positioning System (GPS). Its readings are as shown in Table 1.

4. Network Cell Structure

Figures 2(a), (b), (c) and (d) shows the maps of all the Local Government Areas in Delta North Senatorial District. The cell design structure layout are as seen in fig 3 further explained in tables 1 that shows the feats of each cell site.

Fig 2b: Map of Ika North, North East and South

Fig 2c: Map of Ndokwa West and Ukwauani

Fig 2a: Map of Oshimili North and South, Aniocha North and South

Fig 2d: Map of Ndokwa East and Ndokwa West

From Fig. 3, each base transceiver station (BTS) in the implementation of the robust telephony is connected to its corresponding zone's mobile telephone switching office (MTSO) namely: Agbor, Ogume, Akumazi, Amai, Ogume, Atuma, Ubulu, Otulu, Asaba and Igbuku respectively. Each zone's MTSO is connected to the PSTN in Ubulu-Uku (as the central hub) with an 8XE1 protected link; while the BTSs in each zone is connected to its corresponding MTSOs with the 2XE1 transmission link as in fig 4. Distance between each BTS and its corresponding MTSO lie between 8.6kms to 19.3kms; while the corresponding distances between MTSOs in each corresponding zone and the PSTN at Ubulu-Uku lies between 15.78kms and 102kms as the case may be.

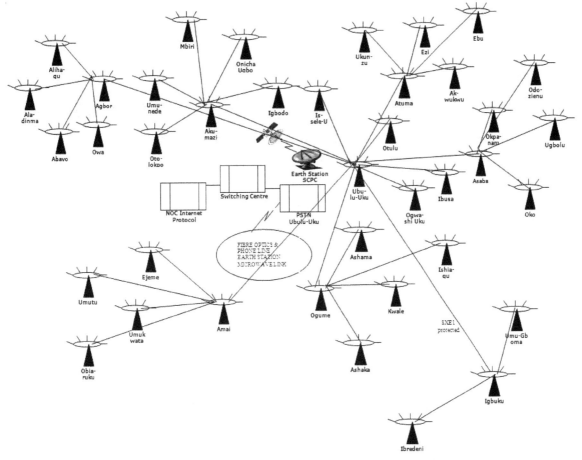

Fig 3: *Cell sites and MTSOs for the Robust Telephony for Delta North*

Table 1: *Selected Cell Sites with Town, Villages and Camps covered in Delta North Senatorial District*

Location	Height	Latitude	Longitude	Areas Covered
Abavo	203	6.180330	6.095506	Abavo, Udomi, Ekuma, Idumuogbo, Ekwuezie, Iyemeni, Igbogili, Azu-Owa, Ojieko, Ete-Erumu, Idumesa, Uteh-Okpu, Oyoko, Obi-Anyima, Obiduhun
Agbor	154	6.089341	6.200879	Agbor, Agbor-Obi, Boji-Boji, Obiliehen, Owanta, Emuru, Aliowa-Agbor, Agbor-Nta, Ottah, Obeihem, Ighobe, Ottah
Aladinma	102	6.307903	6.089131	Aledinma, Owa-Aledinma, Obi-Anyinma, Obi-Iduhun, Obi-Ugbo, Ejeme-Atua, Isomkpe, Ejeme-Uno, Ejeme, Isi-Nkita
Alihagu	120	6.068206	6.240564	Ogan, Alihagu, Owanisua, Alisor, Alisomor, Ozarra, Ahimua, Alijemisi, Omumu, Ozo-Anugogo, Ewuru, Ewotedo, Ewuru, Amahia
Akwukwu	167	6.488999	6.389034	Akwukwu, Onicha-Olona, Ugbo-Obi, Anikoko, Issele-Mkpitime
Akumazi	157	6.380023	6.312250	Akumazi, Uteh-Ogbeje, Anifaka-Eje, Aniwachor, Ali-Ije, Uteh-Alo, Obbio
Anifuma	132	6.310719	6.206131	Anifuma, Uteh-Enugu, Maholo, Uteh-Enugu, Ani-Ugbo, Umuru-Camp, Aniokpor, Ekuku-Agbor, Ete-Erumu
Asaba	187	6.524384	6.154839	Asaba, Okpanam Camp, GRA-Phase, Iyiwudon, Anusai, Urhobo Camp

Location	Height	Latitude	Longitude	Areas Covered
Ashaka	164	6.507700	5.670289	Uchie, Ashaka, Igbuku, Emanweta, Ibrede, Ofagbe, Ellu, Aradhe, Ndoni, Emu-Arade, Emu-Ebendo, Ushie, Avba, Azagba, Emu-Uno, Emu-Obiogo, Emu-Obodeti, Emu-Ebendo
Atuma	102	6.399902	6.209034	Atuma, Issele-Mkpitime, Isa-Issele, Anikoko, Mkpitime
Ebu	205	6.500001	6.779021	Ebu, Eko, Illah, Okpuno, Atawa, Utor, Iyiuku, Ugomuoke, Aniomodue, Ngegu
Ejeme	132	6.387002	6.199012	Ejeme, Ejeme-Aniogor, Ejeme-Uno, Atua
Ezi	203	6.488933	6.600701	Ezi, Achara-Ezi, Umutonu, Ezi-Umuodara, Ukanyi-Egbu, Umutonu
Eziokpor	187	6.285399	5.799824	Eziokpor, Abedei, Umukwata, Abbi, Arogun, Emu, Emu-Uno, Emu-Obiogo, Emu-Obodoeti, Aragba, Umu-Ebu, Emu-Obendo, Umu-Ekumai, Amainge, Ezionum, Umu-Ekukumai, Amai, Ekogboro
Ibredeni	203	6.244598	5.476509	Ibredeni, Asaba-Assa, Ikpede, Onya, Utu-Oku, Etebege, Avba, Ase, Adaiwa, Utu-Ocha, Ivorogbo, Ekpe, Ige
Ibusa	198	6.498932	6.024820	Ibusa, Achala, Aboh-Ogwashi, Achalla-Camp, Ute, Nkpikpa, Aboh, Umuehea,
Igbodo	192	6.305875	6.320010	Igbodo, Ekwuoma, Emuhu, Amurhu, Akuku, Obomkpa, Obbio-Ozili,
Ishiagu	190	6.681719	5.600113	Ishiagu, Isikite, Umute, Nsukwa, Umoni, Ukwu, Abah, Aniofu, Nsukwa,
Issele-Uku	178	6.318023	6.312250	Issele-Uku, Idumuje-Uno, Ukunzu, Ugbodu, Obomkpa, Igbodo
Kwale	213	6.507709	5.690469	Kwale, Utagba, Umu-Osonwu, Osemele, Obetim-Uno, Opaetenasie, Umu-Achiafor, Inyi Creeks, Anuabo, Umu-Edim, Ogbe,
Mbiri	180	6.278341	6.520187	Mbiri, Umunede, Akuku, Otuocha, Mbiri-Iba, Umu-Anatum,
Obiaruku	200	6.200180	5.745092	Obiaruku, Obinomba, Ebedei, Ighobe, Utagba-Uno, Amai, Ikolobie, Obi-Ukwuole,
Odozienu	200	6.355132	6. 400231	Odozienu, Ingene, Anikoko, Aniocha Camp
Ogume	187	6.325875	5.789253	Ogume, Igbe-Ogume, Umu-Osonwu, Obetim-Uno, Ogbagua-Ogume, Obiakamba, Ogbole-Ogume, Umudim-Ogume, Emu-Obiogo, Emu-Obodoeti,
Ogwashi	205	6.483341	6.128288	Ogwashi-Uku, Azagba-Ogwashi, Isa-Asse, Aboh Ogwashi, Nkala-Ofu,
Oko	235	6.681719	6.092131	Oko-Anala, Oko, Oko-Camp, Obiofu, Abala, Odekpe, Otuocha, Akwuebulu,
Okpanam	172	6.680913	6.175504	Okpanam, Anwai, George camp, Azungwu, Abara-Oshimili, Mbalamba
Onicha Ugbo	190	6.388023	6.245220	Onicha-Ugbo, Idumuje-Ugboko, Igbodo, Abukuluku, Uta-Ogbeje, Anifaka-Eje
Otolokpo	151	6.303341	6.200187	Otolokpo, Uteh-Aloh, Uteh-Okpu, Utagba-Ogbe, Utagba
Otulu	241	6.276502	6.178034	Otulu, Ubulu-Okiti, Issele-Azagba, Azagba-Ogwashi, Edo-Ogwashi
Owa	202	6.199802	6.189012	Owa, Owa-Oyibo, Owa-Alero, Owa-Alisor, Okiagbie, Igbanke, Railway Village, Owa-Aladinma, Owa-Ofie, Ugbeka, Ekuku-Agbor, Aledinma, Okiagbie, Owa-Aledinma, Obidi-Ugbo
Ubulu-Uku	198	6.387121	6.203980	Ubulu-Uno, Ubulu-Uku, Edo-Ogwashi, Isa-Ogwashi, Ani-Ugbo, Umuru, Aniokpo, Egbudu, Ali-Ije, Abah-Uno
Ukunzu	167	6.399033	6.589921	Ukunzu, Uburubu, Anioma, Achara-Ezi, Ubodu, Ukala, Ukala-Okpuno, Uku-Anyiegbu, Okuta
Ugbolu	154	6.601129	6.340012	Ugbolu, Egoigwe, Asoko, Ochankie, Iyiwudon
Umudike	190	6.523779	5.819903	Ogbe-Etiti, Onicha, Utagba-Uno, Umudike, Obikwele, Iselegu, Umoni, Inyi, Utchi, Adofi-Camp
Umutu	190	6.392853	5.990824	Umutu, Owahabe, Umuaja, Umu-Isele, Ojeta, Ogbe-Ufie, Obi-Ukwuole, Obi-Ngene, Oliegogo,Umuisele, Ojeta,
Umukwata	162	6.185301	5.799204	Umukwata, Ebedei, Ebedei Uno, Ebedei Waterside, Owa-Abbi, Abbi-Uno, Akuku-Uno, Ogbe-Ufie
Umu-Gboma	198	6.632490	5.676655	Umu-Gboma, Asemoke, Keonokpu, Ononofu, Dioliba, Utu, Okpai, Asaba-Okpai, Abara-Uno, Abo, Issele-Asemoke, Essem

5. Recommendation and Conclusion

From the study, these recommendations were made:

✓ To operate optimally, network requires adequate power supply. Solar energy, compressed wind and gas options may prove to be a better choice. If adequate connection to the national grid exists, it will then serve as backup.

✓ State Governments should investment and deploy such infrastructure to underserved and unserved regions. At completion – handed over to private operators for management and maintenance, with such operators given adequate time to recoup and payoff initial investment via initiating a low-tarrif plan as incentive to trigger interest of dwellers.

✓ An appreciable tax reduction given to operators willing to deploy such network to rural areas – so as encourage participation from them as well as wider coverage within the shortest time.

References

[1] Osuagwu, E., Anyanwu, E and Amaeshi, L., (2003). *Computer-based radar for improved military security surveillance in Nigeria*, Computer Sci. and Application, 9(1), pp.71-81.

[2] Helgert, H., (1991). *Integrated services digital net: Architecture, protocols and standards*, Readings: MA: Addison-Wesley.

[3] Zymanski, W and Zymanski, F., (2004). *Introduction to Computer System*, McGraw Hill Publishers, New York.

[4] Martins, J.A., (1990). *Telecommunications and computer*, Prentice Hall publishers, New Jersey.

[5] Abraham, M., Jajodia, S and Podell, H., (1995) *Information security*, IEEE computer society press, Los Alamitos, CA.

[6] Hafner, K and Lyon, M., (1996). *Where wizards stay late*, Simon Schuster, New York.

[7] Matthews, V.O., Shakunle, O.J and Adetiba, E., (2007). *Hybrid cellular network for mobile networks for rural telecommunications*, Res. in Physical Sci., pp 24-34.

[8] R.E. Schwartz., (1996). *Wireless communications in developing countries: cellular and satellite networks*, Artech house. MA: Boston.

[9] R. Conte., (1994). *Rural telephony: a new appraoch usign mobile satellite communication*, Pacific telemmunication conference, Honolulu: Hawai.

[10] V. Garg, K. Smolik and J.E. Wilkes., (1997). *CDMA Application in wireless nets,* Prentice Hall publications, New Jersey

[11] A. Salamsi and K.S. Gilhousen., (1991). *On system design of CDMA applied to digital cellular and personal network,* Proceedings of IEEE on Vehicular Technical Conference, pp 57-63

[12] R.L Peterson, R.E Ziemer and D.E Borth., (1995). *Introduction to spread spectrum communications*, Prentice Hall publications, Upper Saddle River, NJ.

[13] R.I Pickholtz, D.L Schillings and L.B Milsten., (1982). *"Theory of spread spectrum communications: a tutorial"*, IEEE Transactions on Communications, 30(1), pp 855 – 884.

[14] G.R Cooper and C.D McGillen., (1986). *Modern communications and spread spectrum*, McGraw hill publishers, NJ: Princeton.

[15] J. Rosa., (2007) *"Rural telecommunications via satellite"*, J. of Telecommunications, pp 75 – 81.

[16] R. Conte., (1994). *Rural telephony and telecommunications using mobile satellite communications*, Proceedings of Pacific Telecommunication Conference, Honolulu: Hawaii.

[17] Ojugo, A., Abere, R., Orhionkpaiyo, B., Yoro, E and Eboka, A., (2013). *Technical issues for IP-based telephony in Nigeria*, Int. J. Wireless communications and mobile computing, 1(2), pp 58-67, doi: 10.11648/j.wcmc.20130102.11

[18] J.N. Ofulue., (1985). *Alternate source of energy for telecommunications*, IEEE Proceedings on Communications, pp 87-102.

[19] European Telecommunication Standard Institute (2004). *Electromagnetic compatibility and radio spectrum matters*, Radio sites Engineers equipments and systems, pp 1-60.

[20] STM Wireless., "Rural telephony products" www.stm.com/rural.htm as last accessed Feb 2013.

9

Sideband noise mitigation in a co-located network involving CDMA2000 and WCDMA system

Nosiri Onyebuchi Chikezie[1], Onoh Gregory Nwachukwu[2], Chukwudebe Gloria Azogini[1], Azubogu Austin Chukwuemeka[3]

[1]Department of Electrical & Electronic Engineering, Federal University of Technology Owerri, Nigeria
[2]Department of Electrical & Electronic Engineering, Enugu State University of Science & Technology, Nigeria
[3]Department of Electronic and Computer Engineering, Nnamdi Azikiwe University Awka, Nigeria

Email address:
buchinosiri@gmail.com (Nosiri O. C.), gnonoh@gmail.com (Onoh G. N.), gloriachukwudebe@yahoo.com (Chukwudebe G. A.), autinazu@yahoo.com(Azubogu A. C.)

Abstract: The installation of base station antennas within close frequency range in a co-located scenario constitutes a major interference for radio spectrum engineers. In a co-located setting involving a downlink frequency of CDMA2000 (1960 -1990MHz) and an uplink frequency of WCDMA (1920-1980MHz) as used in the telecommunication industry in Nigeria, the base station receiver is required to receive low amplitude desired signals in the presence of strong transmitting power signals resulting to sideband noise interference. The paper identifies the major mechanism of the sideband noise and proposes the application of a Butterworth Band Pass Filter (BBPF) as a mitigation technique. The technique was developed through the applications of empirical and mathematical analysis conducted in two different scenarios to evaluate the levels of the interference signals on the WCDMA receiver from CDMA2000 transmitter. The first scenario involved a standalone un-collocated WCDMA network while the second scenario involved a co-located network (CDMA2000 and WCDMA). A 52dB required attenuation specification was obtained for the BBPF design.

Keywords: *Sideband Noise, WCDMA, CDMA2000, Co-Location, BBPF*

1. Introduction

Wireless communication technologies having evolved over the years are faced with diverse challenges which the wireless service providers must find ways to navigate for quality service delivery. The desire for wireless service providers to build more cell sites is accelerated by the following factors [1]:

- The need to provide coverage to a geographic region where the service provider has not previously served.
- To cover "dead spot" or areas where existing signals are weak.
- To allow for the reuse of channels or spectrum bandwidth to support a larger number of customers and to meet the higher speed requirements of emerging technologies.

The demand to meet these needs has led to the proliferation of new cell towers which are capital intensive. Possible solution to the proliferation of cell towers is the placement of a number of Radio Frequency (RF) transceiver antennas at close proximity to one another, a concept known as co-location [2].

The benefits of co-location strategy are summarized as follows [3,4]:

- To reduce the proliferation of towers by facilitating sharing thereby maximizing the use of network facilities.
- To co-locate more networks on the same tower to optimize saving and efficient utilization of capital and operational expenditure for site infrastructure and to achieve improved network coverage and capacity.
- To promote fair competition through equal access being granted to the installations and facilities of operators on mutually agreed terms.
- To ensure that the economic advantages derivable from the sharing of facilities are harnessed for the overall benefits of all telecommunication stakeholders.
- To protect the environment from harmful interference.

- To encourage operators to pursue a cost-oriented policy with the added effect of a reduction in the tariffs chargeable to consumers.

Co-location strategy was introduced in Nigeria by the Nigerian Communications Commission (NCC), primarily aimed at reducing capital and operating expenditure of cell sites and also to extend telecommunication services to the un-served and under-served communities in Nigeria [5].

Operators such as MTN, Etisalat, Globalcom, Visaphone, Airtel and Starcomms seeking to expand their network services by building new cell sites are regulated and managed by the NCC licensed service vendors such as the IHS Plc, Swap Technologies, MTI and Helios Towers [6].

One of the essential considerations when analyzing a co-located network is to evaluate if the frequencies of the networks are adjacent to each other, overlap or have close ranges.

However, networks involving the WCDMA (e.g MTN operator) and the CDMA (e.g VISAPHONE operator) at 1.9GHz, when co-located are bound to experience interference due to the close frequency bands of operations between the networks.

The illustrations in Fig. 1 show the frequency spectrum allocation by International Telecommunication Union (ITU) for WCDMA, CDMA2000 and the Universal Mobile Telecommunication System (UMTS) respectively [7]. The frequency band allocation for CDMA 2000 downlink is between 1930-1990MHz and that of WCDMA uplink is between 1920-1980MHz. From Fig 1, it shows that the CDMA2000 transmitter (Tx) frequency band overlaps with the WCDMA receiver band by 50MHz. Therefore, the two frequency bands will interfere in a co-located setting.

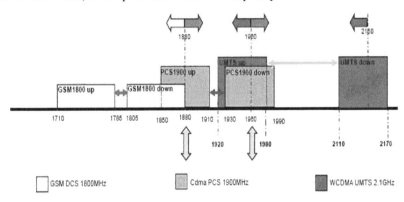

Figure 1. *Spectrum Frequency Band Allocation [7]*

Considering the huge impact of the interference between the two networks if co-located, the regulatory body (NCC) re-banded the frequency between these two networks as shown in Fig. 2. The new CDMA2000 downlink frequency band is between 1960-1990MHz. Fig. 2 shows that the CDMA2000 Tx band overlaps 20MHz on the WCDMA receiver (Rx) band. There is no interference between UMTS base station (Bs) Tx and CDMA2000 Bs Rx because of wide frequency isolation band (see Table 1).

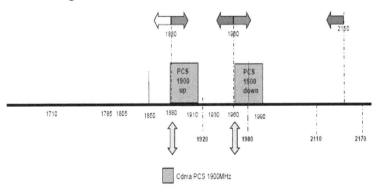

Figure 2. *Spectrum Allocation by NCC (After re-banding)[8]*

Table 1. *Frequency spectrum allocation*

Wireless access technology	Mobile station transmitter (Uplink) (MHz)	Base station Transmitter (downlink)(MHz)	Duplex separation (MHz)
CDMA 2000	1850-1910	1930-1990	80
WCDMA	1920-1980	2110-2170	190

Fig. 3, shows the likelihood of spurious emissions levels on the receiving front end of the WCDMA from the CDMA2000(C2k) in a co-located scenario.

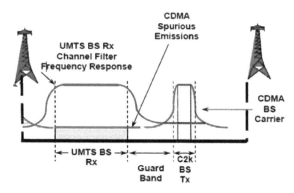

Figure 3. Spurious Emission level [8]

This work was therefore considered necessary because of the salient interference challenges faced between the CDMA2000 Tx band and the WCDMA Rx band at 1.9GHz in a co-located scenario, despite the re-banding by NCC. The co-location strategy investigated in this paper is mainly on passive sharing which involves non-electronic components and facilities such as towers, shelters, electric supply, easements and duct [9].

2. The Mechanism of Sideband Noise

In a co-located network, base station receivers have to receive weak desired signals in the presence of high power transmit signals which may lead to interference[10].

When RF signal is amplified to form the transmit signal, a significant amount of emissions are generated outside the transmit band referred to as sideband noise emission [11]. The emissions are due to the non-linearity and noise generated inside the power amplifier, and may appear as a "skirt" or "shoulders" when observed through the power spectrum at the output of power amplifier. These emissions or undesired noise energy may fall within the pass bands of a co-located receiver, degrading the receiver sensitivity with corresponding rise in the total noise floor level [11]. These undesired noise energy also contribute to the Carrier to Interference (C/I) ratio degradation, reduction in the full utilization of the capacity and the coverage radius, thereby disfranchising the end users from enjoying their hard paid services. Therefore, the undesired spectral components have to be reduced to an acceptable performance level to avoid the introduction of excessive noise in the receiver.

3. Methodology

The first step in recognizing if interference has corrupted a receiver system is to understand the characteristics of the signal that the affected system is intended to receive while the second step requires determining the level of the interfering power that affects the amplitude of the received signal.

In this research, the interference leakage of the CDMA2000 Tx signal power on WCDMA Rx signals was investigated. The test network was set up in Enugu State,

involving a co-located site and an unco-located site respectively. The co-located site, involving CDMA2000 and WCDMA systems with separate antennas placed vertically collinear to each other, situated at the Federal Housing Estate Trans Ekulu Enugu, Enugu. This site has the following characteristics: Visa ID: ENU005, HIS ID:IHS_ENG_007, Network operators: MTN and Visaphone, BTS Local cell ID: 2155, sector ID: 0, Carrier ID:18.

The unco-located site involves a standalone WCDMA, situated at Independent Layout New Haven, Enugu, characterized by the following: Site ID: HENB549, Cell ID: EN0099C, longitude and latitude of 7.5286945 and 6.44658 respectively, the network operator is MTN. All the sites were situated in urban environment.

A detailed experiment was carried out to measure the interference leakage using a *Huawei software M2000 Service Maintenance System CBSS* installed in a laptop. The software provided a window where the sites to be investigated were selected within the sector, carrier and duration. The measurement was carried out remotely from the IHS central office.

3.1. Empirical Analysis and Evaluation for the Un-Collocated Network

The received signal power for the WCDMA was measured, conducted from Huawei site. The Noise Figure (NF) of the WCDMA signals was calculated and obtained as 5dB using "(1)".

$$NF(dB) = MDS(dBm) - KT_0 B_c [12] \qquad (1)$$

Where:
- MDS is the minimum detectable signal (-102dBm) for WCDMA network [12]
- k is the Boltzmann's constant ($1.38x10^{-23} J/k$,)
- T_0 is the ambient temperature in Kelvin ($T_0 = 290k$)
- B_c is the channel bandwidth for the WCDMA (5MHz)

The following parameters were obtained by calculation: the noise floor level and the minimum demodulation C/I ratio.

- **The Noise Floor Level**

Calculating the receiver's noise floor level requires the knowledge of the noise figure (NF). The model in "(2)" was used to calculate the noise floor level.

$$N_{floor} = 10 \frac{KT_0 B_s + NF}{10} (mW) [13] \qquad (2)$$

Where B_s is the signal bandwidth of WCDMA (3.84MHz).

- **The Minimum Demodulation Carrier to Interference (C/I) Ratio**

The C/I ratio in a communication channel characterizes the quality with which information is transferred through the channel. Equation (3) is used to calculate the minimum demodulation of the C/I ratio.

$$\left(\frac{C}{I}\right)_m = S_0 - N_{floor} [13] \qquad (3)$$

Where S_0 is receiver sensitivity.

The summary of the results obtained are shown in Table 2.

3.2. Empirical Analysis and Evaluation for the co- Located Network

In this scenario, the measurements and evaluation of the interfering signal power, noise floor level, minimum Demodulation Carrier to Interference ratio and percentage C/I ratio degradation respectively for the co-located networks were considered. The parameters measured include:

  i.   The CDMA2000 Tx signal power.
  ii.   The WCDMA Rx signal power.

The BTS transmitter power and the BTS received signal power were measured using *Huawei software M2000 Service Maintenance System (CBSS)*. The summary of the measured and calculated results are shown in Table 3.

The following parameters were calculated:

• The Noise Floor Level

The difference in the received signal level for co-located and unco-located networks gives the degraded receiver sensitivity denoted as η. If the receiver sensitivity is degraded by η dB then the interference plus noise power is given by

$$N_{floor} + Interference = 10\frac{KT_0B_s + NF}{10}10\frac{\eta}{10}(mW)[13] \qquad (4)$$

If the interference level is equal to the equivalent noise level of the original signal, the signal sensitivity will be degraded by 3dB [7]. Therefore, it is important to ensure that the noise level of the original signal is always 3dB above the interference noise level to maintain victim's percentage ratio [7].

• The Interfering Signals Power

In the process of proffering solution to this prevailing interference, one of the vital considerations is to evaluate the degree and the impact of the interfering power on the victim receiver channel especially on the system receiver sensitivity and noise floor level. Therefore the interfering power at the receiver input, denoted as (γ) is calculated using "(5)".

$$\gamma = (KT_0B_s + NF) + 10\log_{10}\left\{10\frac{\eta}{10} - 1\right\}dBm[13] \qquad (5)$$

This is expressed in terms of erosion of the receiver sensitivity. By definition, a degradation of receiver sensitivity (dB) is equal to the increase in the total noise plus interference [7]. This denotes that if the noise level at the receiver increases by 1dB, the receiver sensitivity of the

BTS decreases by 1dB accordingly.

• The Minimum Demodulation Carrier to Interference (C/I) Ratio

The minimum demodulation carrier to interference ratio (dB) in a co-located network is obtained using "(6)" and the result shown in Table 3.

$$\left(\frac{C}{I}\right)_m = S_0 - (N_{floor} + Interference)[13] \qquad (6)$$

4. Results and Discussion

Tables 2 and 3 show the detailed summary of the data obtained for the co-located and un-collocated networks.

Table 2. *Summary of data obtained for the un-collocated WCDMA site.*

S/N	Parameter	Obtained values
1	Measured Received signal strength for the WCDMA network	-109.69dBm
2	Calculated Noise floor for the WCDMA network	-103dB
3	Calculated Minimum demodulation C/I (dB) for the WCDMA network	-6.69

Table 3. *Summary of the data obtained for the co-located network.*

S/N	Parameter	Obtained values
1	Measured Received signal strength for the WCDMA network	-110.70dBm
2	Measured Transmitter power for the CDMA2000	41dBm
3	Calculated Noise floor for the WCDMA network	-101.99dB
4	Calculated interfering signal power(γ) for the WCDMA network	-108.80dBm
5	Calculated Minimum demodulation C/I ratio (dB) for the WCDMA network	-8.71
6	Calculated C/I ratio degradation for the WCDMA network	2.01
7	Calculated percentage C/I ratio degradation for the WCDMA network	30
8	Calculated Receiver sensitivity degradation (η) for the WCDMA network	1.01dBm

The behaviour and performance analysis of any system is better explained using graphical representations. The simulated performance results obtained using Matlab 7.0 software tool was shown in Figs. 4-8. The graphs show the various performance effects of interfering power on the system reciever sensitivity and noise floor level. The minimum demodulation C/I ratio after the system is interfered and the percentage C/I ratio degradation effects were also graphically represented.

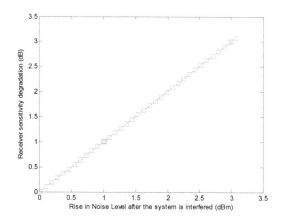

Fig. 4. Interference effect on the system noise level performance

The linear graph in Fig. 4 shows the performance effect of receiver sensitivity degradation caused by the interfering power on the rise noise floor level. A receiver sensitivity degradation of 1.01dBm gave rise to increase in noise floor level by same value.

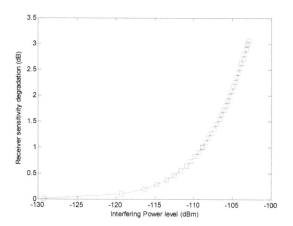

Fig. 5. Effect of rise in interfering power on the system sensitivity.

The exponential performance of the graph of Fig. 5 shows the effects of the CDMA2000 interfering power on the WCDMA receiver sensitivity. Interfering power level of -108.80dBm degrades the sensitivity of the receiver by 1.01dBm. Hence as the interfering power increases, the receiver sensitivity (victim receiver) degrades exponentially.

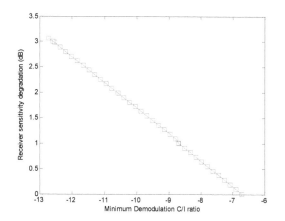

Fig. 6. Minimum demodulation C/I ratio after the system is interfered.

The negative slope in Fig. 6 shows the relationship between the receiver sensitivity degradation of the victim receiver and the minimum demodulation Carrier to Interference ratio. When the receiver sensitivity degrades by 1.01, the minimum demodulation C/I ratio decreases by -8.71.

Fig.7 shows the performance characteristics on the increase in interfering power level and its corresponding effects on the system noise floor. From the graph, increase in interfering power of -108.80dBm gave rise to noise floor level of 1.01dBm. This graph further explains the negative impact of the sideband noise on the channel capacity of the victim receiver. As the noise floor level increases, the overall system capacity reduces, giving rise to poor channel capacity performance. This means that only fewer subscribers may be accommodated.

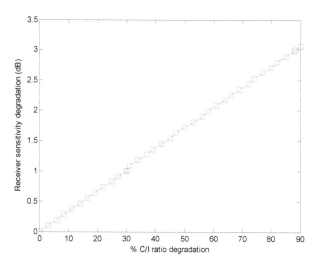

Fig. 7. Rise in system noise floor level after the system is interfered.

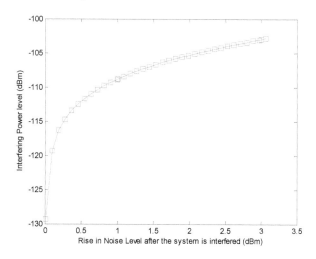

Fig. 8. Percentage C/I ratio degradation after the system is interfered

Fig. 8 clearly explains the percentage degradation performance of the carrier to interference ratio as the receiver sensitivity further degrades. From the graphical representation, 1.01dBm sensitivity degradation resulted to 30% C/I ratio reduction. The degradation of the C/I ratio impacts negatively on the system channel capacity. If not controlled with adequate measures, it could reduce the assigned number of subscribers within the channel.

The summary results from the graph show the following important points:

1. Increase in Interference power of CDMA2000 increases the WCDMA noise level and reduces the sensitivity of the receiver antenna, leading to high call drop rate, hence restraining subscribers from enjoying consistent and high quality services in a WCDMA cell.
2. Interference decreases WCDMA capacity due to the rise in the noise level, hence fewer subscribers will only be accommodated.
3. 1.01 dB degradation gave rise to 1.01dB noise floor and interfering power of 108.80dBm

- **Butterworth Band Pass Filter (BBPF) Design Procedure**

The limits of spurious emission levels by the Three Generation Partnership Project Two (3GPP2) for CDMA far offset from carrier is given as: -13dBm/1MHz (1GHz<f<5GHz)[8].

In order to guarantee that the affected receiver's performance will not degrade, the isolation between the interfering transmitter and affected receiver should be:

$$-13dBm / 1MHz - (-108.80 / 3.84MHz) = 101.80 \approx 102dB \quad (7)$$

The total isolation required to maintain the received signal optimum performance in the co-located scenario is 102dB. The standard antenna-to-antenna isolation specification (dB) by the NCC for personal communication systems (PCS), digital communication systems (DCS) and universal mobile telecommunication systems (UMTS) in a co-located site is 50dB [6]. Therefore a 52dB rejection at 5MHz guard band was obtained as the required specifications for the Butterworth Band Pass Filter (BBPF) offset from the low side edge of the pass band. A guard band of 5MHz pass band was considered in other to offer a faster roll-off space for the BBPF.

- **BBPF Design Specifications**

The BBPF was primarily considered in this work among other infinite impulse response filters (IIR) because of its attributes to the least amount of phase distortion. Every filter design requires creating the filter coefficients to meet specific filtering requirements. The specifications in Hertz are converted to normalized frequencies (ω) using "(8)":

$$\omega = \frac{2\pi f}{f_s} \quad (8)$$

Where:

f is the absolute frequency in Hertz, f_s is the sampling frequency in samples/second and ω is the normalized frequency in π radian/sample. Tables 4 and 5, show the Frequency and Magnitude specifications for the BBPF.

***Table 4.** Frequency Specifications*

Filter parameters	Frequency specifications(KHz)	Normalized frequency (ω) (π radian/samples)
F_{stop1}	1955000	0.4912= ω_{Stop1}
F_{pass1}	1960000	0.4924= ω_{Pass1}
F_{pass2}	1990000	0.5000= ω_{Pass2}
F_{stop2}	1995000	0.5012= ω_{Stop2}
Sampling frequency f_s = 7960000		

***Table 5.** Magnitude Specifications.*

Filter parameters	Magnitude specifications
A_{stop1}	52dB
A_{pass}	0.1dB
A_{stop2}	52dB

Where;

F_{pass1} : Frequency at the edge of the start of the pass band.
F_{pass2} : Frequency at the edge of the end of the pass band.
F_{stop1} : Frequency at the edge of the start of the first stop band,
F_{stop2} : Frequency at the edge of the start of the second Stop band,
A_{stop1}: Attenuation in the first stop band in dB
A_{Pass}: Amount of ripple allowed in the pass band, known as the single pass band gain parameter
A_{stop2}: Attenuation in the second stop band in dB

The output result interprets that any interfering signal that comes out of the filter must be 52dB lower than it went in. In practice, the designed filter should be placed at the CDMA2000 front end to reject the sideband noise falling within the WCDMA receiver pass band. The graphs in Figs. 9 and 10 show the magnitude response and the phase response of the designed filter.

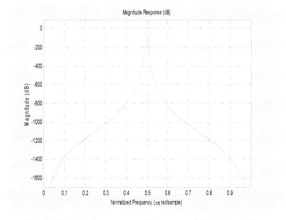

***Fig. 9.** Magnitude response of Butterworth Bandpass filter.*

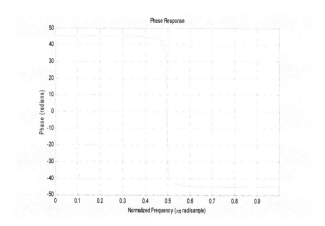

Fig. 10. Phase response of Bandpass filter

monotonically and generates a maximally flat response with no ripples both in the pass band and stop band, this provided a sharp phase linear response with minimal or no distortion.

The graph in Fig. 9 shows a symmetrical magnitude response of the BBPF. Great importance was given to the magnitude response to show how selectively the filter performs as it gets steeper and also the cut-off frequencies. The first cut-off frequency (3-dB point) and the second cut-off frequency were obtained as 0.49215 π rad/sample and 0.50025 π rad/sample respectively.The geometric mean of the upper and lower 3-dB cut-off frequencies was evaluated as 0.4962 π rad/sample. The geometric mean value identifies the point at which the filter achieves its maximum gain.

Phase response is the phase shift of the output relative to its input. Fig.10 shows how the filter response decreases

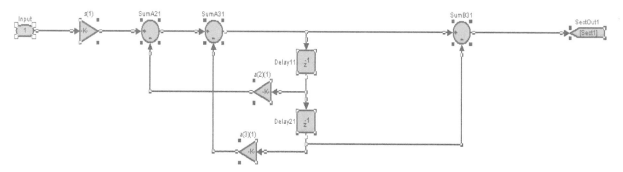

Fig. 11. Simulink block for the Band pass filter

Fig. 11 shows the Simulink block function parameters of the designed BBPF. It shows the first section out of the cascaded 29 sections. The function block parameters represented the gains and the delay samples. Implementation of this filter as a cascade of quadratic factors provides a better control of the stability of the filter.

4. Conclusion

The detailed performance effects of the interfering noise power on the receiver sensitivity, noise floor level and the carrier to interference ratio were clearly evaluated. From the measurements conducted, the interfering power was evaluated as -108.80dBm, which gave rise to 1.01dBm receiver sensitivity degradation and 1.01dBm rise in total system noise floor, about 30% degradation in C/I ratio. A 52dB rejection at 5MHz guard band offset from the low side edge of the pass band was obtained. An application of BBPF was considered necessary as a tool designed to mitigate the challenging effects of sideband noise. In practice, the designed BBPF should be installed at the CDMA2000 front end to reject the sideband noise falling within the WCDMA receiver pass band.

Recommendation for Further Work

With the rapid increase in diversity in wireless

communication systems especially when operating in a dynamic environment, the authors recommend the application of adaptive noise cancellation technique (ANCT) considering the limitations of the BBPF which includes:

a. Poor efficient power handling capability and
b. Low frequency agility response.

References

[1] P. Stox "Wireless co-location on towers and alternate structure" *satellite broadcast TV, linksystem.uk.com.* April, 2010.

[2] Guidelines for siting and sharing of Telecommunication base station infrastructure, "*Rwanda Utilities Regulatory Agency*" RURA/ICT infraDev/Dev/G02/2011. Pp 10-12, 2011.

[3] Guidelines on Collocation and Infrastructure Sharing Issued by the *Nigerian Communications Commission*, pp 1-2, 2010

[4] S. Opara, I. Iteun, "Option for Telecoms operators during recession". A published article pp1-2, June, 2009

[5] F.E. Idachaba, "Telecommunication Cost Reduction in Nigeria through Infrastructure Sharing between Operators". *Pacific Journal of Science and Technology.* 11(1):272-276, 2010.

[6] Nigeria Communications Week "Managed services in Nigeria ", March 11, 2012.

[7] Chen Xinting, "Analysis of co-site interference between different mobile communication system", *wireless technical support department, Huawei technologies Co.* Ltd, pg 9, 2008.

[8] Trino G., Chen L., "interference analysis between co-located networks in Nigeria", *RF design Manager from Starcomms* 2nd version, 2007.

[9] Policy on co-location and infrastructure sharing in SAMOA. *Office of the regulator*, pp 4-8, March, 2012

[10] S. Ahmed, "Interference mitigation in co-located wireless systems". *A PhD thesis, School of Engineering and Science,* Faculty of Health, Engineering and science, Victoria University, pg 3, Dec., 2012.

[11] A. Roussel "Feedforward interference cancellation system applied to the 800MHz CDMA cellular band". *A Master of applied science in Electrical Engineering. Ottawa-Carleton Institute.* Pg 6, 2003.

[12] Chenung, Tze Chiu, , *"Radio Performance"*, *Chapter 4* Virginia Tech: ETD 122298 pp 96-99, 2009

[13] Montegrotto T. "Practical Mechanism to Improve the Compatibility between GSM-R and Public Mobile Networks and Guidance on Practical Co-ordination", *ECC Report* 162, p. 40 May, 2011.

10

KVM vs. LXC: comparing performance and isolation of hardware-assisted virtual routers

Muhammad Siraj Rathore, Markus Hidell, Peter Sjödin

Network Systems Laboratory, School of ICT, KTH Royal Institute of Technology, Stockholm, Sweden

Email address:
siraj@kth.se (M. S. Rathore), mahidell@kth.se (M. Hidell), psj@kth.se (P. Sjödin)

Abstract: Concerns have been raised about the performance of PC-based virtual routers as they do packet processing in software. Furthermore, it becomes challenging to maintain isolation among virtual routers due to resource contention in a shared environment. Hardware vendors recognize this issue and PC hardware with virtualization support (SR-IOV and Intel-VTd) has been introduced in recent years. In this paper, we investigate how such hardware features can be integrated with two different virtualization technologies (LXC and KVM) to enhance performance and isolation of virtual routers on shared environments. We compare LXC and KVM and our results indicate that KVM in combination with hardware support can provide better trade-offs between performance and isolation. We notice that KVM has slightly lower throughput, but has superior isolation properties by providing more explicit control of CPU resources. We demonstrate that KVM allows defining a CPU share for a virtual router, something that is difficult to achieve in LXC, where packet forwarding is done in a kernel shared by all virtual routers.

Keywords: Network Virtualization, Virtual Router (VR), SR-IOV, Virtual Function (VF), SoftIRQ, NAPI

1. Introduction

Network virtualization allows running heterogeneous virtual networks in parallel to support a diverse range of services over a shared substrate. An important building block of network virtualization is router virtualization. One way to enable virtual routers is to use open source virtualization technologies on commodity PC hardware and let each virtual machine act as a router. This is a flexible and low-cost solution. However, there are concerns that PC-based virtual routers could potentially suffer from low performance as packets are processed in software [1][2]. Furthermore, it becomes challenging to maintain isolation among virtual routers due to resource contention in a PC environment [3][4].

It is complicated to provide performance and isolation at the same time in a PC environment. To address this issue, we investigate how virtualization support in PC hardware can be used for virtual routers. Single root I/O virtualization (SR-IOV) [5] is a step in that direction. It provides hardware support to virtualize network interface cards (NICs). SR-IOV divides a single physical PCIe device into multiple PCIe instances, called Virtual Functions (VFs) [5][15]. A VF interface is an Ethernet-like interface that can be used in a virtual router. In addition, SR-IOV offloads packet handling from the host CPU and allows packets to be directly dispatched to virtual routers. This should result in performance improvements. Furthermore, SR-IOV provides dedicated hardware queues (receive and transmit) for each VF, which can be used to isolate traffic streams for different virtual routers.

Different virtualization approaches may have different properties from performance and isolation perspectives. KVM [6] is a full virtualization solution where hardware resources are virtualized through hypervisor software. It is a flexible solution that allows a diversity of virtual machines to run on the same host, but at the potential cost of performance penalties due to overhead. In contrast, LXC [7] is a container-based approach where operating system resources (e.g. files, system libraries, routing tables) are virtualized to create multiple execution environments within the same operating system [2][7]. It is attractive from a performance point of view, but has negative implications for isolation.

In our work we investigate the impact of SR-IOV on performance and isolation of KVM and LXC-based virtual routers. We anticipate a certain degree of performance improvement in both cases due to processing offload.

However, we believe that different virtualization techniques can exploit hardware support differently depending on how packets are processed along the forwarding path. Hence, we examine the forwarding path and modifications made by the hardware. Our hypothesis is that KVM could make use of hardware support more effectively by eliminating a considerable amount of software-based packet processing. In comparison, LXC is already lightweight in nature, and the relative improvement gains could be less significant.

When it comes to isolation, the resource contention among virtual routers may lead to poor isolation properties. For instance, an overloaded virtual router can consume much CPU and starve others. Such behavior can be avoided by defining a guaranteed CPU share for a virtual router. In this regard, Linux kernel introduces a CPU share feature to provide controlled CPU access among different processes [8]. However, CPU share is intended to control CPU for user-space applications (processes) in server environments, and may not be directly applicable for virtual routers. Generally, a forwarding path is a combination of many different components including interrupt processing, kernel and user level devices. It is not trivial to control CPU usage along a forwarding path in a consolidated fashion. In such situations, SR-IOV might be useful. It offloads packet handling by replacing software-based processing modules along the forwarding path. We investigate how CPU share can be combined with hardware-assisted forwarding paths in KVM and LXC-based virtual routers. Furthermore, we analyze how CPU guarantees can be enforced and what the effects would be on isolation between virtual routers. Finally, we identify an approach that provides a suitable trade-off between performance and isolation after hardware support.

The rest of this paper is organized as follows: Section 2 surveys related work on virtual router platforms. Section 3 presents the packet forwarding architecture for KVM and LXC-based virtual routers. Thereafter, section 4 describes our performance and isolation measurements, results and discussion. Finally, Section 5 concludes the paper.

2. Related Work and Contributions

There are several studies in the literature where Xen is proposed to enable virtual routers [9][10][11]. The work presented in [10] compares performance of two different versions of Xen (3.1 and 3.2). It shows that guest domain packet forwarding is improved in version 3.2, but that isolation is weakened when more virtual routers are added. Another work [11] suggests an architecture using Xen and Click. It achieves encouraging performance and isolation results, but requires dedicated NICs to be allocated to each virtual router, something that might be difficult to realize in practice. Others introduce customized components to improve forwarding performance [12], but observe degraded isolation (in terms of packet loss) at high network load.

We conclude from previous work that software-based solutions have difficulties providing high performance and strong isolation at the same time. We therefore propose a hardware-assisted platform for router virtualization. There are some examples where SR-IOV is proposed to improve performance of virtual machines [13][14] in a server setting. The focus of these studies is not on virtual routers and therefore no relevant results are available (e.g. packet forwarding rate, latency etc).

In our previous work, we study PC-based virtual routers [2][15]. The first study [2] compares two container-based approaches (i.e. OpenVZ and LXC) from a performance perspective. Our results show that LXC achieves better performance than OpenVZ. However, the work does not consider hardware assistance for virtualization. The other work [15] focuses on hardware assistance (i.e. SR-IOV) to enable virtual routers in LXC environment. We investigate how SR-IOV can be used to enable parallel forwarding paths over a multi-core platform. We dedicate a CPU core to each virtual router in order to improve performance and isolation. The focus of the current paper is different, since it aims to compare two different virtualization techniques. In addition, we evaluate a more challenging and practical scenario where a CPU core is shared between virtual routers. We anticipate that such resource sharing may result in degraded isolation properties for a shared kernel environment (i.e. LXC). In this regard, a KVM-based approach might be a better alternative, since it uses a different packet forwarding architecture.

To our knowledge, there is no previous work on comparing KVM and LXC for hardware-assisted virtual routers. These techniques are part of the mainstream Linux kernel and hence readily available. A solution based on them should be more up to date and adoptable. On the other hand, Xen based solutions in existing literature are not completely in line with main stream kernel. For instance, Xen uses custom process scheduler and memory management system.

We evaluate the CPU share feature of Linux kernel to control CPU usage of parallel running virtual routers, something not investigated before. In addition, we study how non-virtualized components of the underlying host (e.g. NAPI will be discussed later) can impact the behavior of virtual routers.

3. Virtual Routers Forwarding Architecture

In this section we explore the packet forwarding path for KVM and LXC-based virtual routers. First we discuss various software-based devices to enable virtual routers. Then we investigate how these devices can be replaced with hardware-assisted devices. We analyze impact of hardware support on forwarding paths from performance and isolation perspectives.

3.1. KVM-based Virtual Routers

Figure 1 shows the packet forwarding path for a KVM-based virtual router. When a packet is received on a physical network interface (NIC1) it is copied to the main memory of the host system, and an interrupt is generated to notify the CPU. The Linux packet reception API (NAPI [16]) handles this interrupt. It adds the network interface to a queue (the NAPI poll list for interfaces with incoming traffic) and disables interrupts for more incoming packets on that interface. An RX SoftIRQ (software interrupt/kernel thread) is scheduled to process the packet. Through the SoftIRQ, NAPI serves network interfaces (i.e. packet processing) from the NAPI poll list in a round robin fashion.

In a virtualized environment, the first processing task is to identify the virtual interface (VIF) within the virtual router that should process the incoming packet. A VIF is an Ethernet-like interface with a unique MAC address. A virtual router may contain one or more VIFs. The VIF is identified based on the destination MAC address in the Ethernet header. This process is known as physical-virtual *device mapping* (Figure 1).

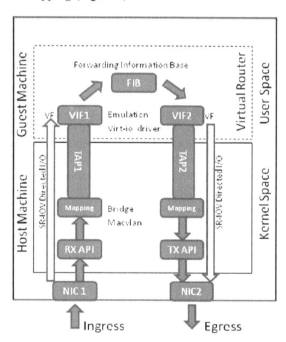

Figure 1. *Forwarding path- KVM-based virtual router*

There are various options for how to perform device mapping. The most common is Linux bridging. It is a software implementation of a bridge function that can switch packets between any pair of interfaces based on MAC addresses. However, Linux bridging has a considerable amount of overhead for redirecting packets between virtual and physical interfaces. Similar solutions are virtual switch [17] and Open vSwitch [18]. An attractive solution from a performance perspective is to replace the software bridge with macvlan devices. A macvlan device maintains a static MAC address table to

provide physical to virtual address mapping and thus incurs less processing overhead [2].

After the device mapping, the packet should be made available to the virtual router on the VIF. However, according to the KVM architecture, a guest machine runs in user space with its own memory management system [6]. This requires a copy operation to move the packet from kernel to the user memory that belongs to the VIF. Furthermore, a CPU context switch is also required from kernel to user mode. Hence, a packet cannot be immediately delivered to the VIF after device mapping. Instead, it must be queued in kernel space. We use the tap device for this purpose (Figure 1). The tap device is a layer 2 network device that consists of two interfaces, one in kernel space and one in user space. The two interfaces are connected in such a way that data written at one end is available for reading at the other end.

At this point there are several alternatives for a VIF. For instance, a virtual interface can be an emulated network device using Qemu. An emulated device can be attractive as it uses standard network device drivers without any changes (e.g. Intel e1000). However, emulation is a CPU-intensive task and leads towards high performance penalties. Another option is the virtio para-virtualized network device. It is optimized to reduce virtualization overhead but requires modifications in the guest kernel. Yet another option is the macvtap device. It integrates the functionality of macvlan (device mapping), tap device and VIF in a single device, which makes it an attractive choice.

Once the packet is available on the ingress VIF (VIF1 in Figure 1) of the virtual router, the forwarding decision is taken and next hop is determined. The packet is placed on the outgoing virtual interface VIF2. After that, the tap interface and the mapping device are used in order to place the packet on the outgoing physical interface (NIC2).

We conclude that the software-based approach demands much packet processing in order to deliver a packet to a virtual router, which could lead to performance penalties. In contrast, a hardware-assisted approach makes it possible to directly deliver a packet to a virtual interface inside a virtual router without any major intervention from the system CPU. With such an approach, when a packet is received on a NIC, it is passed to a hardware switch inside the NIC. A destination MAC address lookup is performed to determine the VF. After that, the packet is placed on a hardware queue reserved for that VF. The next step is to transfer the packet from NIC to the VF's memory areas in the virtual router (i.e. the guest machine). The guest machine has its own memory addresses separated from host memory addresses, so guest memory addresses need to be translated to host memory addresses in order to perform a DMA operation. Hardware support for such address translation is integrated inside the CPU chipset [19] (Intel-VTd, directed I/O), which makes it possible to directly transfer a packet to virtual router memory (something that would otherwise require software intervention). This should improve forwarding performance.

We see in Figure 1 that SR-IOV replaces many software-based devices (e.g. bridge, macvlan, and tap) and therefore appears promising for considerable processing offload. We also observe other architectural benefits; SR-IOV replaces the entire kernel level packet processing required in the software-based solution. Packets are processed only in user space. This should improve cache locality and reduce the amount of CPU context switches (between kernel and user space) compared to a software-based architecture. In addition, it should be more straight-forward to control a CPU in such an architecture, by for example defining CPU share for user space processes. As a result, we expect to achieve better isolation between virtual routers.

3.2. LXC-based Virtual Routers

In the case of LXC, packet reception and device mapping are similar to that of KVM. However, in contrast to KVM, LXC performs packet forwarding in kernel space, as shown in Figure 2. It does not require any packet copying or context switching operations, and the packet is immediately available to a virtual router after device mapping. This should result in better performance, compared to KVM.

We have two choices for VIFs. The first is the virtual Ethernet (veth) device. It is an Ethernet-like device, which can be used in combination with the bridge device in order to access the host's physical devices. Alternatively we can use the macvlan device, which combines the functionality of virtual interfaces and device mapping.

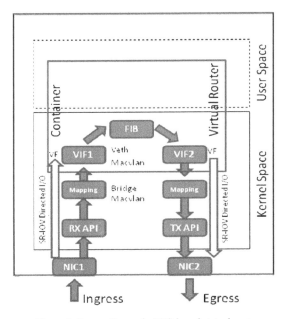

Figure 2. *Forwarding path- LXC-based virtual router*

We see in Figure 2 that SR-IOV can be used in the same way as for KVM to offload packet processing onto the hardware. However, in contrast to KVM, there are not many differences between software and hardware based solutions from an architectural point of view, since both solutions perform packet processing in kernel space. The kernel space processing might be an advantage from a performance point of view, but there are potential drawbacks from a resource management perspective. The reason is that packet handling inside the kernel is mainly done through SoftIRQ processing. SoftIRQ serves all (virtual) interfaces in a round robin fashion with equal priority, even though the interfaces may belong to different virtual routers. Accordingly, it is not possible to control the CPU usage of individual forwarding paths, something that we expect could lead to poor isolation properties.

4. Experimental Evaluation

In this section we evaluate performance and isolation of KVM and LXC-based virtual routers. First, we measure performance both for software and hardware based solutions. For KVM, we compare macvtap and SR-IOV whereas macvlan and SR-IOV are compared in LXC. We investigate the level of performance gains that can be achieved using hardware assistance. As the next step we increase the number of virtual router running in parallel and measure the effect on aggregated performance.

For the isolation study, we consider two SR-IOV-based virtual routers running in parallel with different performance requirements. We overload one of the virtual routers and study the impact on the performance of the other virtual router. We evaluate two different scenarios by varying performance requirements and offered loads.

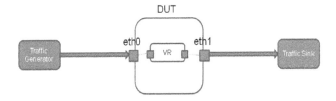

Figure 3. *Experimental test bed*

The experimental setup is shown in Figure 3. We use three Linux machines where the first, Traffic Generator, is used to generate network load using pktgen [20]. The load is fed into the device under test (DUT). The DUT has two physical interfaces and a virtual router that forward packets from one interface to the other. The virtual router is configured with two virtual interfaces as shown in Figure 3; one virtual interface is connected to the physical ingress interface while the other virtual interface is connected to the physical egress interface. The network load is received on a third machine, Traffic Sink. All performance measurements are taken at Traffic Sink using pktgen receiver side utility [21].

The hardware used for DUT is Intel i7 Quad Core 3.4 GHz processor (Intel VT-d supported, chipset Intel Q-67 Express) and 4GB of RAM, running Linux kernel net-next 3.2-rc1. We use a single CPU core in all experiments unless otherwise stated. The machine is also equipped with one 1 Gbps dual-port NIC with an Intel 82576 GbE controller. On each port a maximum of eight SR-IOV devices are supported. It means that we can run up to eight parallel virtual routers using the configuration shown in Figure 3.

4.1. Performance Results

We offer network load (100 UDP Flows) on DUT and gradually increase the load until line rate i.e. 1488 kilo packets per second (kpps) using 64 byte packets. As a baseline, we relate the performance of the virtual router to the performance of regular "IP forwarding" in a non-virtualized Linux-based router.

The throughput results are presented in Figure 4. It can be seen that only the baseline IP forwarder is able to achieve line rate. The rest of the configurations are below line rate. SR-IOV has the highest rate compared to the other software-based approaches. It is interesting to note that the difference between SR-IOV (KVM) and SR-IOV (LXC) is marginal. The former achieves 1250 kpps whereas SR-IOV (LXC) obtains 1300 kpps. It indicates that full virtualization with proper hardware support can achieve performance comparable to that of lightweight containers.

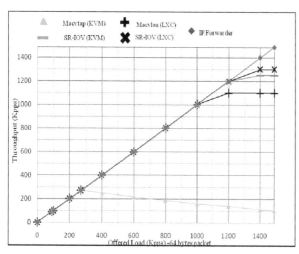

Figure 4. *Throughput- LXC vs. KVM based virtual router*

A clear performance difference can be observed among the software-based approaches. The macvlan (LXC) achieves around 11 times higher throughput than macvtap (KVM). This is remarkable, so we investigate more closely the poor performance of macvtap. With macvtap, there is a significant amount of packet drop on the tap interface (kernel side) at high packet rates. The amount of packet drop increases when the load is increased. This results in more throughput degradation. For instance, we see in Figure 4 that throughput is around 280 kpps for macvtap (KVM) at an offered load of 280 kpps. However, when load is increased up to the line rate, throughput degrades to 100 kpps.

We further investigate the reasons behind this large packet drop by measuring how CPU utilization varies with offered load. In addition to total CPU utilization, we also measure CPU consumption in kernel and user space. We see in Figure 5 that macvtap-kernel CPU usage increases

very quickly with offered load. This behavior points towards an architectural bottleneck of the KVM software-based setup: In the software-based approach, packets are switched between kernel and user space during forwarding and this switching becomes a bottleneck at high load. The packet handling is done through SoftIRQs inside the kernel, which runs at higher CPU priority than user space processes. At high offered load, the CPU is occupied with SoftIRQ processing most of the time. This results in starvation of user space processes and the virtual router is unable to process its incoming queue. As a result, packets are simply dropped after RX SoftIRQ handling, without further processing. This is clearly a waste of CPU resources and results in throughput degradation. The SR-IOV (KVM) eliminates this bottleneck by offloading the kernel side packet handling to hardware. This results in much higher throughput.

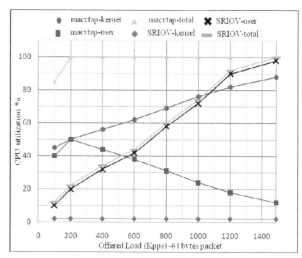

Figure 5. *CPU utilization of KVM based virtual router*

In addition to throughput, we measure latency at maximum offered load (i.e. line rate). The results follow the same pattern as for the throughput measurements. We see in Figure 6 that SR-IOV is very effective for KVM and produces comparable results to SR-IOV (LXC).

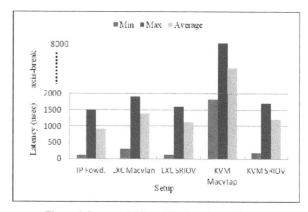

Figure 6. *Latency- LXC vs. KVM based virtual router*

As a next step, we gradually increase the number of virtual routers running in parallel. We offer load at line rate and measure the aggregated throughput. We see in Figure 7 that throughput for the macvtap case drops to zero with five virtual routers. It shows complete starvation of user space tap devices as a result of extensive SoftIRQ processing. The performance is quite reasonable for the rest of the configurations. It appears that SR-IOV (LXC) scales better than SR-IOV (KVM). The performance difference becomes more pronounced as the number of virtual routers increases. The SR-IOV (LXC) achieves 1230 kpps whereas SR-IOV (KVM) obtains 949 kpps for eight parallel virtual routers. The difference is considerable; still, SR-IOV (KVM) is achieving reasonable performance at high-load conditions (i.e. 64-byte packets).

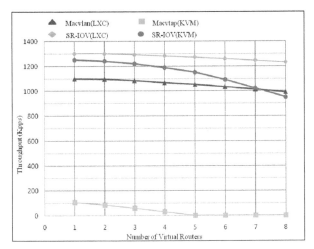

Figure 7. *Throughput vs. no. of virtual routers (1 CPU core)*

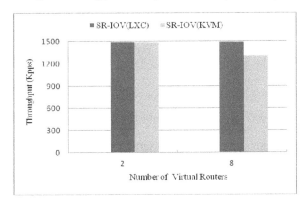

Figure 8. *Throughput vs. no. of virtual routers (2 CPU cores)*

As the next step, we investigate how performance scales while introducing another CPU core. We configure a CPU core to process all traffic belonging to one virtual router while the other core services a second virtual router. It can be seen in Figure 8 that line rate (1488 kpps) is achieved both for SR-IOV (LXC) and SR-IOV (KVM). However, some performance drop can be seen while adding more virtual routers for SR-IOV (KVM). Apart from this, performance is still quite high and we observe performance scalability for both cases. It can also be noticed that we are approaching line rate and adding more CPU cores probably would not increase throughput much. In this regard, it might be interesting to test performance scalability over a

10Gbps network.

4.2. Isolation Results

For the isolation experiments we focus on SR-IOV (KVM) and SR-IOV (LXC). We consider two identical virtual routers VR1 and VR2, with two VFs on each virtual router (Figure 9). We offer network load on eth0. The two virtual routers are responsible for processing packets in parallel and for forwarding them onto the same outgoing interface eth1.

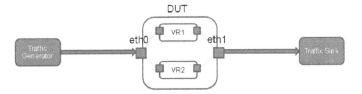

Figure 9. *Isolation test setup*

In this setup we use a single CPU core for both virtual routers, in order to explore possible CPU contention under stress and then investigate different ways of resolving it. We consider two different scenarios of router overload:

4.2.1. Scenario I: VR1 overloaded, VR2 fixed at 400 kpps

In this scenario we assume that the performance requirement for VR2 is 400 kpps. The VR1 is free to attain as much throughput as possible. However, the objective is that the load conditions on VR1 should not affect the performance of VR2. We offer a network load of 400 kpps towards each virtual router. The aggregated offered load on the DUT is 800 kpps. At this point, we gradually overload VR1 and study the impact on the performance of VR2. The offered load for VR1 is increased up to 1088 kpps (when a line rate of 1488 kpps on eth0 is reached) while it remains at 400 kpps for VR2.

The results are shown in Table 1 and Table 2 for LXC and KVM respectively. We can see that isolation is maintained both for LXC and KVM. The performance for VR2 remains at 400 kpps no matter the load conditions on VR1.

We believe that the high degree of isolation of LXC comes from the NAPI RX API in the host kernel. In our setup, we have incoming traffic on two VFs belonging to two different virtual routers. This means that the host kernel is responsible for handling incoming traffic on two interfaces in parallel. The NAPI algorithm maintains fairness among network interfaces that share a CPU [16]. The network interfaces are served in a round robin fashion during a RX SoftIRQ. The NAPI processes only a certain number of packets for an interface before it switches to serve the other interface. In this way it is not possible for an interface to monopolize the CPU. As a result, in an overload situation the excessive packets are simply dropped. This may result in some throughput degradation but provides isolation between interfaces.

The strong isolation properties of KVM may come from

Linux completely fair process scheduler (CFS) [8]. KVM-based virtual routers (which are simple user space processes as discussed in Section 3) are scheduled using CFS. The CFS maintains a mechanism (i.e. red block tree) to impose certain degree of fairness in CPU time allocation among processes. A process that receives less CPU time is given priority over those that have consumed more CPU time. As a result all running processes receive a fair amount of CPU time.

Table 1. *LXC: VR1 overloaded and VR2 fixed at 400 kpps*

Offered load (kpps)			CPU%	Throughput (kpps)		
VR1	VR2	Total	Total	VR1	VR2	Total
400	400	800	61	400	400	800
600	400	1000	75	600	400	1000
800	400	1200	85	705	400	1105
1000	400	1400	100	715	400	1115
1088	400	1488	100	720	400	1120

Table 2. *KVM: VR1 overloaded and VR2 fixed at 400 kpps*

Offered load (kpps)			CPU%			Throughput (kpps)		
VR1	VR2	Tot	VR1	VR2	Tot	VR1	VR2	Tot
400	400	800	33	33	66	400	400	800
600	400	1000	49	33	82	600	400	1000
800	400	1200	60	33	93	800	400	1200
1000	400	1400	67	33	100	836	400	1236
1088	400	1488	67	33	100	842	400	1242

4.2.2. Scenario II: VR2 overloaded, VR1 fixed at 800 kpps

We see in scenario I that isolation is achieved thanks to built-in fairness policies in Linux kernel. However, for other scenarios where for instance one virtual router should be given priority over other, fairness policies might be less suitable.

In order to test our hypothesis we make some changes to scenario I. We still assume a performance requirement of 400 kpps for VR2. In addition, we consider VR1 with a performance requirement of 800 kpps. However, here we overload VR2 instead. The offered load for VR2 is increased from 400 kpps to 688 kpps whereas a constant load of 800 kpps is offered on VR1. Ideally, the overload conditions on VR2 should not degrade VR1 performance.

The results are shown in Table 3 and Table 4 for LXC and KVM respectively. We notice that both LXC and KVM yield poor isolation. In LXC, we observe that VR1 performance decreases from 719 kpps to 667 kpps whereas

it increases from 400 kpps to 633 kpps for VR2. In KVM, we notice that VR1 performance decreases from 800 kpps to 682 kpps whereas it increases from 400 kpps to 567 kpps for VR2. The poor isolation is related to the lack of sufficient CPU resource for VR1. For instance we see in Table 4 that 61% CPU is required to support 800 kpps. However, when the load increases on VR2 the CPU usage for VR1 drops to 53%, as a result of CPU contention between VR1 and VR2. As a result, we observe performance degradation for VR1.

In order to achieve the required isolation, we should give priority to VR1 over VR2 in terms of CPU time. However, this is hard in LXC for two reasons. First, it is difficult to control CPU usage for each virtual router in a shared kernel. This is the reason why we are unable to present such data in Table 1 and Table 3. Secondly, LXC uses NAPI for packet processing in host kernel. The fairness policy of NAPI algorithm provides equal CPU time to the virtual routers. There is no easy way to adapt NAPI's behavior to our requirements. In the first scenario, our performance requirements match with NAPI behavior and we achieve isolation. However, in the second scenario, it is not the case and we observe poor isolation. We also notice that the CPU share feature has no control over shared-kernel forwarding paths as it is intended for user space processes.

In contrast to LXC, the KVM virtual router is a user space process and hence priority can be given to any virtual router using CPU share. When CPU share is configured, it changes CFS from a "fair" to a "proportional weight" scheduler. We allocate a CPU share to VR1 that is two times to the CPU share of VR2. It allows VR1 to obtain at least 66.66% of CPU time whereas VR2 is allowed to get at least 33.33%. The behavior of CPU share is work-conservative, which means that a VR is privileged to consume any unused share (or a portion of share) of the other VR(s). We repeat our experiment with these settings and the results are shown in Table 5. We see that both virtual routers achieve the required performance even in overload situations. VR1 is ensured its required CPU share regardless the load conditions on VR2. We also notice that VR1 consumes less than its allocated share and that an additional share is used by VR2 without causing any isolation problems.

Table 3. *LXC: VR2 overloaded and VR1 fixed at 800 kpps*

Offered load (kpps)			CPU%	Throughput (kpps)		
VR1	VR2	Total	Total	VR1	VR2	Total
800	400	1200	87	719	400	1119
800	500	1300	100	708	500	1208
800	600	1400	100	687	600	1287
800	688	1488	100	667	633	1300

Table 4. KVM: VR2 overloaded and VR1 fixed at 800 kpps

Offered load (kpps)			CPU%			Throughput (kpps)		
VR1	VR2	Tot	VR1	VR2	Tot	VR1	VR2	Tot
800	400	1200	61	33	94	800	400	1200
800	500	1300	59	41	100	778	460	1238
800	600	1400	56	44	100	739	502	1241
800	688	1488	53	47	100	682	567	1249

Table 5. KVM CPU share: VR2 overloaded and VR1 fixed at 800

Offered load (kpps)			CPU%			Throughput (kpps)		
VR1	VR2	Tot	VR1	VR2	Tot	VR1	VR2	Tot
800	400	1200	60	33	93	800	400	1200
800	500	1300	61	39	100	800	441	1241
800	600	1400	60	40	100	800	443	1243
800	688	1488	61	39	100	800	440	1240

5. Conclusions and Future Work

In this paper we compare KVM and LXC as means to enable virtual routers. We investigate the level of performance that can be gained using hardware support for virtualization. The results show that hardware support is especially effective for full virtualization (KVM). We find that switching packets between kernel and user space is a potential bottleneck for the software-based KVM approach at high offered loads. The hardware assistance makes it possible to perform the entire packet processing in user space. This alleviates the bottleneck and we can see a significant performance improvement (Figure 4). In comparison, the behavior of LXC is somewhat different. We see some performance gain when moving from a software-based to a hardware-assisted approach. The gain is expected as we offload some packet processing (e.g. device mapping) onto hardware. However, in contrast to KVM, the hardware assistance does not constitute any major architectural changes, since packet processing is still done in the shared kernel. The shared kernel makes it hard to restrict the CPU allocation for a particular virtual router, which leads to very limited possibilities for isolating virtual routers from each other. In contrast, the KVM hardware-assisted architecture achieves a much higher degree of isolation between virtual routers. It is sufficient to control the CPU share for a virtual router process running in user space.

For future work, we plan to enable multiple virtual networks with QoS guarantees for different types of services on a shared substrate.

References

[1] J. Whiteaker, F. Schneider, R. Teixeira, "Explaining packet delays under virtualization," ACM SIGCOMM Computer Communication Review, Vol. 41 Number 1, January 2011.

[2] S. Rathore, M. Hidell, P. Sjödin, "Performance Evaluation of Open Virtual Routers," IEEE GlobeCom workshop on future Internet, Miami USA, December 2010.

[3] S. Rathore, M. Hidell, P. Sjödin, "Data Plane Optimization in OpenVirtual Routers," IFIP Networking, Valencia Spain, May 2011.

[4] G. Somani, S. Chaudhary, "Application performance isolation in virtualization," IEEE International Conference on Cloud Computing, Bangalore India, September 2009.

[5] PCI-SIG: PCI-SIG Single Root I/O Virtualization Specifications, http://www.pcisig.com/specifications/iov/single_root/

[6] A.Kivity, Y.Kamay, D.Laor, "KVM: Linux virtual machine monitor," Proceedings of Linux Symposium, Ottawa Canada, June 2007.

[7] Linux Namespaces, http://lxc.sourceforge.net/index.php/about/kernel-namespaces/

[8] Linux process scheduler, https://www.kernel.org/doc/Documentation/scheduler/sched-design-CFS.txt

[9] N. Egi, A. Greenhalgh, M. Handley, M. Hoerdt, L. Mathy, and T. Schooley, "Evaluating Xen for virtual routers," IEEE ICCCN workshop on Performance Modeling and Evaluation in Computer and Telecommunication Networks (PMECT07), Honolulu USA, August 2007.

[10] F. Anhalt, P. Primet, "Analysis and experimental evaluation of data plane virtualization with Xen," IEEE International Conference on Networking and Services (ICNS), Valencia Spain, April 2009.

[11] Greenhalgh, M. Handley, L. Mathy, N. Egi, M. Hoerdt, and F. Huici, "Fairness issues in software virtual routers," ACM SIGCOMM PRESTO Workshop, Seattle USA, August 2008.

[12] S. Bhatia et al. "Hosting virtual networks on commodity hardware," Georgia Tech. University, Tech. Report, GT-CS-07-10, January 2008.

[13] J. Liu, "Evaluating Standard-Based Self-Virtualizing Devices: A Performance Study on 10 GbE NICs with SR-IOV support," IEEE International Symposium on parallel and distributed processing, Atlanta USA, 2010.

[14] Y. Dong, X. Yang, X. Li, J. Li, K. Tian, H. Guan, "High Performance Network Virtualization with SR-IOV," IEEE International Symposium on High Performance Computer Architecture, Banglore India, 2010.

[15] S. Rathore, M. Hidell, P. Sjödin, "PC-based Router Virtualization with Hardware Support," IEEE International Conference on Advanced Information Networking and Applications (AINA), Fukuoka Japan, March 2012.

[16] J.H.Salim, R.Olsson, A.Kuznetsov, "Beyond softnet," Proceedings of the 5th Annual Linux Showcase &

Conference (ALS 2001), Oakland USA, 2001.

[17] Ben Pfaff, Justin Petit, Teemu Koponen, Keith Amidon, Martin Casado, Scott Shenker, "Extending Networking into the virtualization layer," ACM SIGCOMM HotNets Workshop, New York USA, September 2009.

[18] The Openvswitch Project, http://openvswitch.org/

[19] Intel   Virtualization   Technology,
http://www.intel.com/technology/itj/2006/v10i3/2-io/7-concl usion.htm

[20] R. Olsson, "pktgen the Linux packet Generator," Proceedings of the Linux Symposium, Vol.2 pp. 11-24, Ottawa Canada, July 2005.

[21] D. Turull, "Open source traffic analyzer," Master's thesis, KTH Information and Communication Technology, 2010. http://tslab.ssvl.kth.se/pktgen/docs/DanielTurull-thesis.pdf

11

Defending WSNs against jamming attacks

Abdulaziz Rashid Alazemi

Computer Engineering Department, Kuwait University, Kuwait

Email address:

fortinbras222@hotmail.com (A. R. Alazemi)

Abstract: Wireless sensor networks WSNs consist of a group of distributed monitor nodes working autonomously together cooperatively to achieve a common goal. Generally they face many threats, threatening the security and integrity of such networks. Jamming attacks are one of the most common attacks used against WSNs. In this paper we discuss the jamming attack and defense mechanisms proposed by two papers and suggest improvements on those four approaches.

Keywords: Jamming Attacks, Network Security, Wireless Sensor Networks

1. Introduction

WSNs are generally heterogeneous distributed wireless radio networks, with little to no manual operation. Sensors or nodes use low power frequencies to conserve energy and prolong battery life. Thus, their transmission range is often low and relay on other sensors to route their messages. For this reason, multi-hop communications [1] are used as an alternative for single hop communications as they consume less power. The word node and sensor are used interchangeable in the text, but they reference the same thing. Signal propagation problems stem from long-distance wireless communication were the signal might be jammed, lost, blocked, due to physical objects, or its power level drops before reaching its destination. That said WSN usually operate in an autonomous fashion, with minimal intervention. They are usually scalable networks with high fault tolerance aspects, as the sensors, gateway sensors, or even the links making the network may fail at any time during operation.

WSNs consist of inexpensive sensors or nodes, geographically scattered to cover an area in which the WSN is deployed for monitoring. Communications are done solely by radio frequency, which is inherently insecure. The sensors are cheap component off the shelf units built using the prevalent Micro Electro Mechanical Systems MEMS technologies. They consist of very simple designs; they consist of a battery, a small flash memory ROM (using a primitive OS like TinyOS), a microcontroller, a radio transceiver with an antenna and a transducer to acquire readings. The transducer is a device that converts energy from one domain to another [2]. these sensors is their low power consumption necessary to insure long battery life, as battery power is usually not replenshable. Another characteristic is the robustness of these sensors as they are placed in industrial, military, and hazardous zones. Other importance characteristic is their long life expectancy as they monitor areas for a long time with minimal intervention and maintenance, see Fig.1.

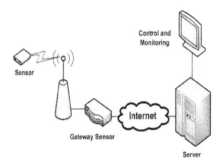

Figure 1. *WSN Architecture.*

The development of WSNs can be traced back to the Cold War, were the SOund Surveillance System (SOSUS) was developed and deployed in the United States as a surveillance system. SOSUS consisted of sensors deployed at the bottom of the sea to monitor Soviet submarines and possible attacks. late 1990's that WSNs got huge commercial breakthrough. The huge advancements in digital consumer electronics have made low-cost, low power, multifunctional radio sensors -the elements that make up the WSN- became a commodity [3]. Protocols especially designed for WSN were developed like ZigBee/IEEE 802.15.4 and ISA 100 [4]. Protocols were specified to address the heterogeneity and diversity of sensors and other components that make up the WSN and intercommunications between these components.

Counteracting this issue, the Institute of Electrical and Electronics Engineers IEEE and the National Institute of Standards and Technology NIST in 1993 worked on IEEE 1451, the Standard for Smart Sensor Networks.

1.1. Applications of WSNs

WSN applications are varied, due to the advent of radio networks [5]. Monitoring is an application for WSNs, surveillance of secure or territorial areas were motion sensors are deployed to locate any source of motion. Pollution control, were sensors monitor air, water pollution, greenhouse effects, and waste control systems, used in heavily populated cities. Disaster alert systems, like forest fire, earthquakes, and tornados, were many weather and geological sensors are used to forecast natural disasters. Major applications of WSNs are in industrial monitoring, to monitor mechanical ware of large industrial machines, give day-to-day status checks.

Structural monitoring is an application were WSNs dominate, were sensors measure the exact structural integrity of large structures like bridges, dams and skyscrapers, these sensors measure pressure, bending, heat and wind speeds, readings that are vital to control and safety engineers. WSNs are also used in planetary and space exploration, were satellites are in practice the sensors. As more satellites are deployed and together, they form a WSN that monitors interplanetary movements and anti-matter presence [6]. Medical monitoring is another new field that WSNs are used in, were a micro sensor is attached to a device that is implanted inside an organ, thus providing real time monitoring of the device and the organ.

2. Related Work

WSN are used in many applications and in some applications, security compromises may lead to threat to national security, commercial lose or even environmental damage [4]. WSNs use radio frequency for communication between its different parts, this is by default a shared medium, security becomes a serious issue [7]. There are many attacks against WSN, but jamming attacks are the most common and widely used attacks that threaten WSNs. Jamming attacks are relatively cheap and easy to implement than other attack types, for example synch flooding. Usually we can defend against jamming attacks using two main methods, spread spectrum techniques and authentications as in [8].

In [9] the authors represented SPREAD (Second-generation Protocol Resiliency Enabled by Adaptive Diversification), a technique to resist smart jammers. By smart jammers, we mean jammers that act more like reactive jammers, discussed later. These jammers only jam the payload of the delivered packet, knowing exactly what protocol is in use. This serious attack will have a high rate of corrupting the data carried by the packets and can effectively reduce throughput to zero. Furthermore, these smart jammers utilize cross-layer attacks, they jam certain layers. Smart jammers reduce the power needed to jam the entire

channel but they should carefully time their attacks to target protocol specific information.

The authors also mention other attempts to defend against jamming; the spectrum spread techs, and how it is not practical for WSNs as it was developed for voice communications. They mention that there are jam attack patterns that target specific layer protocols, the Wolfpack program [10]. This ultimately leads to SPREAD, a technique that avoids smart jammers that target specific protocols, by using a parallel collection of network protocol stacks and switching between them.

In [11] the authors represented a novel mapping service to detect jamming attacks. JAM (Jamming Area Mapping) is a service that provides quick and accurate jamming attack response, which alerts the WSN for a possible jamming attack in effect. As geographic information is important for most WSNs, knowing where exactly the jamming is and what sensors does it effect, certainly will help in mitigating and leveraging its effects. As jammers often, attack specific areas like the gateway sensor area or critical proxy areas. Finding where the jamming are coming from and what sensors are currently cut-off, is very essential in the next step which is avoiding or challenging the jammer. The authors suggested that cost of other solutions like spread spectrum techniques [12] is high, and only practical in military WSN, were security compromise is not an option.

3. Security Issues

Security of WSNs is an important research area, as it plays a major role defending from malicious hackers and possibly terrorists. WSN are used in many applications and some applications security compromise may lead to threat to national security, commercial lose or even environmental damage, making security breaches not an option. Security issues include confidentiality, meaning that authorized personnel shall not access the data; this is hard due to the use of radio waves. In addition, another issue is data integrity that means that the data is not tampered with when received by the other side. Service availability is another security issue, availability means that authorized access of data and other WSN resources is made ready when requested or demanded. Attackers range from the hackers, for blackmailing and monetary gain, or industry spies, gathering confidential business insight, or espionage spies, for confidential and top secret information.

Since WSNs use radio frequency for communication between its different parts, this is by default a shared medium, security becomes a serious issue. In addition, the constraints such as limited processing and limited memory capabilities of the sensors and their dependency on battery power alone make it more difficult to keep these networks secure and safe. WSNs are especially venerable to Denial of Service Attacks DoS, since we have limited processing capabilities and the dependency on battery power for the scattered sensors and the inability to secure the shared medium used for communications.

3.1. Types Attackers

Attacks on WSNs come from two groups, insiders, or outsiders. Outsiders usually sniff the packets sent over the channels; they can tamper with the data or jam the signals all together. Insiders are capable of damage that is more malicious. As they have certain keys the nodes use for their communications. Insiders with keys can take hold of one of the sensors, inject their own programming to harm the whole WSN or even take parts out like in a blackout only to insert their own nodes in order to take hold of the entire WSN.

Aside from this categorization of attackers, we will classify attackers into four kinds as found in [15]. First kind is the Passerby; they are not determined, very little knowledge about the WSN but with common tools that disturb the flow of the WSN work. More malicious are the Vandals, they intend to damage the WSN on purpose, with significantly more resources and inside knowledge. Even more malicious are Hacker, extremely determined and fueled by curiosity or financial interests. Hackers are skilled network intruders, they tend to cause damage spontaneously and to show off or just for self-recognition. Another kind is the Raiders, motivated by personal, economical, or political gains. They are determined, usually very well knowledgeable with lots of insiders' information and backed up by organizations or governments. Finally, we have the Terrorist or Foreign Powers; they cause international security damage by breaking in or hindering of critical military or civil systems. Politically motivated, in most cases will die for their cause, highly knowledgeable. Terrorists are usually well funded with money and work force; this in effect allows them to cause the most harm among all the kinds of attacks. Security policies were mainly driven by the amount of damage possible from the kind of attacks that may attack the WSN.

Usually attackers with high computational powers as PCs or laptops do much harm than malicious attackers using injected sensors. These attacks usually have virtually unlimited processing powers and can easily out number or use brute force to enter the WSN. Furthermore, they can send their own programming into the entire network through fake packets in effect overriding the original programming of the sensors. In addition, when the attacker is a group of computers, they can launch what it is called a sandwich attack. Sandwich attacks are attacks that try to take control of more than a single node in the network at different time. Allowing attackers to acquire authentication keys and group keys used for exchanging secure messages around, in effect taking control of the entire WSN. In addition, attackers can cause target localization intrusions, when WSNs are used for intruder detection and motion, attackers can mislead the sensors by faking, or misinterpreting their signals thus giving wrong readings making it possible to go under the radar under the area WSN is covering.

3.2. Layers of Threats

Let us dissect the security threats facing WSNs by the network layers [2]. First the physical layer, we have the jamming and tampering, were the sending and receiving frequencies are jammed or distorted. Then link layer, we have exhausting floods, were attackers flood links, and make packets drop. We also have collision attacks, it is a technique to make packet drop and get reordered more frequently thus disturbing the data transmission in the WSNs. Another attack unique to WSNs is the denial of sleep attacks [14], in these attacks external intruders keeps the sensors busy with fake communications or even sending fake or empty packets in order to let the sensor in the on mode as much as possible. Preventing sensors to go to sleep mode will ultimately drain the battery, once the battery is drained the sensor goes down, thus making the network go into a partitioning process and blacking out parts of it.

In the network layer, we have the malicious sinkhole and wormhole attacks were packets are drawn out of the network to different destinations the attacker wants, usually their own databases. The wormhole attack may also include selective forwarding attacks, were packets are forwarded by the wormhole to different locations maliciously. Transport layer, we have the synchronization flooding attacks we channels are flooded with fake packets that require a full TCP communication. Finally in the application layer, we have the clone attack were a node is taken over by a clone node with the same key and identity. In addition, a popular attack is the DoS attacks, since WSNs are inherently weak to it, an attacker can easily deny the service by sending fake empty messages continuously on the same receiving frequency of the sensors. Other application layer attacks are Deluge or reprogramming attacks [9]. These attacks often done by professional insiders were they send their own programming using authenticated messages to certain sensors, the program is like a virus it replicate itself through the network resulting in complete take over by the attackers. A summary of the attacks is shown in table 1, which is illustrated using the Open Systems Interface standard network layers OSI.

Table 1. OSI Layers with the corresponding threat and defense mechanism.

Network Layer	Attack	Defense
Physical	Jamming, Tampering	Spectrum spread, Authentication
Link	Collision, Exhausting flood, Denial-of-sleep	Authentication, Error Correction Codes, Anti-Replaying
Network	Sinkhole, Wormhole, Selective Forwarding	Authentication, Keying Techniques
Transport	Synchronization flooding	Authentication, Synchronization Cookies
Application	Clone, Deluge, DoS	Authentication, Anti-Replay

4. Jamming Attacks

Jamming in the physical layer [13] is usually done by distorting the sending and receiving frequencies using heavy noise levels. They are easily accomplished by either by-passing link layer protocols or emitting a signal targeted that jams a certain channel. Most WSNs are made of commodity sensors and components, thus these technologies are easily targeted by attackers. That means attackers are can easily gain access to communications channels used between sensors since they already know technology and accompanied protocol. Jamming attacks target the shared medium and even with the technological advancements in security of this shared medium, it is still difficult to defend against jamming attacks. Attacker usually prevent legitimate data from reaching its target, or even make packet collide thus no legitimate packets can be delivered over the channels [14].

4.1. Types of Jamming Attacks

We have four types of jamming attacks. The first one is constant jamming, were attackers emits a radio signal, this is done using waveform generators that continuously send radio signals or sends random bits to channels' link layer protocols, in effect jamming the channels. The link layer protocols allow sensors to send data when only if the channel is idle, when jammed constantly, sensors can't get any data through effectively. Major back draw of constant jamming is the nearly unlimited resources the attacker must have. The second type of jamming is deceptive jamming; the attackers won't send random data but will send a stream of real packets of data. This stream will be continuous so that receiving sensors will never go to sleep or ever be able to send its legitimate packets, because the stream of fake packets has no gap between those packets. Requiring again near unlimited resources, also attackers must sniff packets streams in order to replicate them.

Third type is random jamming; this can be constant jamming attack or deceptive jamming attack with power conservation taken into account. Attackers jam the signal for a set period, later they go to sleep, to conserve power. Usually attackers attack at certain send patterns for the sensors, then go to sleep when sensors aren't sending. Requiring far less power and processing resources, but careful timings are required to get effective jamming results. The last type of jamming is reactive jamming attacks. This is the hardest jamming attack to detect and hardest to implement. Unlike all previous jamming types, which are active, this type is reactive, meaning that jamming only start with legitimate sensor sending data out into channels. Blocking channels only when data is about to be sent. This type of jamming however requires precise sniffing and complex pattern recognition in order to occupy the whole channels effectively. In addition, this type of jamming attacks strikes a balance between power consumption and effectiveness, making it an efficient choice for attackers. Check Fig 2.

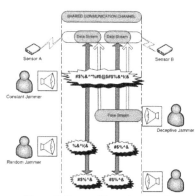

Figure 2. Types of Jamming Attacks.

4.2. Defense Mechanisms Against Jamming

First of all we must be able to detect the jamming signals from the legitimate. It is both difficult and imperative to distinguish between the legitimate signals and jamming signals. The first method to detect jamming is signal strength, whenever jamming is present; the signal levels are usually high when compared with signal levels without jamming. The other way to detect jamming is use the packet delivery ratio, in the presence of jamming packet delivery ratios is below average or even drops to zero if the jammer is completely blocking the signal. Thus, we can determine that the channel is being jammed and that the sensors should switch to another channel.

After detecting the presence of jamming, now we take action. Many security techniques were developed for protection against jamming. Mainly they fall into two main categories, evasion strategies, and competition strategies. In evasion strategies, sensors under attack from a jamming signal will evade the jammer by changing their broadcast channels or by physically moving away from the jamming. First evasive technique is called Channel Surfing; a technique that allows sensors to change their broadcast frequencies in presence of jamming. Change in frequency is on demand and done after the sensing of jamming on certain channels. In other evasive techniques, sensors physically move out of range from jamming thus allowing the jamming signals to be very weak and die out before affecting the WSN. The other strategy is the competition strategy; here sensors compete against the jamming attack. Sensors after detecting the jamming signal; will try to compete with the jamming. While competing, the sensors will use stricter ECCs that effectively lowers the data rates in each packet, but more successfully decoded packets will be received.

5. Selected Papers

5.1. Combined Approach for Distinguishing Different Types of Jamming Attacks Against Wireless Networks

The authors have described a novel method not just only to detect jamming attacks but distinguish which type of jamming attacks [16]. The work discusses how jamming attacks are one of the most prevalent attacks launched against WSNs. Main reason is that jamming attacks are one of the most effective Denial of Service DoS attack [17], as it completely deny any communications done between sensors in a jammed area. Another reason is that jamming, as the authors presumed, from an attacker perspective, is easy to implement and launch, using commodity Radio Frequency RF devices. The authors stated that detection of the jamming attack isn't always easy, as many attributes must be considered. Attributes hinder clear detection of jamming include low signal to noise ratios SNR, this is where outside interference with the channel can reduce its quality. Low SNR ratio however, isn't deliberate as opposed to jamming attacks. Other attributes hindering jamming detection is caused by low battery power that can give off low signal transmission strengths. In addition, sensors mobility can have the same effect on the signal strength. State of the art mechanism to detect jamming is Signal Strength Consistency Checks [18], as stated by the authors. This method can effectively distinguish jamming signals from other interference and noise. The authors proposed approach uses this technique to detect jamming. The authors presented a novel method that uses the Packet Delivery Ratio PDR, and Packet Sent Ration PSR, in their statistical method to detect and distinguish the jamming attack. Method described in there paper works upon the MAC layer, as this layer is responsible for the delivery of data frames in the shared medium, which are radio channels.

The basic protocol applied in the MAC layer is the Carrier Sense Multiple Access CSMA. The authors stated that CSMA/CD, which is with Collision Detection is not suitable for WSN [19], as multiple transmissions are possible but what matter is the receiver ready to receive the transmission or not. For two situations, hidden station problem and exposed station problem. In hidden station problem, it is a situation where three sensors in the same vicinity try to communicate with each other. For example, if sensors A, B, and C, are in the same vicinity, if A is communicating with B, then if C tries to communicate with B, its transmission is using a different channel, which is between C and B. Under CSMA/CD, channel between C and B is idle, so C starts to send its packet stream over the channel. However since channel is part of the shared medium, and sensors are in the same vicinity, the stream from C corrupts the stream sent from A. Other problem is exposed channel problem, in the same settings as before, if A is communicating with B again and C is wishing to transmit to a nearby sensor D. D is out of range from A and B, but C is in there transmission range still. C will sense that the channel is busy and will never transmit to D until A finishes. However, C will be idly waiting for A to finish while in fact the channel from C to D is idle. For these reasons, CSMA/CD is not suitable and CSMA/CA, which is CSMA over Collision Avoidance, is used instead. In this protocol, the sender before sending the sending sensor sends a handshake message, Request to Send RTS, and then waits for an acknowledgment, which is Clear To Send CTS message from the receiving sensor, and then communication is established. Even with CSMA/CA protocol in use,

jamming can corrupt the handshake or even send fake packets to occupy the channels.

To distinguish jamming attacks, authors used the fact that jamming can both corrupt the protocol messages or handshaking or jammers send fake useless packets to fill channels. Jamming signals usually ignores the protocol messaging to get ACKs and just send junk messages over medium. The authors for their approach used three parameters the Signal Strength SS, PSR and PDR. SS parameter is used to detect jamming, it is used with threshold parameters for distinguishing whether jamming is disturbing channels or not. PSR determines how idle channels are, as if more packets were sent from total packets; it indicates that no constant or deceptive jamming is occurring in channels. Other parameter, PDR, indicates that jamming occurred actually, as all types exemplify extremely low rates, low successful transmissions. The case of zero PDR is when reactive jamming is done against the WSN, as almost no packets can go through the channel without being distorted halfway, this means that when the receiver will run its cycle redundancy checks CRCs, it finds that packets are corrupted, and won't acknowledge it back. In addition, when PSR drops to nearly zero value, with PDR nearing zero, it indicates constant jamming. Differences in PSR and the huge drop in PDR will result in detection of jamming.

For distinguishing four types of jamming, the authors used two main thresholds. Two thresholds are experimental threshold statistically set. Threshold1 is for the signal strength value, and the PDR, used to detect jamming. Threshold2 is about PSR, when it falls to a certain level, it is used to distinguish which jamming attack is the channel is under. Approach has two phases, in the first phase checks if PDR is below threshold1, if yes; it checks the signal strength, if it is above this threshold, threshold1. If no then no jamming had occurred, just interference or noise. If yes then jamming is detected, and the channel is under one of the four type of jamming. Second phase is to check the PSR with threshold2; if PSR is above threshold2 then it is either reactive jamming or random jamming. If PSR is below threshold2 then it's either deceptive jamming or constant jamming. The authors used statistical models to distinguish between four types; usually they proclaimed that with reactive jamming PSR is very high, above 60% for random jamming, which saves power for the jammer, PSR is below 60% but above threshold2. If PSR is, zero then its deceptive jamming. Finally, it is under threshold2 but not zero then its constant jamming. After detection of jamming, the authors suggest that the sensors communicate in another secure channel with different frequency, already set up in advance. The final stage of the method then, after detecting the jamming and its type, is to utilize the secret channel, on the different secret frequency not under attack. This means that the sensors under the jamming will negotiate what to do next. This technique is similar to channel surfing [20] for avoiding the jamming attack. The method does not specify what methods to use, but since the sensors under attack know exactly which type of jamming. They are freely communicating in this secret channel to initiate a solution. Fig.4 illustrates how the method can both detect the jamming attack and distinguish its type.

For the experiment the authors used the Network Simulator NS2, simulating all four jamming types. The results from the simulation runs showed that effectively all jamming types would drop the PDR value to nearly under 10%. This is expected, as one of the main aims of jamming is to corrupt packets in transmits. However, the PSR value shows huge differences as random and reactive will keep it above 70%, while deceptive and constant drop it under 10%. The method used to detect SS, the signal strength consistency checks, a well know and widely used technique for signal strength measurement. The SS is especially useful in the first phase, as it will help distinguish jamming from other factors contributing to low PDR and PSR. These other factors are low batteries, as they will drop the signal power, and moving out of range in case of mobile sensors, they will affect PDR value and drop it. In this case, the SS is measured, if it is lower than the threshold1, then this interference is not due to jamming, as usually jamming have high signal strengths, even higher than normal. This is called interference or noise distortions, and not considered jamming. As only, as the authored assumed in the experiments, if the SS is above threshold1, then the drop of PDR is deliberate and it is the result of jamming attacks.

The thresholds used in the experiments were based on statistical models of average signal strengths, PSR, and PDR rates in normal situations. Threshold2 and PSR are used to determine the type of jamming. The authors divided their method into two main algorithms, the PDR detection, and the PSR differentiation. The differentiation that occurs after detection does not really specifies a special avoidance plan. The method after detection and differentiation of the jamming, sensors just send jamming detection messages with the type of jamming, then use a suitable method to counteract this jamming type efficiently. The authors have not specified or suggested any jamming countermeasure to be used in their experiments though.

5.1.1. Outcomes, Limitations and Improvements

The authors came up with a novel method based on experimental and statistical results. Detection process doesn't need extra hardware; it utilizes SS and PDR only. PDR and PSR can be collected locally in the WSN. Distinguishing process between the different types helps in counter measurements selection. As in defense strategies, competition strategies specifically are more effective if the jamming mode is known. The method is simple, does not require any further modification to the sensors hardware. Even protocol modifications are not needed. Statistical data for the PDR and PSR needed however. The experiment showed clear and consistent results. The algorithm based on the approach, used simple switching statements can lead to identifying the jamming type. Authors showed how CSMA/CA is vulnerable against jamming as the jammer can still bypass the protocol, or utilize the protocol to occupy the channel with

fake packets. The jamming attacks will leave the sensors in constant waiting or receiving modes according to the protocol CSMA/CA.

The method also distinguishes between interference disturbances in the signal and jamming. Interference comes from moving out of range or low batteries giving low signal strengths. This hinders PDR but isn't necessarily from jamming attacks. This allows multiple defensive techniques to be used depending on the type of jamming attack. The method also helps in protecting against multiple jamming attacks against multiple parts of the WSN. The method detection of the type of jamming helps against jammers that switch their modes after detection are caught, and suitable counter measurements are taking against them accordingly.

The assumption that was used is that jammers will constantly use a single model of jamming is an oversimplification. This is not entirely true, as jammers tend frequently to avoid detection by continually switching their modes. Switching the jamming mode is found in [21]. The statistical model applied is not entirely true, not for all WSNs, as the thresholds are approximation values from the experiment only. The approach is evasive in nature, as once the jamming is detected; the authors suggest using a secret channel to send messages to confirm the jamming attack and its type. This means that sensors avoid the jamming by using predefined secret channels to conform only of the jamming attack. Without proper usage of an effective method, the jamming problem will still affect the WSN. In addition, other defensive technique leaves the distinguishing of the jamming type rather useless as once jamming is detected no need really to know the type, in some defensive techniques. This leads to the truncation of the second phase of the approach as sensors are now using a different channel with different frequency.

Possible hopping to another channel to convey jamming attack conformation may not be possible either. As most jammers jam entire spectrums, at high costs though. On the other hand, attackers will try to retune their frequencies to that new channel. Therefore, the retune will reduce the possibility of using those secret channels. Furthermore, threshold calculations, PSR, and PDR, is not easily managed in large WSNs with scattered sensors. As larger areas are jammed and more sensors involved, the harder it is to get reasonable PSR and PDR readings indicating the average values. The authors did not specify any means to acquire PSR or PDR, other than theoretically, but not physically between the sensors under the jamming attack. Compared to SPREAD, this method tries to differentiate the jamming type, while this method the authors after detection, try to differentiate the type of the jamming but this means that more damage is done before it distinguishing. In SPREAD, immediate actions are taken to hop to another protocol or to change the protocol settings. This method of detection and differentiation is very slow in effectively taking action against the jamming. As only, after the damage is quite apparent can the sensors counter act. The sensors only send messages acknowledging that a jamming attack is affecting the WSN and its type only, as the authors have not specified any defense counter measurement.

Improvements to this method include the use of PDR and PSR values that are predefined depending on the size of WSN and the average message exchanged between the sensors. This eliminates the stochastic values overhead taken during operation. Furthermore, the use of channel surfing techniques after the detection and identification of the jamming attack. This will be used instead of just exchanging messages regarding that the jamming over the secret channels. Finally, if the defensive technique used to defend against the jamming attack is not for a specific type, we can simply skip the second phase and apply the technique after the detection phase. Another improvement is to use a mapping protocol like in [11], to map an area of the jamming attack, or possibly locate the jammer, as the method was developed under the assumption that the sensors are mobile.

5.2. A Defense Technique for Jamming Attacks in Wireless Sensor Networks Based on SI

In this paper [22], the authors have presented a novel jamming detection and evasion method based on Swarm Intelligence SI inspired by biological systems. Their technique utilizes autonomous agents to detect possible jamming attacks, and further mitigate its effects. Authors stated that other techniques detected jamming attacks through modifying used protocol or by using new MAC layer protocols; this is not always a practical solution to the jamming attacks problem. The authors mentioned the work in [21], and also how it employs a variant or alternative MAC layer protocol to defend against stealthy jammers. These techniques are all evasive techniques, similar to the SI method proposed. The jamming attack model used is the pulse jamming attack. This type of jamming attack is done through sending discrete pulses that destroy parts of the frames making up the packets, in affect corrupting them.

The proposed detection and evasion method is based on SI techniques. SI [23] is based on biological behaviors of social insects. SI is an Artificial Intelligence AI technique, used in cellular robotics for coordination, team work, and monitoring. SI consists of simple agents that try to solve complex problems together, like bees and ants. The authors stated that the agents they employ act much like ants, which get useful information from following trails of chemicals, and certain body movements. In the sense that the work done by ants is not supervised by any other ant, instead their combination of exchanging information, knowledge and work partitioning, achieves their ultimate goal without the need of any supervision. The ants in the colony do all their work in this manner, communication without supervision. In the same manner, the authors have proposed a similar technique called Swarm Based Defense Technique SBDT.

The technique utilizes AI, namely the Swarm Intelligence, a type of biological inspired algorithms. The method is based on intelligent agents, like ants in an ant colony, working together to achieve a common goal. The authors claimed that such a method is adaptable, as the intelligent agents gather enough information about WSN status dy-

namically. Agents gather both topology and traffic information that help in generating an overall view of the current status of WSN's channels. Agents collect non-local information that helps other sensor to be updated about the current channels availability. Agent use stochastic components, they act like pheromone table for the swarm agents, and they are autonomous. The gathered information helps in finding the best routes in which the packets will face less congested traffic, or avoid deliberate jamming attacks. The pheromone P, borrowed from the ant metaphor, will act as the information or probability provided to guide the agents in choosing which channels are safe to use.

SBDT, is based on four main principles, these principles organize autonomous work of agents. First principle is positive feedback, this feedback is the information regarding that the channels aren't under any jamming and still running without deliberate interference. Second principle is negative feedback, this feedback is alert information, resulting that jamming and its interferences are found on the travelled path, these channels are under heavy interference from jamming pulses, thus this path should be avoided. Third principle is randomness, the authors haven't stated what this principle is exactly, but it's a factor that help agents in selecting next hops, this factor helps reduce overhead of maintaining channels, and reduces updating data of channels very quickly. Randomness, is used when channels information of the next possible hop is available, thus agents depending on factor choose a channel. Last principle is multiple interactions; this is how agents communicative together. Multiple interactions between agents, means that agents traversed WSN will effactually transmit their gathered information to the other agents that traverse WSN right after them. This leads to acquiring prior knowledge about the channel status for other agents. agents use channel hopping, similar to what is found in [20] but not exactly the same, the hopping is used to evade the jammer. The hopping is based on a pair-wise key shared key K, this secret key will generate an encryption sequence factor E, and this is used to create a pseudorandom channel sequence, in the following equation (1):

$$Chnext = E(i) \mod Chcurrent, i \geq 0 \qquad (1)$$

Furthermore, the MAC layer will provide the packet fragmentation over the channels. The sending sensor will transmit its fragments on a certain channel after filling its transmit FIFO queue then it issues the transmission command. The authors listed an equation that calculates the time to fragment and issue the sending of such packet fragment. As each fragment will be send on its channel, and using the secret sequence, the sender will hop from channel to channel for each packet. The authors used Pulse jammers as the used model for jamming. Pulse jammers, are jammers that use a single channel and send random pulse to destroy such fragments, as whole packets cannot be detected quickly as fragments are scattered over many channels on a secret sequence. The method has two main agents, the forward ants FA, and the backward ants BA. The FA agents are the agents that start from a starting point; the author did not state exactly where, and traverses the whole WSN's channels collecting any information about possible jammers. When they reach the end of the network, they are transformed into new BA agents and their information is inherited and carried over. Finally, the BA agents will retrace back the same path their FA counterparts have travelled and update the current information inherited from the FA agents.

When the BA agents reach the source of which the FA agents started from, all of their information is verified and the current status of the channels is detected, then BA agents turn again to FA agents and inherit the information and redo the process in a new iteration. The agents will unicast or broadcast depending on the availability of the channels. Agents try to go through channels that we do not have any previous information about, newly created channels. As sensor, mobility is considered in the experiments, and sensor random movements create and destroy dynamically these channels. Channels which we have already recent information are not likely chosen, this as the authored said would lessen the overhead of maintaining channel information. When the agents face channels were all information is recently acquired, then the agents apply the randomness principle, this means randomly choosing their next channel. For choosing the channels randomly, much like in the biological system where the pheromones are used by ants to guide their way, the author used a channel probability equation. This equation depends on the available information about channel from the sender and the receiver. Also in the channel, probability equations are two relative weight values. Also in the equation is a variable called λ, this variable's value depends on the jamming pulses generated in each channel by some jammers. The variable λ value change depends on its previous value, if the change is positive; this means a negative feedback is sensed by the FA. If the change is negative, this leads to positive to the FA, which means that the jamming is lessened for some reason on that channel. Each channel neighbor or end points have also a probability equation that the feedback will update. Since the sensors are mobile, this means that channels are dynamic and are not fixed, and the sensors movements are random and not predefined or limited. Fig. 5 shows how the agents spread from a starting point sensor A, until they reach sensors F and G, which are situated at the end of the WSN. The FA agents will turn to BA agents upon reaching the end point of the WSN. The BA agents will traverse back the same path taken by the FA agents. Likewise, the BA agents turn to FA when reaching sensor A again, to start a new iteration. Channel information is always updated with each iteration.

The authored did their experiment using the Network Simulator NS2 tool. The MAC layer used is the IEEE 802.15.4; supporting the direct sequence spectrum spread DSSS, implemented in the hardware. The authored have compared their method SBDT with DEEJAM. The main attributes are the aggregated throughput, packet drop rate PDR, and the packet dropped during transmission. The results showed that SBDT usually have higher PDR than DEEJAM, much lower packet dropped rate, and higher aggregate throughput. The main reasons for these improvements are due to the positive and negative feedbacks and the lessened maintenance overhead. The feedbacks can give a message to the routing of the packets quickly to avoid channels with bad quality, for some reasons. Therefore, channel status can quickly be acquired and used to avoid channels under jamming attacks and reduce the traffic through them by omitted the jammed channels. The other contributing factor is the channel maintenance criteria. The agents, FA and the BA, will always choose the channels that they do not have recent information about their status. In case all channels information is known, they choose randomly using a probabilistic equation. This behavior will reduce the overhead of channel maintenance. In addition, it will also speed up the convergence process of having a clear view of the channels status quickly. Using he channel status the WSN can detect the quality of each channel faster, checking if any channel is under jamming or not.

These factors helped SBDT also achieve a much faster detection and convergence. In affect lessening the damage caused by jamming attacks much faster. The authored also used mobile sensors were topology is not fixed. The mobility of the sensors in the experiment is random. This in affect made channel creation and destruction completely dynamic. The other also used in their experiment, two different attackers. The attackers were placed randomly in the path of the channels in range. This random model of jamming however is not entirely true. As most deliberate jamming attacks usually target pivotal channels, those that link several parts of the WSN.

5.2.1. Outcomes, Limitations and Improvements

The authors proposed method based on swarm intelligence, to our knowledge, this utilization of agent-based method is novel. The probability equations used help reduce the overhead of maintaining or updating the channel values to soon or very quickly that will not be relevant as the channels are dynamic in nature. The agents work independently and try to listen in each channel to the traffic, thus if multiple jamming at multiple channels is present, such agents will detect each jammer individually. Randomness and the stochastic model also give the agents freedom and allow the method to adapt to changes in the WSN. The method uses channel hopping, by default, to spread the fragments over the channels using a pair-wise keying method. However, when jamming affects certain channels, they are omitted and skipped, meaning not used. This helps fragments escape the jamming on such channels. Finally, the brilliant aspect of this method is the fast convergence speed of the method, as the FA agents with the BA agents in a single iteration can give a very accurate view of the current status of the channels and how much traffic on each, every iteration updates theses values.

The authors did not take into consideration that limited resources available for each sensor. Nowhere in the paper is ever mentioned that the mobile sensors have limited memory or power constraints. As the SBDT's agents are continuously traversing the WSN, power is drained and memory is used. In addition, the authors used sensors that are DSSS capable; this is of most used hardware technique to avoid jamming [24]. This means that channel-hopping technique avoids the jamming, but the SBDT is used further to detect where the jamming attacks are and what channels to be avoided. The paper did not specify how or where the agents start, or what is the end point where the FA agents are turned to BA agents. Taking into consideration this is a mobile WSN, and the channels are not fixed. How will the autonomous agents know which channels exist and which that does not anymore? In addition, FA agents al go through different paths depending on the current topology, then the BA must retrace those route that may not exist anymore. The method also has no coordination regarding how the information will be used, when a FA or BA agent dismiss its information for each sensor, how will each sensor get the information relevant to its use? Sensors in this mobile setup will eventually need information regarding the whole WSN.

The paper used for jamming, the pulse jamming model, this model specifically jams only a single frequency of the whole frequency band allowed for the sensors. Under this model alone, the SBDT will detect affected frequencies and omit them. Other jamming models like constant jamming, where the jamming affects an entire frequency band. In addition, deceptive jamming model is not considered, when the jammer sends valid but empty packet fragments, how can the FA agents, and the BA agents detect such a jamming attack? The λ variable is a heuristic value dependent on the number of pulses sent on that channel. This means that deceptive jammers will not be easily detected as they append fake packet fragments into the channel. The injected packet fragments are continuous; therefore, it does not appear as a pulse but as a stream filling the channel. Also in the experiment, the authors considered the jamming attacks to be done randomly; this meant that the jamming attacks do not target certain channels. The agents FA and BA agents eventually gather information after at least the first iteration. The question is will this information be relevant, in case of a random jammer, jamming can be done randomly at random times and channels. Will information gathered by the agents be relevant to this random pattern? Also the computational and memory overhead is not considered in this method. In addition, the authors suggested that jamming attacks are mitigated by routing the traffic to other channels not under jamming attacks. The only defense mechanism in the SBDT method is when certain channels are found to be under jamming attacks is to switch their traffic to other unaffected channels. This is not entirely true all the time, as in the case where the jammed channels are the only path to certain parts of the WSN. Alternatively, maybe the jammers have jammed certain bridging channels, which resulted in partitioning the WSN. Furthermore, how can we, in these cases then routing the traffic to other unaffected channels? Routing to other unaffected channels will not certainly help all the time.

Improvements to the SBDT are to use better jamming

detection. Jamming or random and deceptive jamming can be caught by using pattern tables like in [16], by using the localizing approach found in [25] to locate the jamming and try to avoid it. For further reducing the agents overhead, sensor may have their individual agents. Each sensor has a FA and its BA counterpart. The heuristic model should be replaced with a stochastic model based on routes this sensor use, and the patterns known for jamming attacks. Comparing the SBDT with DEEJAM, it seems apparent that SBDT is much more costly, in terms of agents' maintenance and coordination. When compared with SPREAD, we think that SBDT reaction to jamming is not as fast. For example, SPREAD, when jamming is detected, immediately the jamming is mitigated by using a new protocol or changes its settings. SBDT, on the other hand, must wait for the agents to traverse the entire WSN, and then verify the information, and then jammed channels are not used or omitted. When compared with JAM, this method gathers information about dynamic channels, which are quickly destroyed and created. Whereas JAM, maps an area where the jammer is launching attacks against WSN. Mapping the area of jamming is more helpful than information about the channels themselves.

5.3. SPREAD: Foiling Smart Jammers Using Multi-layer Agility

The authors represented SPREAD (Second-generation Protocol Resiliency Enabled by Adaptive Diversification) [9], a technique to resist smart jammers. By smart jammers, the authors mean jammers that act more like reactive jammers. To be more precise, these jammers only jam the payload of the delivered packet, meaning they know exactly what protocol is in use. This serious attack will have a high rate of corrupting the data carried by the packets and can effectively reduce the throughput to zero. Furthermore, these smart jammers utilize cross-layer attacks, meaning they jam certain layers to hinder the other layers from communicating with this layer. Smart jammers reduce the power needed to jam the entire channel but they should carefully time their attacks to target protocol specific information.

The authors also mention another attempt to defend against jamming attacks, the spectrum spread, and how it is not practical for WSNs as it was developed for voice communications. They mention that there are jam attack patterns that target specific layer protocols, the Wolfpack program [10]. This ultimately leads to SPREAD, a technique that avoids smart jammers that target specific protocols, by using a parallel collection of network protocol stacks and switching between them.

The technique has two major parts, the core, and the layers. The core is the central control part of the protocol; it uses the data collected from the layers to analyze when and how to hop to another protocol sequence in case of jamming on the current protocol. The layers are physical layer, Medium Access Control MAC layer, Data Link layer, and the Transport layer. The layers use certain set of protocols, and report to the core the variables readings, variables like network congestion, channel status, PDR and the energy levels used.

The core, depending on these reading will decide to initiate the hopping sequence to other protocols in case of a smart jamming found targeting the used protocols. Two technique of hopping are used, the Inter-protocol hopping and the Intra-protocol hopping. The inter-protocol hopping means that several instances of interleaved protocols are used and the network uses them redundantly, taking into consideration that a smart jammer cannot possibly target all of the protocols at once. The other type, intra-protocol hopping, means that a single protocol is running but the core cryptographically and dynamically changes the packet size, coding rates, and the transmission rates.

The authors applied their technique against Extended Inter-frame Space EIFS attacks, were the jammers jams the period the channel is supposedly idle. This attack will reduce the throughput to zero as sensors will always check for the channel to be idle and simply find it busy. Then for fine-tuning SPREAD, the authors used game theory [26], to demonstrate how SPREAD can be tuned to fend off the jamming attacks more effectively.

5.3.1. Outcomes, Limitations and Improvements

The authors have presented a novel method to defend against smart jammers. SPREAD helps to mitigate and leverage the damage cause by jammers that target protocol critical information. The authors assume that it is more efficient that other methods like spectrum spread or using more than a single channel for communications. SPREAD framework utilizes the same hardware, only needs fundamental protocol changes and implementations. They also presented their work in a game theoretical framework, which illustrates how SPREAD is tuned to lower the effects of the jamming.

The authors suggested that most jammers are smart jammers. This is not typically the case, as most jammers use nearly unlimited power supplies to effectively cutoff the channels. Most jammers target many protocols simultaneously, in case the WSN used different protocols. Another effective defense against jamming is channel surfing that dynamically change the frequency instead of the protocol. It is less complex and more effective, as most WSN employ ZigBee and IEEE 802.15.4 compatible. These protocols complaint sensors support multi-channel frequencies. The authors suggest that the jammer only use different jamming rates also. This is not entirely true as jammers can be deceptive, that mean sending jamming signals right after the packet payload to trick the receiver into continuously receiving junk data. In addition, reactive jammers target the payloads, even if the WSN used multiple protocols and used redundancy in sending the packets, these jammers will corrupt most of the data payloads leaving the protocol data unaffected.

5.4. JAM: A Jammed-Area Mapping Service for Sensor Networks

In [11] the authors represented a novel mapping service to detect jamming attacks. JAM (Jamming Area Mapping) is a service that provides quick and accurate jamming attack response, which alerts the WSN for a possible jamming attack in effect. As geographic information is important for most WSNs, knowing where exactly the jamming is and what sensors does it effect, certainly will help in mitigating and leveraging its effects. As jammers often, attack specific areas like the gateway sensor area or critical proxy areas. Finding where the jamming are coming from and what sensors are currently cut-off, is very essential in the next step which is avoiding or challenging the jammer. The authors suggested that cost of other solutions like spread spectrum techniques is high, and only practical in military WSN, were security compromise is not an option.

The authors described JAM, as a service that provides feedback to the routing directories, thus warning the WSN of the jammer current activities. It also provides preemptive warnings to entry of individuals or sensors or even vehicles to this area as the enemy or jammer is currently active in. Finally, the mapped area aids the WSN in deploying its strategies against the jammer effectively. For example, the WSN can switch off the sensors in the jammed area, reallocate the sensors, if they are mobile or capable of moving, or reroute all packets to avoid that area. Power management strategies can also be used to foil the jammer possible damage. The authors clearly claim that JAM provides a much cheaper and convenient solution to jamming attacks, as it aids the WSN to take critical actions and imply simple solutions to jamming.

JAM is a protocol applied by all the sensors in the WSN. The protocol works as follows; the jammed sensors inside the jamming area send 'Jammed' messages to sensors outside of the jammed area. Then the sensors outside will cooperatively map an area of the possible jamming attack. The JAM paradigm utilize a loose group semantics were sensors do not wait for acknowledgment ACKs messages and eager eavesdropping were only needed sensors must be notified. Many assumptions are presumed in the sensors and in the jamming attack itself. JAM works in two main modules, the jam detection module, and the mapping module that follows. In the jamming detection, the authors assumed that the sensors could override the MAC layer carrier sense multiple access CSMA policies in relaying the 'Jammed' message. This is critical, as the channels are most likely to be filled all the time by the jammer. This overriding with sent 'Jammed' messages blindly to the jammed sensor's nearest neighbor. Many factors will lead the sensor to detect it is under jamming, some of the main factors are, low signal-to-noise ratio SNR, repeated collisions, and protocol violations. The jammed sensors will send 'Jammed' messages to their nearest neighbors to instantiate the next phase. In addition, when the jamming affects wares off, they send an 'UnJammed message to their previous neighbors to update the status.

The next phase is the mapping phase, here an edge sensor, the sensor that receives the actual 'Jammed' message, starts this phase upon receiving the message. When an edge sensor receives this message, the edge sensor instantiates a random group ID, in case it knows no other group. This group ID, with it assigns a normalized direction vector pointing to its jammed neighbor. It also includes its ID, along with the group ID and the normalized vector in a message called BUILD. Finally, this edge sensor now becomes a mapping sensor or node, and sends the BUILD message to its neighbors. Neighbors receiving this message will do the same but will no further propagate the message to their neighbors. These messages contain also a sequence number to prevent duplicates from being processed. When receiving more than one BUILD message, sensors check the group IDs, and check their direction vectors for coalescing compatible groups together.

Edge sensors are those that did not receive any BUILD message from their neighbors, thus they send Probe messages to detect other possible groups nearby. A bridge sensor or node is the sensor that upon checking the normalized vectors between two different groups, it appears they are very much pointing to the same area, and then there are coalesced. When coalescing, the group ID that is higher is used for the new group. Ideally, we want to get a mapping with a single group containing all the sensors surrounding the jammed area. The convergence of the protocol is, as the authors assume, achieved within seconds of the jamming detection. The use of loose group semantics and eager eavesdropping assure quick knowledge diffusion. In addition, the one-hop back flooding helps in reducing redundant information being exchanged. Probing and coalescing help in gathering compatible groups together to quick build a map of the current jammed area. The authors used extensive simulation to demonstrate JAM, using GloMoSim simulator [27].

5.4.1. Outcomes, Limitations and Improvements

The authors presented a protocol to leverage and mitigate the effects of jamming in a novel and descriptive way. The approach does not require extra hardware added to the sensors, thus implementing JAM is cost effective. The approach uses many heuristics to determine whether the channel is jammed or not. The convergence of the protocol is very high; this means that at the end we get a single map or group surrounding the jammed area. The normalized vectors used, not only help in pointing where the jammed area boundaries is, but also help in reducing the sent Probe messages to the neighbors that are on the edges of this jammed area, left and right only. Furthermore, during the protocol no messages are acknowledged, this reduce the complexity, timeouts are used instead with reception events. The use of probabilistic uniqueness in randomly selecting group IDs prevent any synchronization needed among the sensors. The simulation was very extensive and considered many parameters in different situations.

The authors assumed that most sensors are not capably of spread spectrum techniques, as most ZigBee and IEEE 802.15.4 compatible sensors are capable of Frequency Hopping Spread Spectrum FHSS and Direct Sequence Spread Spectrum DSS. The authors assumed that not all or huge parts of the WSN are jammed, this allows JAM to detect the area and maps it. In addition, they do not allow multiple jammers to jam nearby areas or even merging their areas together in an ad hoc fashion. The critical assumption that the authors assume is that the MAC layer, CSMA is overridden in the protocol 'Jammed' message to start the second phase. The authors suggested modification if necessary. This is not always possible with hardware implementations of the MAC layer CSMA in some sensors, thus no override is possible. Also high sensor counts create message explosions from the excessive back-flooding messages. The overall protocol complexity is also noticeable, as in some cases, no single group, or map is achieved. The protocol assumes that all sensors know their location, IDs of their neighbors. They assume that also the network only use a single channel for all communications. The protocol is only compatible with static WSN, as mobile sensors may never reach any convergence in computational times.

6. Conclusion

WSNs are a rapidly growing field, with many opportunities and challenges. Strict architectural, economical, and technological aspects of such networks give it its unique characteristics' and traits. As more dependent we grow on WSNs, we cannot afford to compromise the availability and security of such networks. Since WSN hardware and software have many limitations, it allowed security issues to rise to the surface. In this paper, we have discussed the jamming attacks and sinkhole attacks. We discussed their main aspects and types, and how attackers utilize such techniques to launch their attacks.

We have discussed through two major papers that proposed techniques to defend against jamming attacks. We discussed and criticized the papers in terms of their positive contributions and limitations. Furthermore, we suggested improvements to such shortcomings. The improvements we suggested stemmed from other papers who suggested other defense mechanisms. The future work lies in further studying more techniques that try to generally improve the overall security levels and standards of WSN. Finally, we hope that WSN of the future, are designed and realized with security in mind as WSNs today lack such focus on security. As more secure WSN will be in the future, more possibilities and applications are sure to use WSNs.

Acknowledgements

The author would like to thank Dr. Abraham Al-Rashid, Computer Engineering Dept. Kuwait University for his support and careful follow up with this project.

References

[1] Kai Xing, Fang Liu, Xiuzhen Cheng, David H.C. Du

"Real-time Detection of Clone Attacks in Wireless Sensor Networks" The 28th International Conference on Distributed Computing Systems ICDCS, IEEE 2008.

[2]   Lewis, Frank L; "Wireless Sensor Networks," Smart Environments: Technologies, Protocols, and Applications, pp. 11-46, © 2005 John Wiley & Sons, Inc.

[3]   I.F. Akyildiz, W. Su, Y. Sankarasubramaniam, E. Cayirci, "Wireless sensor networks: a survey", Computer Networks, Volume 38, Issue 4, pp. 393-422, 15 March 200.

[4]   Lewis, Frank L; "Wireless Sensor Networks," Smart Environments: Technologies, Protocols, and Applications, pp. 11-46, © 2005 John Wiley & Sons, Inc.

[5]   Sukumar Ghosh, Distributed Systems: An Algorithmic Approach, 2006 CRC Press.

[6]   Pottie, G.J.; "Wireless sensor networks," Information Theory Workshop, pp.139-140, 22-26 June 1998.

[7]   Wenyuan Xu; Ke Ma; Trappe, W.; Yanyong Zhang; "Jamming sensor networks: attack and defense strategies," Network, IEEE, vol.20, no.3, pp. 41- 47, May-June 2006.

[8]   Alzaid, Hani; Park, Dong Gook; Nieto, Juan Gonzalez; Boyd, Colin; Foo, Ernest; "A Forward and Backward Secure Key Management in Wireless Sensor Networks for PCS/SCADA," Sensor Systems and Software, Lecture Notes of the Institute for Computer Sciences, Social Informatics and Telecommunications Engineering, pp. 66-82, Vol.24, © 2010, Springer Berlin Heidelberg.

[9]   Xin Liu, Guevara Noubir, Ravi Sundaram, San Tan, "SPREAD: Foiling Smart Jammers using Multi-layer Agility" IEEE Communications Society subject matter experts for publication in the IEEE INFOCOM 2007 proceedings, IEEE 2007.

[10] DARPA, http://ww.darpa.mil/ato/programs/WolfPack/index.htm.

[11] Anthony D.Wood, John A. Stankovic, and Sang H. Son "JAM: A Jammed-Area Mapping Service for Sensor Networks" Proceedings of the 24th IEEE International Real-Time Systems Symposium (RTSS'03), IEEE, 2003.

[12] R. L. Pickholtz, D. L. Schilling, and L. B.Milstein "Theory of spread spectrum communications, a tutorial" IEEE Transactions on Communications, IEEE, May 1982.

[13] Onel, Tolga; Onur, Ertan; Ersoy, Cem; Delic, Hakan; Byrnes, Jim; "Wireless Sensor Networks For Security: Issues and Challenges," Advances in Sensing with Security Applications, NATO Security through Science Series, pp. 95-119, Vol. 2, © 2006 Springer Netherlands.

[14] AusCERT, "AA-2004.02 — Denial of Service Vulnerability in IEEE 802.11 Wireless Devices," http://www.auscert.org

[15] W. Xu, Timothy Wood, Wade Trappe, Yanyong Zhang "Channel Surfing and Spatial Retreats: Defenses against Wireless Denial of Service," In the Proceedings of Wireless Security Workshop, ACM, pp. 80–89, 2004

[16] Wang, Le; Wyglinski, Alexander M.; , "A Combined Approach for Distinguishing Different Types of Jamming Attacks against Wireless Networks," In the Proceedings of the Conference on Communications, Computers and Signal Processing Pacific Rim, pp.809-814, 23-26 IEEE, Aug. 2011.

[17] Wood A. D., Stankovic J. A., "A Taxonomy for denial-of service attacks in wireless sensor networks," in Handbook of Sensor Networks: Compact Wireless and Wired Sensing Systems, edited by Mohammad Ilyas and Imad Mahgoub, CRC Press LLC, 2005.

[18] Faraz Ahsan, Ali Zahir, Sajjad Mohsin, Khalid Hussain "Survey On Survival Approaches In Wireless Network Against Jamming Attack" Journal of Theoretical and Applied Information Technology, Vol. 30 No.1, JATIT & LLS, 15th August 2011.

[19] Y. Hu, A. Perrig, and D. Johnson "Ariadne: A secure on demand routing protocol for ad hoc networks". In 8th ACM International Conference on Mobile Computing and Networking, pages 12-23, September 2002.

[20] Wenyuan Xu, Wade Trappe, Yanyong Zhang "Channel Surfing: Defending Wireless Sensor Networks from Interference" In Proceedings of the 6th international conference on Information processing in sensor networks IPSN'07, April 25-27, Massachusetts, USA. ACM, 2007.

[21] Wood, A.D.; Stankovic, J.A.; Gang Zhou; "DEEJAM: Defeating Energy-Efficient Jamming in IEEE 802.15.4-based Wireless Networks," fourth Annual IEEE Communications Society Conference on Sensor, Mesh and Ad Hoc Communications and Networks SECON '07. PP.60-69, 18-21 June 2007.

[22] Periyanayagi, S.; Sumathy, V.; Kulandaivel, R.; "A Defense Technique for Jamming Attacks in Wireless Sensor Networks Based on SI," International Conference on Process Automation, Control and Computing PACC, pp.1-5, 20-22 July 2011.

[23] Swarm Intelligence, http://www.sce.carleton.ca/netmanage/tony/swarm.html

[24] Aristides Mpitziopoulos, Damianos Gavalas, Grammati Pantziou, and Charalampos Konstantopoulos "Defending Wireless Sensor Networks From Jamming Attacks" The 18th Annual IEEE International Symposium on Personal, Indoor and Mobile Radio Communications (PIMRC'07), IEEE, 2007.

[25] Zhenhua Liu, Hongbo Liu, Wenyuan Xu, and Yingying Che, "Exploiting Jamming-Caused Neighbor Changes for Jammer Lcalization", IEEE Transactions on Parallel and Distributed Systems, Vol. 23, No. 3, March 2012.

[26] Hai-Yan Shi, Wan-Liang Wang, Ngai-Ming Kwok and Sheng-Yong Chen "Game Theory for Wireless Sensor Networks: A Survey", Sensors 2012, doi:10.3390/s120709055, July 2012.

[27] X. Zeng, R. Bagrodia, and M. Gerla "GloMoSim: A library for parallel simulation of large-scale wireless networks" Workshop on Parallel and Distributed Simulation, pages 154–161, 1998.

12

Analysis of the automatic transmission fault diagnosis based on Bayesian network sensitivity

Luo Jin[*], **Ma Qihua, Luo Yiping**

Faculty of Automotive Engineering, Shanghai University of Engineering and Science, Shanghai, China

Email address:
84276661@qq.com (Luo Jin), mqh0386@163.com (Ma Qihua), lyp777@sina.com (Luo Yiping)

Abstract: Automobile automatic transmission is a complex system integrates machine, electricity, liquid, in terms of its fault diagnosis has the characteristics of uncertainty and high complex correlation, thus put forward the Bayesian network based on the theory of probability and theory of the fault tree for fault diagnosis methods of automobile automatic transmission, it has a certain value for application. Take a failure automatic transmission as an example, analyzes the causes of the automatic transmission fault and the relationship between the affecting factors, according to the properties of the automatic transmission fault establishes a Bayesian network model, and through calculating sensitivity to determine the probability of a certain causes to the automatic transmission fault. The results show that the method is operable in automatic transmission fault diagnosis, diagnosis results are accurate and credible.

Keywords: Automatic Transmission, Bayesian Network, Sensitivity

1. Introduction

Automatic transmission with a lot of advantage, such as good transmission ratio, improve the comfort and trafficability of the vehicle，reduce driving operation fatigue strength and reduce the emissions, is widely used in modern cars. At the same time, the automatic transmission is the complex machinery and equipment with a collection of mechanical, electronic control, and hydraulic transmission, its fault identification and diagnosis is difficult because the complexity of its structure and working principle. Therefore, it's have great significance to establish an operable, accurate and rapid diagnostic methods to ensure transmission can work reliably.

Since the automatic transmission was born to now, automatic transmission fault diagnosis technology has been an important part of automatic transmission technology research and development. At present, foreigner studying an online automatic transmission fault diagnosis system that diagnosis range is more broader, its principle is based on the entire model of the automatic transmission, make full use of existing sensor information of automatic transmission, as much as possible to obtain meaningful information for faults. Jin-Oh Hahn[1] proposed a algorithm based on state observer to evaluate the oil pressure of the shift actuator, it's have great significance for the improvement of the automatic

transmission control technology and study of the online fault diagnosis system. In domestic, Hu Ning[2] et al proposed function analysis methods of the automatic transmission fault, according to the function of the of each components, to study the component failure impact on the automatic transmission. Qin Guihe[3] proposed that use exists in measurement data and local power transmission system model and redundant information between the structure logical relationship to detect and diagnose the faults of the sensors, actuators and AMT parts. Wei Shaoyuan[4] proposed to diagnose automotive automatic transmission fault based on fuzzy theory and put forward to set up a system of automatic transmission fault diagnosis. Overall, the current domestic research on fault diagnosis of the automatic transmission also stay on how to solve some targeted specific example.

This paper make a detailed analysis of the automatic transmission fault diagnosis, apply for Bayesian network in the field of automatic transmission failure, and by calculating the sensitivity with examples to verify the feasibility of Bayesian networks for automatic transmission failure. On the basis of Bayesian networks, establish a network diagram of the automatic transmission fault, According to the specific fault reasoning procedure and sensitivity calculation to achieve automatic transmission fault diagnosis, to solve the uncertain reasoning problems in the process of found fault reason. A comprehensive of preliminary testing, failure analysis, probability calculation and fault recognition, thus forming a complete automatic transmission fault diagnosis system based on Bayesian network. The results show that Bayesian network have realized the optimization of judgment steps, it has good effect.

2. Introduction to Bayesian Networks

Bayesian networks is a probabilistic network, which is based on probabilistic reasoning graphical network, and the basis of this network is Bayesian probability formula. Bayesian network use directed graph to describe the relationship between probability theory, it is suitable for the uncertainty and probabilistic things, applied to related issues what conditionally dependent on a variety of control factors, is a kind of probability inference technology, can use probability theory to deal with the uncertainty produced by the knowledge relationship because of conditions correlation.

In simple terms, the Bayesian network diagram is a kind of cycle directed graph and related parameters properties, it's consist of the model structure and related parameters. In Bayesian network, the node represent the variable, the contact between them represented by directed arc, usually indicates the cause node to result node. So that can draw a causal relationship between each node and to describe the degree of variable may affect another variable by probability. In probabilistic reasoning, the random variable is used to represent events or things, through the random variable instance into various instances, it can be modeling a series of events or the existing state of things. Based on the theory of Bayesian probability, it can be calculated the joint probability under the a certain condition. Fig. 1 is a simple but typical example of Bayesian network (omitting the conditional probability values).

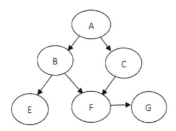

Fig 1. Sample of Bayesian network structure

Assuming that there are n variables, to remember $X_1, X_2, \cdots, X_n$, to contain a combination of variable distribution $P(X_1, X_2, \cdots, X_n)$, then decomposed this joint distribution with the chain rule, see equation (1) :

$$P(X_1, X_2, \cdots, X_n) = \prod_{i=1}^{n} P(X_i | X_1, X_2, \cdots, X_{i-1})$$ (1)

Assuming the existence of $\pi(X_i) \in \{X_1, X_2, \cdots, X_{i-1}\}$ an

arbitrary X_i, and to set the conditions under $\pi(X_i)$, X_i and $\{X_1, X_2, \cdots, X_n\}$ in the other variables are independent of each other, then formula (1) can be rewritten as formula (2) :

$$P(X_1, X_2, \cdots, X_n) = \prod_{i=1}^{n} P(X_i | \pi(X_i))$$ (2)

A and B are provided as two random variables, the probability of A occurrence is $P(A)$, $(P(A) > 0)$, the probability of B occurrence is $P(B)$, $(P(B) > 0)$, the probability of B occurrence conditions for A occurrence is $P(B|A)$, there is a multiplication formula, see equation (3) :

$$P(AB) = P(B|A)P(A)$$ (3)

Let $A_1, A_2, \cdots, A_n$ constitute an integral event, and have $P(A_i) > 0$, the total probability of the event B has a formula, see equation (4) :

$$P(B) = \sum_{i=1}^{n} P(A_i)P(B|A_i)$$ (4)

Equation (2) ~ (4) can be calculated the joint distribution function between the variables of each factor in the accident.

3. The Construction of a Bayesian Network Diagram Automatic Transmission Fault

Different manufacturers and different car manufacturers with different types of automatic transmission. Though the structure of different types of automatic transmission are different, the basic function and working principle of various components are basically the same. In general, the automatic transmission is consist of external operating system, hydraulic transmission device, mechanical structure, hydraulic control system and circuit control system.

Automatic transmission fault caused by many reasons, but the occurrence probability of reasons is different. Need to find a greater probability caused by fault reasons from the symptoms, therefore, put the automatic transmission fault classification is very important. Automatic transmission fault generally have the following symptoms, respectively: no forward gears and reverse gears, big shift impact, shift up too late, can't shift gears, hang power engine shut down immediately, automatic transmission oil temperature is too high, appearing vibration in the process of driving, the transmission gear jumping, and so on. Removing some of the new cars because its quality problems causes the fault of the automatic transmission, but with the decline in the status of automatic transmission technology and life shortening, there will be produce a series of fault, if the fault not resolved quickly often cause serious damage, however the common failure usually manifested through certain phenomenon, so it's important as soon as possible to find out the fault reason through the phenomenon. Assuming that automatic transmission malfunction is the parent node, its associated child nodes respectively hydraulic torque converter, mechanical drive mechanism, pressure of automatic transmission oil, pressure of the governor, pressure of the throttle, circuit control system, locking clutch, and so on. The above factors listed in sequence into Bayesian network diagram (see fig. 2). Analysis is as follows: (1) influence most is the oil, followed is the circuit, then the mechanical fault and hydraulic torque converter fault; (2) the focus of fault diagnosis is the oil and circuit.

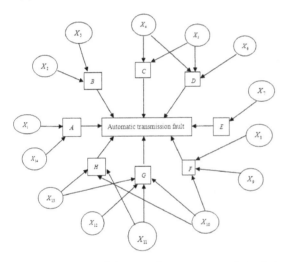

Fig 2. *Bayesian network diagram of automatic transmission fault*

A: no forward gears and reverse gears; B: big shift impact; C: can't shift gears; D: shift up too late; E: hang power engine shut down immediately; F: automatic transmission oil temperature is too high; G: appearing vibration in the process of driving; H: the transmission gear jumping; X_1: automatic transmission light-leaking; X_2: Energy storage shock absorber of failure; X_3: Work oil pressure is high; X_4: throttle pressure is high; X_5: the governor oil pressure too low; X_6: circuit control system failure; X_7: hydraulic torque converter failure; X_8: cooler pipeline or blockage in a one-way valve; X_9: the oil sump level is too low; X_{10}: clutch skid; X_{11}: shift institution failure; X_{12}: engine failure; X_{13}: electronic components failure; X_{14}: the main serious leaks oil

The figure 2 is a directed acyclic graph, actually it's a Bayesian network, the fault diagnostic process can be seen as a probabilistic Bayesian network inference process, and thus can be use Bayesian network as the representation and solving methods of the uncertain problems in fault diagnosis process, so based on the Bayesian network graph can also understand that: solving object is automatic transmission fault, the fault produced by the oil circuit, circuit, mechanical fault, fault of hydraulic torque converter; its failure of automatic transmission phenomenon is bad working conditions of oil, circuit control system, governor,

locking clutch, hydraulic torque converter.

Using Bayesian network for automatic transmission failure analysis in accordance with people's normal thinking. When the fault happens, firstly, you will be to determine the fault symptoms, and then judge the character of failure and types, according to the information to choose the most suitable solution finally. By Bayesian network probabilistic inference, the fault will be sorted according to the probability of risk events, determine the priority for troubleshooting.

In a word, the Bayesian figure can be described as full of information, it can play a big role in the automatic transmission fault diagnosis.

4. The calculation of Sensitivity Analysis Method Based on Bayesian Network

4.1. The Definition of Parameter Sensitivity Analysis

To introduce the method of sensitivity analysis for the fault diagnosis of complex systems, the priority task is to determine the related issues of parameter sensitivity and node sensitivity, let's introduce relevant definitions below.

Definition 1 Parametric sensitivity: In a Bayesian network, assuming that there is a parameter $\theta = P(b_i|\pi)$ on node B, the value of target node A is a, value of evidence is e, then the sensitivity of parameters of θ can be expressed as follows:

$$I(\theta) = \frac{1}{rs}\sum_{a,e}\frac{\partial P(a|e)(\theta)}{\partial \theta} \qquad (5)$$

Where r and s represent the number of values of A and e.

Definition 2 Node sensitivity: Suppose there are two node B and node A in a Bayesian network and there are edges from B to A, the parameter $\theta = P(b_i|\pi)$ for node B, the value of target node A is a, value of evidence is e, then the sensitivity of node B with respect to the node A can be expressed as follows:

$$EI(B \to A) = \frac{1}{rt}\sum_{j=1}^{r}\sum_{i=1}^{t}I(\theta_{ij}) \qquad (6)$$

Where r and t represent the number of values of A and B. A more important node, we call it is a high sensitivity node.

Due to the particularity of fault diagnosis, we can simplify the following formula (5):

$$I(\theta) = \frac{\partial P(A=a)(\theta)}{\partial \theta} \qquad (7)$$

The corresponding formula (6) simplified as follows:

$$EI(B \to A) = \frac{1}{t} \sum_{i=1}^{t} I(\theta_i) \qquad (8)$$

By the simplify formula (8), we can see, the higher sensitivity of the parent node, the bigger influence of state and state parameters to child nodes stay normal working, therefore, when the parent node produce problems, it is easy to cause the abnormal state of relatively child nodes.

4.2. Process of SA_FD Algorithm [5]

As shown in Figure 2, when a child node have multiple parent nodes, according to the formula (8), we can determine which node should bear the greatest responsibility for the child node produce fault, thus in the process of fault diagnosis system, avoid finding fault blindly, improve the efficiency of fault diagnosis.

Put forward a kind of fault diagnosis algorithm combining with Bayesian network sensitivity analysis: SA_FD (sensitivity analysis fault diagnosis), the whole process of diagnosis algorithm is summarized as follows:

Step 1 Abnormal events occur through the system, the system state of the parent node;

Step 2 Suppose through step1 to find the parent node are $X_1, X_2, \cdots, X_n$, then according to its sensitivity size detect abnormal node;

Step 3 Judge the abnormal nodes if it have parents, if it have then carried out step 4, if the abnormal modes don't have parent node, so the faulty node is the abnormal nodes;

Step 4 Repeat step 3 until find out the fault node.

5. Examples of Application

Figure 2 is about Bayesian network diagram of automatic transmission malfunction, select one of the G child node as the research object, the network parameters see literature [6], when computing child node have more than one parent node appear failure, the size of its parent node CPT (Conditional Probability Distribution) can be obtained according to variant $P(X_i|H) = \frac{P(HX_i)}{P(H)}$ of the formula(3), here only show CPT change size of each parent node, the calculation results as shown in tab. 1.

Tab 1. Change of the parent node CPT

Child node	Parent node	Change of the CPT
G	X_{10}、X_{11}、X_{12}、X_{13}	$X_{10} > X_{12} > X_{11} > X_{13}$

SA_FD algorithm is not only need to consider the change degree of parent node state, also need to consider the influence degree of parent node's state parameter for the state of "normal" child nodes, the influence degree is obtained through sensitivity function. According to the formula (8), we can calculate the sensitivity of each node relative to the its child nodes. The calculation results are shown in tab. 2.

Tab 2. Sensitivity of nodes

Child node	Parent node				Sensitivity			
G	X_{10}	X_{11}	X_{12}	X_{13}	0.79	0.32	0.41	0.28

Derived from Table 2, when there is too much vibration in the process of driving, the probability is bigger caused by the X_{10} "clutch skid".

For other associated nodes, can be treated using similar methods, respectively obtain the sensitivity of a certain parent node to its child nodes, thus gain the conclusion that this child node (fault symptoms) occurs when the occur probability of a certain parent node (fault reason) is bigger.

6. Conclusion

In this paper, sensitivity analysis based on Bayesian networks is applied to fault diagnosis in the field of automotive automatic transmission, when a fault occurs, quantitatively calculated the sensitivity between the child nodes and the parent nodes, thus obtained a conclusion that the bigger probability what a certain cause lead to the automatic transmission fault, improve the accuracy of diagnostic results, saving a lot of time and effort at same time. However, the study objects are limited to static Bayesian network model in this paper, remains to be further put forward a more efficient and precise fault diagnosis method for the dynamic Bayesian network model.

References

[1] Jin-Oh Hahn, Jae-Woong Hur, Young Man Cho and Kyo Lee. Robust Observer-Based Monitoring of A Hydraulic Actuator In A Vehicle Power Transmission Control System. Proc. Of the 40th IEEE Conference on Decision and Control (pp. 522-528)

[2] Hu Ning, Hu Jia, The function of the automatic transmission failure analysis. The technology of automotive, 2003(8):39-41.

[3] Qin Guihe, Ge Anlin, Lei Yulong. On-line fault diagnosis system of automated mechanical transmission. Journal of agricultural mechanization, 2001(32)5:75−77.

[4] Wei Shaoyuan, Wang Dongmei, Zhang Zhongyang. Automotive automatic transmission based on the theory of fuzzy fault diagnosis system. Mechanical design and manufacturing, 2011 (1):230-232.

[5] Yuan Zheng. Bayesian network sensitivity analysis method and its application in complex system fault diagnosis. [Master's degree thesis]. Hefei university of technology,2012.

[6] Liu Xuejun, Liu Cunxiang, Zhang Erli. Bayesnets application in vehicle gearbox fault analysis. Journal of Nanning College for Vocational Technology. 2013(18) 6:90-92.

13

Enhancement of transient and steady state responses of voltage by using DVR

Jyothilal Nayak Bharothu, D Sunder Singh

Associate professor of Electrical & Electronics Engineering , Sri Vasavi Institute of Engineering & Technology, Nandamuru, A.P.; India

Email address:
nayakeee@gmail.com(J. N. Bharothu), dsundersingh@yahoo.com (D S. Singh)

Abstract: Voltage sag is one of the most important power quality problems challenging the utility industry voltage sags can be compensated for by voltage and power injection into the distribution system. Voltage sags are an important power quality problem and Dynamic Voltage Restorer is known as an effective device to mitigate voltage sags .The dynamic voltage restorer has become popular as a cost effective solution for the protection of sensitive loads from voltage sags. Simulations of the dynamic voltage restorer have been proposed at low voltage level; and give an opportunity to protect high power sensitive loads from voltage sags. This paper reports simulation test results obtained on a low voltage level using a dynamic voltage restorer. Dynamic voltage restorer was designed to protect a load from voltage sag. The proposed Dynamic Voltage Restorer obtains good transient and steady state responses.

Keywords: Voltage Stability, Transient, Voltage Sag, Simulation, DVR Etc

1. Introduction

Significant deviations from the nominal voltage are a problem for sensitive consumers in the grid system. Interruptions are generally considered to be the worst case with the load disconnected from the supply. The number of interruptions, though expensive, can be minimized with parallel feeders and are less likely to occur with the transition from over headlines to cables in the LV and MV distribution system.

Voltage sags are characterized by a reduction in voltage, but the load is still connected to the supply. Sags are in most cases considered less critical compared to interruptions, but they typically occur more frequently. Voltage sags have in several cases been reported as a threat to sensitive equipment and have resulted in shutdowns, loss of production and hence a major cost burden.

Sags are so far almost impossible to avoid, because of the finite clearing time of the faults causing the voltage sags and the wide propagation of sags from the Equipment can be made more tolerant of sags either via more intelligent control of the equipment or by storing more energy in equipment. Instead of modifying each component in for instance a factory to be very tolerant to voltage sags, a better solution might be to install one dynamic voltage restorer to mitigate voltage sags. A DVR can eliminate most sag and minimize the risk of load tripping at very deep sags.

The control of a DVR is not a straight – forward because of the requirements of fast response, large variation in the type of sags to be compensated and variation in the type of connected load. The DVR must also be able to distinguish between background power quality problems and the voltage sags with a phase jump and illustrates different control strategies for a DVR and some DVR limitations, which should be included in the control strategy. Two control methods are proposed with the ability to protect the load from a sudden phase shift caused by voltage sag with phase jump. Simulations and measurements illustrate how symmetrical voltage sags with phase jump successfully can be compensated.

Dynamic voltage restorer (DVR) was originally proposed to compensate voltage disturbances on distribution systems. The restoration is based on injecting ac voltages in series with the incoming three- phase net work, the purpose of which is to improve voltage quality by adjustment in voltage magnitude, wave –shape and phase shift. These are important voltage attributes as they can affect the performance of the load equipment. The ideal restoration is Compensation, which enables the voltage seen by the load to be unchanged in The face of upstream disturbances. Also, voltage restoration involves energy injection into the distribution systems and this determines the capacity of the energy storage device required in the restoration scheme.

If the voltage restoration is to be realized by in-phase voltage injection, minimum –magnitude injected voltage is achieved. However, in-phase injection does not take into consideration of minimizing the energy required to achieve voltage restoration. By employing reactive power to implement compensation, voltage injection with an appropriate phase advance with respect to source side voltage can reduce the consumption of energy, from the perspective of the DVR energy storage device. The increased reactive power is generated electronically from within the voltage source inverter (VSI). Thus the energy saving voltage   restoration methods means that the capacity of the energy storage device can be reduced. In other words for a given DVR this has fixed energy storage capacity, Reduced energy injection means increased ride-through ability for the compensated load. Voltage injection with a phase advance, however, does require a larger injected voltage, and the resulting voltage shift can cause voltage waveform discontinuity, inaccurate voltage zero crossing, and load power swing.

Although the energy saving concept was mentioned, no systematic analysis was given to explore such possibility and the associated problems were not addressed. Balanced voltage sag is considered first. The voltage injection control strategy proposed is to achieve energy saving subject to voltage injection limit placed on the DVR. The results are then extended to include unbalanced voltage sags. Simulation results are given to show the efficacy of the proposed method.

In other words for a given DVR this has fixed energy storage capacity, Reduced energy injection means increased ride-through ability for the compensated load. Voltage injection with a phase advance, however, does require a larger injected voltage, and the resulting voltage shift can cause voltage waveform discontinuity, inaccurate voltage zero crossing, and load power swing.

1.1. Literature Survey

Power quality has a significant influence on high–technology equipments related to communication, advanced control, automation, precise manufacturing technique and on-line service. For example, voltage sag can have a bad influence on the products of semiconductor fabrication with considerable financial losses; power quality problems include transients, sags, interruptions and other distortions to the sinusoidal waveform. One of the most important power quality issues is voltage sag that is a sudden short duration reduction in voltage magnitude between 10 and 90% completed to nominal voltage. Voltage sag is deemed as a momentary decrease in the RMS voltage, with duration ranging from half a cycle to up to one minute. Deep voltage sags, even of relatively short duration, can have significant costs because of the proliferation of voltage sensitive computer-based and variable speed drive loads. The fraction of load that is sensitive to low voltage is expected to grow rapidly in coming decades. Studies have shown that transmission faults, while relatively rare, can cause widespread sags that may constitute major process interruptions for very long distances from the faulted point.

Distribution faults are considerably more common but the resulting sags are more limited in geographic extent. The majority of voltage sags are within 40% of the nominal voltage. Therefore, by designing drives and other critical loads capable of riding through sags with magnitude of up to 40%, interruption of processes can be reduced significantly. The DVR can correct sags resulting from faults in either transmission or distribution system.

2. Introduction to Power Quality

The term "power quality" has barren used to describe the extent of variations of the voltages, current and frequency on the power systems. Most power apparatus made over a decade back could operate normally with relatively wide variations of three parameters. However, equipment added to power system in recent years generally is not tolerant to these variations for two main reasons. First one is the design tolerances have been going down in a competitive market and second one is the increasing use of sophisticated electronic controls. Hence, system disturbances, which were tolerated earlier, may now cause interruption to industrial power system with a resulting loss of production and this could be substantial with greater stress on productivity and quality now. Especially for a developing country like India, power quality is of prime importance considering the need for the energy conservation. It is a paradox that some of the energy conserving devices themselves is the reasons for some of the power quality problems.

2.1. Power Quality Problems

Power quality refers to ac power supply with rather perfect sine wave, constant magnitude and constant frequency.

The power quality problems originate primarily from industries; transmissions and distribution systems. The real power quality problems can be put under the following main categories:

- Voltage dips or sags
- Voltage swells
- Voltage fluctuations and flicker
- Interruptions
- Harmonics
- Unbalances
- Transients
- Commutation notches
- Electric noise

PQ variations fall into three categories: 1) transient disturbances: these include uni-polar transients, oscillatory transients, localized faults and other events typically lasting less than 10ms. These disturbances are expected to change shape as they propagate through the power system. Hence the recorder/monitor location, relative to the point of origin of disturbances is critical.2) momentary disturbances: these are voltage increases (swells) or voltage decreases (sags) lasting more than 10 ms, but less than 3 sec.3) steady state disturbances: these are voltage deviations lasting more than 3 sec.

With the revolution in technology, the trend in the industrial world is towards completely automated process control. Hence, PQ problem results in failure or mal-operation of electronic equipment, sensitive microprocessors personal computers and all electric gadgets. This has driven the utilities to concentrate their research efforts on improvement of power quality.

2.2. What is Power Quality

Maintaining the near sinusoidal waveform of power distribution bus voltage at rated voltage magnitudes and frequency termed as "power quality". There can be various definitions

(1) As meant by consumers: the relative absence of utility related voltage variations, particularly, the absence of outages, sags, surges and harmonics as measured at the point of service.

(2) As meant by manufactures: power quality means equipment compatibility at the supply of electrical power.

2.3. Aspects of Power Quality

2.3.1. Voltage Fluctuations and Flicker
These cause variations in the intensity of illumination. The frequency of variation is in the range of 10hz.The electric furnaces are the major source of voltage fluctuation that is caused by the rapid variation in the reactive power drawl. The adverse effects are irritation at home and offices, and interface with certain equipment.

2.3.2. Voltage DIP or SAG
A dip or sag in the voltage is referred to as a decrease in RMS voltage or current at the power frequency followed by a rapid return to the initial value. The voltage dips and sags are caused by lightning strikes, ground faults, energizing heavy loads, etc.   The problems caused due to the voltage dips and sags include loss of conduction, loss of data and damage to the equipment, etc.

2.3.3. Voltage Swells
An increase in RMS voltage or current at the power frequency for duration from 0.5 cycles to 1 minute is known as voltage swell. The cause include drop of large loads and Unsymmetrical faults. The voltage swells may result in loss of production damage or loss of life of equipment, etc.

2.3.4. Interruptions
An interruption is a type of short duration variations of voltage resulting in complete loss of voltage (<10%) on one or more phase conductors for a time period greater than one minute. The causes of interruptions are lightning. Wind, tree and animal contact, equipment failure, accidents involving power lines, generator failures, etc.

2.3.5. Harmonics
The Harmonics are caused by converters, rectifier loads, switched mode power supplies, etc. These harmonics distort the voltage and currents. The harmonics cause heating in machines, noise, mal-operation of control and protection equipment, etc.

2.3.6. Unbalances
The unbalances could be due to the differences in either amplitude or phase or both. The unbalance is caused due to the traction and other unbalanced loads. The ill effects of unbalance include generation of harmonics, when non-linear loads are supplied by unbalanced power supply, and damaging effect on machines due to the flow of negative sequence currents.

2.3.7. Transients
Lightning, power line feeder switching, capacitor bank switching and system faults Cause the transients. The transients may stress equipment insulation or even cause damage to Equipment.

2.3.8. Commutation Notches
Commutation notches are caused by converters, rectifiers, switched mode power supplies, etc. The commutation notches may stress insulation, because heating in machines, noise, and mal-operation of control and protection equipment.

2.3.9. Electric noise
The cause of electric noise in the supply voltage and currents include radar and radio Coupling, arcing utility equipment, industrial arc equipment, converters and switching circuits. It can cause interference with communication systems, damage to equipment, etc.

2.3.10. Geo Magnetic Disturbances
Results from low frequency magnetic fields, typically below 5 Hz, associated with solar flares. The slowly varying field induces low frequency current in long transmission lines. These currents when pass through transformer winding, saturate the magnetic core and the voltage waveforms at the transformer may be distorted.

2.4. Interest in Power Quality

The fact that power quality has become an issue recently does not mean that it was not important in the past. Utilities all over the world have for decades worked on the improvement of what is now known as power quality. And actually, even the term has been in use for a rather long time already. The oldest mentioning of the term "power quality" known to the author was in a paper published in 1968.The paper detailed a study by the U.S navy after specifications for the power required by electronic equipment. That paper gives a remarkably good overview of the power quality field, including the use of monitoring equipment and even the suggested use of a static transfer switch. Several publications appeared soon after, which used the term power quality in relation to airborne power systems. Already in 1970"high power quality" is being mentioned as one of the aims of industrial power system design, together with "safety"," reliable service" and "low initial and operating costs". At about the same time the term "voltage quality" was used in the Scandinavian Countries and in the Soviet Union, mainly with reference to slow variations in the voltage magnitude.

The recent increased interest in power quality can be explained in a number of ways. The main explanations given are summarized below. Of course it is hard to say which of these came first; some explanations for the interest in power quality given below will by others be classified as consequences of the increased interest in power quality.

2.5. Overview of Power Quality Phenomenon

We saw in the previous section that power quality is concerned with deviations of the voltage from its ideal waveform (voltage quality)and the deviations of the current from its ideal waveform(current quality).such a deviation is called a 'power quality phenomenon' or a 'power quality disturbance'.

Power quality phenomena can be divided into two types, which need to be treated in a different way.

- A characteristic of voltage or current (e.g., frequency or power factor) is never exactly equal to its nominal or desired value. The small deviations from the nominal or desired value are called "voltage variations" or "current variations". A property of any variation is that it has a value at any moment in time. e.g. The frequency is never exactly equal to 50hz or 60hz; the power factor is never exactly unity .Monitoring of a variation thus has to take place continuously.
- Occasionally the voltage or current deviates significantly from its nominal or ideal wave shape .These sudden deviations are called "events". Example are a sudden drop to zero of the voltage due to the operation of a circuit breaker (a voltage event) and a heavily distorted over current due to switching of a non_ loaded transformer (a current event).monitoring of events take place by using a triggering mechanism where recording of voltage and /or current starts the moment a threshold is exceeded.

The classification of a phenomenon in one of these two types is not unique .it may depend on the kind of the problem due to phenomenon.

2.5.1. Voltage and Current Variations

- Voltage magnitude variation

- Voltage frequency variation
- Current magnitude variation
- Current phase variation
- Voltage and current unbalance
- Voltage fluctuation
- Harmonic voltage distortion
- Harmonic current distortion
- Inter harmonic voltage and current components
- Periodic voltage notching
- Mains signaling voltage
  - ➢ Ripple control signals
  - ➢ Power line carrier signals
  - ➢ Mains marking signals
- High frequency voltage noise

2.5.2. Events

Events are phenomenon which only happen every once in a while. An interruption of the supply voltage is the best known example. They are
- interruptions
- under voltages
- voltage magnitude steps
- over voltages
- fast voltage events
- phase angle jumps and three phase unbalance

2.5.2.1. Interruptions

A voltage interruption / supply interruption / interruption is a condition in which the voltage at the supply terminals is close to zero. Close to zero is defined as lower than 1% of the declared voltage. Interruptions can also be subdivided based on their duration, thus based on the way of restoring the supply.
- Automatic switching
- Manual switching
- Repair or replacement of the faulted component

2.5.2.2. Under Voltages

Under voltages of various duration are known under different names. Short-duration under voltages are called "voltage sags" or "voltage sags" or "voltage dips". Long duration under voltage is normally simply referred to as "under voltage".

Voltage sag is a reduction in the supply voltage magnitude followed by a voltage recovery after a short period of time. When a voltage magnitude reduction of finite duration can be called a voltage sag or voltage dip. According to IEC, a supply voltage dip is a sudden reduction in the supply voltage to a value between 90% and 1% of the declared voltage, followed by a recovery between 10ms and 1 min later.

2.5.2.3. Voltage Magnitude Steps

Load switching, transformer tap changers, and switching actions in the system can lead to a sudden change in the voltage magnitude. Such a voltage magnitude step is called a "rapid voltage change" or "voltage change".

2.5.2.4. Over Voltages

Just like with under voltage, over voltage events are given different names based on their duration. Over voltages of very short duration, and high magnitude, are called "transient over- voltages", "voltage spikes", or sometimes "voltage surges". The letter term is rather confusing as it sometimes used to refer over voltages with duration between about 1 cycle and 1 min. The latter event is more correctly called voltage swell or "temporary power frequency over voltages". Longer duration over voltages is simply referred to as "over voltages". Long and short over voltages originates from, among others, lightning strokes, switching operations, sudden load reduction, single-phase short circuits and nonlinearities.

A resonance between the nonlinear magnetizing reactance of a transformer and a capacitance can lead to a large over voltage of long duration. This phenomenon is called Ferro resonance, and it can lead to serious damage to power system equipment.

2.5.2.5. Fast Voltage Events

Voltage events with a very short duration typically one cycle system frequency or less referred to as "transients"/transient voltages / voltage transients/wave shape faults. The term transient is not fully correct, as it should only be used for the transition between two steady states. Events due to switching actions could under that definition be called transients. Events due to lightning strikes could not be called transients. Even very short duration voltage sags (due to fuse clearing) are referred to as voltage transients, or also "notches". Fast voltage events can be divided into impulsive transients (mainly due to lightning) and oscillatory transients (mainly due to switching actions).

2.5.2.6. Phase Angle Jumps and Three Phase Unbalance

We will see that voltage sag is often associated with a phase angle jump and some three phase unbalance. An interesting thought is whether or not a jump in phase angle without drop in voltage magnitude should be called voltage sag. Such an event could occur when one of two parallel feeders is taken out of operation.

3. Voltage Sags- Characterization

3.1. Introduction

Voltage sags are short duration reductions in RMS voltage, caused by short circuits, overloads and starting of large motors. The interest in voltage sags is mainly due to the problems they cause on several types of equipment adjustable-speed drives, process-control equipment and computers are notorious for their sensitivity. Some pieces of equipment trip when the RMS voltage drops below 90% for longer than one or two cycles.

3.1.2. Voltage Sag Magnitude

3.1.2.1. Monitoring

The magnitude of voltage sag can be determined in a number of ways. Most existing monitors obtain the sag magnitude from the RMS voltages. But this situation might well change in the future. There are several magnitude of the fundamental (power frequency) component of the voltage and the peak voltage over each cycle or half cycle. As long as the voltage is sinusoidal, it does not matter whether RMS voltage, fundamental voltage, or peak voltage is used to obtain the sag magnitude. But especially during voltage sag this is often not the case.

RMS VOLTAGE:

As voltage sags are initially recorded as sampled points in time, the RMS voltage will have to be calculated from the sampled time-domain voltages. This is done by using the following equation:

$$V_{rms} = \sqrt{1/N \sum_{i=1}^{N} (v_i)2} \qquad (3.1)$$

Where N is the number of samples per cycle and v1 are the sampled voltages in time domain.

Fundamental voltage component:

The advantage of using the fundamental component of the voltage is that the phase-angle jump can be determined in the same way. The fundamental voltage component as a function of time may be calculated as

$$V_{fund}(t) = \frac{2}{T} \int_{t-T}^{t} v(\tau)e^{j\omega_0 \tau} d\tau \qquad (3.2)$$

Where $\omega_0 = 2\Pi/T$ and T one cycle of the fundamental frequency. Note that this results in a complex voltage as a function of time. The absolute value of this complex voltage is the voltage magnitude as a function of time; its argument can be used to obtain the phase angle jump. In a similar way we can obtain magnitude and phase angle of a harmonic voltage component as function of time. This so-called "time-frequency analysis" is a well-developed area within digital signal processing with a large application potential in power engineering.

Peak Voltage:

The peak voltage as a function of time can be obtained by using the following expression.

$$V_{peak} = max \mid v(t-\tau) \mid \qquad 0< \quad <<T \qquad (3.3)$$

Where V (t) the sampled voltage waveform and T an integer multiple of one half cycle.

One cycle voltage sag: the voltage is low in one phase for about one cycle and recovers rather fast after that. The other phases show some transient phenomenon, but no clear sag or swell.

Obtaining one sag magnitude:

Until now, we have calculated the sag or as the

fundamental voltage component obtained over a certain window. There are various ways of obtaining one value for the sag magnitude as a function of time. Most monitors take the lowest value. Thinking about equipment sensitivity, this corresponds to the assumption that the equipment trips instantaneously constant RMS value during the deep part of the sags, using the lowest value appears an acceptable assumption.

3.2. Theoretical Calculations

To quantify sag magnitude in radial systems, the voltage divider model shown in fig.1 can be used. This might appear a rather simplified model, especially for transmission systems.

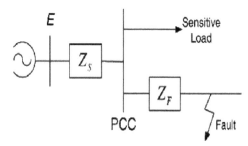

Fig. 1. *Voltage divider for voltage sag*

In fig.1 we see two impedances: Z_s is the source impedance at the point-of-common coupling; and Z_f is the impedance between the point-of-common coupling and the fault. The point-of-common coupling is the point from which both the fault and the load are fed. In other words: it is the place where the load current branches off from the fault current. We will often abbreviate "point-of-common coupling" as PCC. In the voltage at the PCC and thus the voltage at the equipment terminals, can be found form

$$V_{sag} = \frac{Z_F}{Z_S + Z_F} E \qquad (3.4)$$

In the remainder of this chapter, we will assume that the pre-event voltage is exactly 1pu, thus E=1. This results in the following expression for the sag magnitude.

$$V_{sag} = \frac{Z_F}{Z_F + Z_S} E \qquad (3.5)$$

Any fault impedance should be included in the feeder impedance Z_f. We see from that the sag becomes deeper for faults electrically closer to the customer= (when Z_F becomes smaller), and for systems with a smaller fault level (when Z_S becomes larger)

Note that a single phase model has been used here, whereas in reality the system is three phase. That means that this equation strictly speaking only holds for three phase faults.

Above equation can be used to calculate the sag magnitude as a function of the distance to the fault. Therefore we have to write Z_f=z x∈, where z the impedance of the feeder per unit length and ∈ the distance

between the fault and the PCC, leading to

$$V_{sag} = \frac{z\varepsilon}{Z_S + z\varepsilon} \qquad (3.6)$$

Influence of cross section: overhead lines of different cross section have different impedance, and lines and cables also have different impedance. It is thus to be expected that the cross section of the line or cable influences the sag magnitude as well.

Faults behind transformers:
The impedance between the fault and the PCC not only consists of lines or cables but also of power transformers. As transformers have rather large impedance, among other to limit the fault level on the low voltage side, the presence of transformer between the fault and the PCC will lead to relatively shallow sags.

Fault levels:
Often the source impedance at a certain bus is not immediately available, but instead the fault levels. One can of course translate the fault level into source impedance and use to calculate the sag magnitude. But one may calculate the sag magnitude directly if the fault levels at the PCC and at the fault position are known. Let S_{flt} be fault level at the fault position and S_{pc} at the point-of-common coupling. For a rated voltage V_n the relations between fault level and source impedance are as follows:

$$S_{FLT} = \frac{V^2_n}{Z_S + Z_F} \qquad (3.7)$$

$$S_{pcc} = \frac{V^2_n}{Z_S} \qquad (3.8)$$

With the voltage at the PCC can be written as

$$V_{sag} = 1 - \frac{S_{FLT}}{S_{PCC}} \qquad (3.9)$$

We use to calculate the magnitude of sags behind transformers.

Critical Distance

Equations give the voltage magnitude as a function of the distance to the fault. From this equation we can obtain the distance at which a fault will lead to sag of a certain magnitude. If we assume equal X/R ratio of source and feeder, we obtain

$$E \text{ critical} = \frac{Z_S}{Z} \times \frac{V}{1-V} \qquad (3.10)$$

We refer to this distance as the critical distance for a voltage V. supposes that a piece of equipment trips when the voltage drops below a certain level the critical voltage). The definition of critical distance is such that each fault within the critical distance will cause the equipment to trip.

If we assume further that the number of fault is proportional to the line length within the critical distance, we would except that the number of sags below a level V is proportional to V/ (1-V) another assumption is needed to be infinitely long without branching off. Of course this is not the case in reality. Still this equation has been compared with a number of large power quality surveys.

4.1. Introduction to DVR

Dynamic voltage restorer is a series connected device for mitigating voltage sag and swell. The first DVR was installed by the then Westinghouse in 1996. Since, then a lot of installations have taken place worldwide along with wide spread research in different aspects of DVR and control philosophies.

To improve power quality a custom power device Dynamic Voltage Restorer (DVR) can be used to eliminate voltage sags and swells. DVR is an inverter based voltage sag compensator. DVR protects the precision manufacturing processes and sophisticate sensitive electronic equipments from the voltage fluctuations and power outages. DVR offers sub cycle protection, restores the quality of electric power delivered to the sensitive load. The DVR regulates voltage within acceptable tolerances and meet the critical sensitive power quality needs. The DVR has been developed by Westinghouse for advance distribution. DVR injects a set of three single-phase voltages of an appropriate magnitude and duration in series with the supply voltage in synchronism via booster transformer to restore the power quality.

The injection voltages are of controllable amplitude and phase angle. The reactive power exchange between the DVR and distribution system without capacitors or inductors. The DVR is a series conditioner based on a pulse width modulated voltage source inverter, which is generating or absorbing real or reactive power independently. The ideal restoration is compensation which enables the voltage seen by the load to be unchanged in the face of upstream disturbances. The DVR injects the independent voltages to restore the line voltage to sensitive loads from sags caused by unsymmetrical line-to-ground, line-to-line, double-line-to-ground and symmetrical three phase faults. The output voltage waveform of DVR is highly regulated and clean. The DVR provides harmonic compensation and mitigates voltage transients.

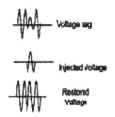

Fig. 2. *Voltage sag, DVR voltage, final load voltage*

4.2. Operating Principle of DVR

The DVR is designed to inject the missing voltage into the distribution line. Its basic idea is to dynamically inject a voltage u_c (t) as shown in Figure. The upper part of Figure

shows a simplified single-phase equivalent circuit of a distribution feeder with a DVR, where the supply voltage u_s (t), the DVR-injection voltage u_c(t) and the load voltage u_L(t) are in series. So, the DVR is considered to be an external voltage source where the amplitude, the frequency and the phase shift of $u, (f)$ can be controlled. The purpose is to maintain the amplitude of the load voltage fixed and prevent phase jumps.

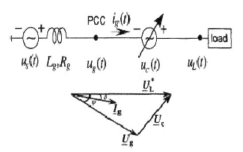

Fig. 3. *DVR operational principal: up) simplified equivalent circuit, down) Phasor diagram.*

A phasor diagram for a voltage dip with a phase-jump case is shown in Figure (down). From Figure, the load voltage is deduced:

$$u_L (t) = u_s(t) + u_c(t)$$

If the supply voltage u_s (t) has dropped due to a voltage dip or increased due to a voltage swells, the DVR compensating voltage u_c (t) should be controlled so that the load voltage remains the same as during no-disturbance conditions. Thus, the instantaneous amplitude of u_c (t) is controlled such as to eliminate any detrimental effects of a system fault to the load voltage as long as the disturbance does not cause the circuit breaker to open.

4.3. Block Diagram of DVR

A schematic diagram of the DVR incorporated into a distribution network is shown in Fig. V_s are the source voltage, V_1 is the incoming supply voltage before compensation, V_2 is the load voltage after compensation, V_{DVR} is the series injected voltage of the DVR, and I is the line current.

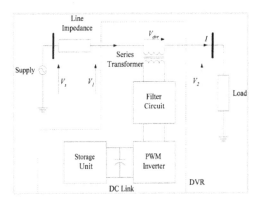

Fig. 4. *Typical schematic of a power distribution system compensated by a DVR.*

The restorer typically consists of an injection transformer, the secondary winding of which is connected in series with the distribution line, a voltage-sourced PWM Inverter Bridge is connected to the primary of the injection transformer (actually three single phase transformers) and an energy storage device is connected at the dc-link of the inverter bridge. The primary winding (connected in series with the line) of the transformer must be designed to carry the full line current. During the transient period at the occurrence of voltage sag, DVR booster transformer can experience a flux linkage that is up to twice its normal steady state value. In order to overcome the transformer from saturating a rating of flux i.e. twice that of steady state limit is used. An alternative method is to limit the flux linkage during the transient switch-on period, thus preventing saturation. The primary voltage rating of the transformer is the maximum voltage the DVR can inject into the line. The DVR rating (per phase), is the maximum injection voltage times the primary current. The bridges are independently controllable to allow each phase to be compensated separately. The inverter bridge output is filtered in order to mitigate the switching frequency harmonics generated in the inverter. The series injected voltage of the DVR V_{DVR} is synthesized by modulating pulse widths of the inverter-bridge switches. While online, the DVR can get heated-up due to switching and conduction losses in semiconductor switches. Therefore, it is necessary to provide proper means of heat sinking in order to operate the DVR safely and to increase the life-span of semiconductor switches.

The injection of an appropriate V_{DVR} in the face of an up-stream voltage disturbance requires a certain amount of real and reactive power supply from the DVR. The reactive power requirement is generated by the inverter. To inject the active power, energy storage is needed to increase the ride through capability of DVR. Ride through capability is the time that DVR can restore or sustain the output voltage to 100% of DVR rating. The available energy storage devices are batteries, capacitor banks, super capacitors, superconducting magnetic energy storage (SMES), flywheels, and he1 cells. The required active power can also be obtained from the grid through an AC/DC rectifier. In this case, a capacitor should be used to link the AC/DC rectifier with the converter. The capacitor size depends on the required active power to be injected and the voltage drop at its terminal during the discharge period.

The capacitor size is characterized as a time constant T, defined as the ratio between the stored energy at rated dc voltage and the rated apparent power of the converter as:

$$T = \frac{1}{2} \frac{CU^2_{dcN}}{S_N}$$

Where U_{dcN} is the nominal dc-converter voltage and S_N is its rated apparent power. The capacitor cost is approximately proportional to the square of its terminal

voltage.

Another source of stored energy is batteries. They are much cheaper than capacitors per KJ. As the internal impedance of batteries is rather high, the amount of power that can be extracted from them is small, so that the cost per kW is still rather high. Batteries are more suitable for minute-scale events like interruptions. The disadvantage of using batteries is the required maintenance and their adverse environmental effects.

The flywheel has been used for a long time, as cheap short-term energy storage, particularly in machine applications. As the energy stored by the flywheel is directly proportional to the square of the operational speed, the higher speed and composite flywheel can provide a ride-through capability for voltage dips and momentary interruptions.

In SMES, the electricity is stored by circulating a current in a superconducting coil. Because no conversion of energy to other forms is involved (like the flywheel), the round-trip efficiency can be rather high. Choosing the suitable energy storage for DVR applications depends on the expected total cost and the designed rated power.

Startup:

At the moment, capacitors are the most commonly used energy storage for the DVR because they provide a fast discharging response and have no moving parts. To startup the DVR with a capacitor bank as energy storage, the capacitor bank should be charged to the rated dc voltage. Charging of the capacitor bank can be realized by exploiting a separate energy supply such as an electric generator operated by a diesel engine. This solution is considered a good solution for the DVR in the respect of being always ready and not affected by grid disturbances, but it would be very expensive and slow in response. Two other techniques can be implemented to charge the dc capacitor bank. One of them is based on using a shunt diode rectifier, which is placed either on the load side or the grid side. The other exploits the VSC of the DVR to charge its capacitor bank with a proper control algorithm. After charging the dc capacitor bank, the DVR is transferred to the idle state where the dc voltage is blocked by the valves of the VSC and the bypass switch. When a voltage dip is detected in the grid, the bypass switch is turned off and the DVR transfers to the compensating state.

4.4. Topologies of DVR

Different types of topologies for DVR are discussed in the literature, which are classified according to

4.4.1. Location of a DVR

a) DVR-medium voltage three wire system

b) DVR-low voltage four wire system

In both systems the main purpose is to inject a synchronous voltage during faults. The difference between a Low voltage connection and a Medium voltage connection is the flow of zero sequence currents and the generation of zero sequence voltages.

DVR-medium voltage three wire system   DVR-Low voltage four wire system

Fig. 5. *DVR with & low voltage*

4.4.2. Converter Type

Voltage sourced converters compensate for the missing voltage requires a boost inductance (L) and a line filter to damp the generated switching harmonics.   Choosing a more complicated converter topology can help achieve low DVR impedance, low current ripple and low switching frequency, but the penalty is higher complexity.

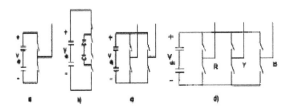

Fig. 6. *Different types of series converters for a DVR*

a)   Two level (half bridge)
b)   Three level (half bridge)
c)   Three level (full bridge)
d)   Series converter DVR with single 3-phase inverter block

4.4.3. Energy Source

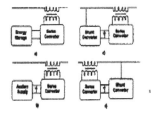

Different types of connection to source power to the DVR

Fig. 7 *different types of connection*

a)   Energy storage
b)   Auxiliary supply
c)   Front connected shunt converter
d)   Back connected shunt converter

DVR needs the active power to restore the voltage waveform, which can be supplied from a Energy storage or auxiliary supply or from line connected shunt converter. The different topologies are illustrated in above figure.

4.4.4. Transformerless DVR

For the transformer less DVR scheme, the largest injection voltage is determined solely by DC- link voltage. Two topologies for transformer less DVR are shown below.

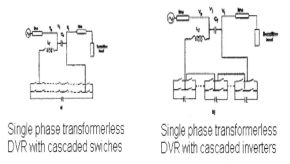

Single phase transformerless   Single phase transformerless
DVR with cascaded swiches   DVR with cascaded inverters

Fig. 8. *single phase Transformer less DVR*

The injection transformer used in the conventional DVR is expensive, occupies more space and contribute towards losses. From the design, operation maintenance point of view, the transformer is an added complexity to the restorer. The transformer less DVR can satisfactorily mitigate the voltage sag problems. The design is promising, less costly restorer and compact in size to conventional DVR. Transformer less DVR posses' superior voltage regulation property with less loss.

4.4.5. Filter

The filtering scheme in the Dynamic Voltage Restorer can be placed either line side or inverter side of the booster transformer. Due to the inverter switching operations harmonics are produced which should be filtered before the load. Using a properly designed filter, harmonic voltages can be attenuated.

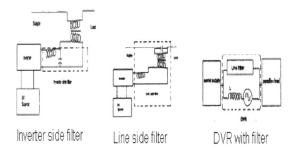

Inverter side filter   Line side filter   DVR with filter

Fig. 9 *Different filters*

4.4.6. Dynamic SAG Correcter (Dy.SC)

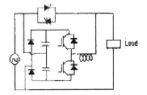

Dynamic Sag Corrector

Fig. 10. *Dynamic sag corrector*

The series parallel connected Dynamic Sag Corrector provides statistically significant protection at greatly reduced cost. Above figure shows the schematics of a Dy.SC which is performing the DVR operation i.e.

eliminating voltage excursions. Dy.SC units have been shown to be smaller in size and lower cost, while providing a high level protection. Correction for sag down to 50% as well as protection from momentary loss of power has been demonstrated. Operating efficiency greater the 99%, response time is less than 1/8 cycles, long operation life of at least ten years. The Dy.SC is rated from 1.5kVA one phase to 2000kVA three phase and features a single stage power conversion circuit with minimal stored energy.

4.5. Features of DVR

- Extremely fast response less than 1/4th cycle
- Expandable modular design
- Low maintenance
- High reliability
- Protection is available up to 120MVA

4.6. Compensation Techniques of DVR

After the detection of the voltage dip, the DVR starts *to* inject the correct three phase voltage to keep the load voltage unchanged. The response time of the control unit dedicates the transient time of the DVR. During the dip, the DVR works at a steady state, injecting the missing voltage in the grid until the grid recovers to its initial state before the dip. The operation of the DVR in steady state can be obtained via different compensation strategies. Such compensation strategies are illustrated in this section.

4.6.1. Voltage Difference Compensation

This is also called as pre-sag compensation. With the voltage difference compensation strategy, the injected voltage U_i is calculated by subtracting the grid voltage U_g from the reference of the load voltage U^*_L

$$\underline{U}_i = \underline{U}^*_L - \underline{U}_g$$

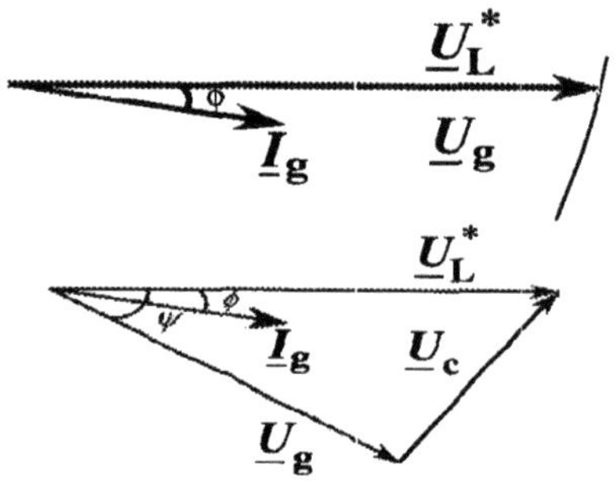

Fig. 11. Vector diagram of voltage difference compensation: up (before dip, bottom) after dip.

The vector diagram of voltage-difference compensation is shown in Figure Provided that the energy storage and the rated voltage of the DVR are large enough, the grid voltage can be placed anywhere inside the circle whose radius is the magnitude of the reference load voltage. Both the load voltage magnitude and phase angle are exactly the same as before the dip occurs. The disadvantage of this strategy is

that it necessitates a big size of the energy storage, particularly if the DVR is sized to compensate for deeper dips.

4.6.2. In-Phase Compensation

With in-phase compensation strategy, the injected voltage is placed in phase with the grid voltage during the dip. The magnitude of the injected voltage is calculated by subtracting the magnitude of the grid voltage from the reference load voltage;

$$U_i = U^*_L - U_g$$

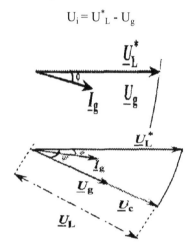

Fig. 12. Vector diagram of in-phase compensation: up) before dip, bottom) after dip.

If the grid voltage during the dip has a phase-angle jump, the load voltage will have a phase-angle jump as well. In other words, by this strategy only the magnitude of the load voltage is restored as depicted in Figure. The restored load voltage *CL* has the same phase jump as the grid voltage. In-phase voltage compensation results in maximum voltage boost. The magnitude of the injected voltage is minimized since there is no phase difference: between the input and the output voltage. According to the fact that in-phase compensation introduces a phase-angle jump to the load voltage, this phase angle jump may cause waveform discontinuity and inaccurate zero-crossings, which may cause, for instance, nuisance tripping of adjustable speed derives. In-phase compensation minimizes the magnitude of the injected voltage, or equivalently the rated voltage of the DVR, but its puts more demand on the size of the energy storage compared to voltage difference compensation. This is true in case of voltage dips with phase angle jump. If the voltage dip is not associated with a phase angle jump, the two methods (voltage difference and In-phase compensation) are equivalent.

4.6.3. ADVANCED Phase Compensation

It is also called as Energy optimal compensation. The injected active power can be: minimized by injecting a voltage into the grid with a phase advance with respect to the grid voltage during the dip. Zero injected active power can be realized when the injected voltage is orthogonal to the load current. This case is not always possible or it will result in partial compensation of the load voltage.

The vector diagram of the advanced phase compensation is depicted in Figure. Where the injected voltage $\underline{U}_C$ has a phase advance angle of β with respect to the grid voltage. The reduction in active power is subsidized by an increase in the reactive power, which is generated internally by the VSC of the DVR. This strategy demands lower energy storage compared with the voltage difference and in-phase compensation strategies. On the other hand, it introduces a phase-angle jump that will cause waveform discontinuity and inaccurate zero-crossings.

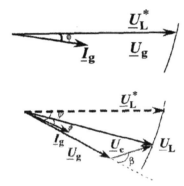

Fig. 13. *Vector diagram of advanced phase compensation up a) before dip, bottom b) after dip.*

It is worth noting that the magnitude of the injected voltage with the advanced phase compensation is larger than the one in case of in-phase compensation. As the phase advancement angle needs to be determined in real time, the control system is complex. There could be significant increase in the reactive power supplied by the DVR depending in the advance angle and nature of sag.

4.6.4. Progressive Advanced Phase Compensation

To overcome the problems associated with the phase angle jump introduced by the advanced phase compensation, the progressive advanced phase compensation has been proposed, but it has not been implemented yet. The strategy is based on making the phase advance of the injected voltage progressively in such a way that the load will not sense a rapid phase-angle jump. This is done by dividing the phase advance angle β(in above Figure) into small steps. The phase advance caused by this strategy should be removed after the distribution system recovers to its initial state before the dip. This is done through small backward steps. The response time of the DVR is longer, compared with other strategies, because of the delay introduced by the progressive phase advance.

4.7. Design Criteria and Rated Power Calculations

4.7.1. Design Criteria

The design of the DVR is affected by the load, the supply characteristics and by the expected voltage-clip characteristics. When designing for a DVR for certain application, the following items should be considered:
- Maximum load power and power factor: The load

size strongly affects the current rating of the voltage-source converter and the injection transformer as well as the amount of energy storage needed.
- Maximum depth and duration of voltage dips to be corrected: These characteristics, together with the load size, dictate the necessary storage capacity of the energy storage device. The maximum depth determines the voltage rating of the voltage-source converter and the injection transformer. The maximum depth and duration of voltage dips to be corrected is determined by the statistics of the voltage dips at the DVR location and by the acceptable number of equipment trips.
- Maximum allowed voltage drop of the DVR during the standby mode: This affects the control mode during normal operation and indirectly the reaction speed at the beginning of a voltage dip.
- Coupling of the step-down transformer: (Y/Δ or Y/Y…..etc.) at input and output sides of the DVR.
- Parameters of the step-down transformers: Coupling of the step-down transformer *CIA* or *YIY,. . .*etc.) at input and output sides of the DVR.
- Harmonic requirements of the load and of the system: These affect the harmonic filtering needed for the DVR and also influence the choice of charging method for the capacitors.
- At the first instance when designing a DVR, some assumption could be made to simplify the analysis, such as:
  - ➤ Ideal switches
  - ➤ DC-side capacitors are large enough to maintain a ripple free DC bus voltage, even for unbalanced input voltage.
  - ➤ Series transformer and output filter components are ideal.

In order to design a DVR, the concept of "boost rating" is introduced to define the maximum voltage that the DVR is capable to inject into the power line with respect to the nominal distribution system voltage. This boost ratio is defined as

$$B = \frac{U_C}{U_{sl}/\sqrt{3}}$$

Here, U_{sl}, is the nominal line-to-line voltage of the supply.

4.7.2. Rated Power Calculations

The DVR function, in case of voltage dips, is to exchange real power between the power system and the energy storage. Reactive power exchanged between the DVR and the distribution system is internally generated by the DVR without any ac passive reactive components. The real power injected by the DVR is an important feature to precede its design process. To calculate the active and the reactive power, a factor & is defined to indicate the reduction of the positive sequence voltage with respect to the nominal voltage of the load. For a certain supply voltage U, and required load voltage UL, the DVR injected voltage is written as:

$$\underline{U}_c = \underline{U}_L - \underline{U}_s = (1 - \underline{M}_F)\,\underline{U}_L$$

The range of the modulus of $\underline{M}_F$ is defined by the maximum variation of $\underline{U}_s$, for which the DVR is designed. So, in normal operation $\underline{M}_F$ will be unity and $\underline{U}_c$ is zero.

$$\underline{M}_F = \underline{U}_s / \underline{U}_L$$

Considering the fact that the DVR current should be designed to be the same as the rated load current, the apparent power required by the DVR is then calculated in terms of the apparent load power, $\underline{S}_L$ and $\underline{M}_F$ by the following formula.

$$\underline{S}_c = \underline{S}_L\,(1 - \underline{M}_F)$$

Consequently, the active and reactive powers are calculated by separating $\underline{S}_c$ into its real and imaginary parts as

$$\mathbf{P_c} = S_L\left\{\cos(\phi_L) - |\underline{M}_F|\cos(\phi_S)\right\}$$

$$\mathbf{Q_c} = S_L\left\{\sin(\phi_L) - |\underline{M}_F|\sin(\phi_S)\right\}$$

Where $\cos(\Phi_S)$ is the source power factor and $\cos(\Phi_L)$ is the load power factor.

$$\mathbf{P_c} = P_L\left\{1 - \frac{U_S\cos(\phi_L + \psi)}{\cos(\phi_L)}\right\}$$

In the above equation, the load voltage is assumed to be constant and equal to 1pu. So, the required active power of a DVR depends on the magnitude and the phase-angle jump of the supply voltage as well as the load power factor.

4.8. Control Strategies

Different control techniques to control the DVR have been reported in literature. Both the feed forward and feedback techniques have been implemented in the control unit of the DVR applications. Feed forward control technique does not sense the load voltage and it calculates the injected voltage based on the difference between the pre-dip and during-dip voltages. The load voltage is sensed in case of using feedback technique and the difference between the voltage reference of the load and actual load voltage is injected. Feedback structures however have inherent delay in responding to power system disturbances. To optimize the dynamic performance, a direct feed forward control should be used. The feed forward technique has been applied by measuring the supply voltages and computing the appropriate compensation voltages. The required voltage reference is obtained by synchronization with the phase voltage, without performing symmetrical component decomposition. The proposed

control algorithm is implemented in a DSP as illustrated in below figure. The algorithm consists of the following steps: 1) calculation of the phase angle of the synchronization signal; 2) generating the reference voltages, and 3) computing the modulating signal. The proposed technique involves the calculation of the RMS of the fundamental voltage, which requires at least one half periods at the grid frequency. The switching frequency used in this *DVR* application is 20 kHz, which is considered to be a very high frequency. Higher switching frequencies lead to smaller current ripple which allows for a reduction in filter size.

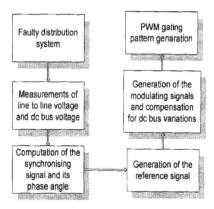

Fig. 14. *Feed forward control algorithm*

Vector control has been proposed to control the injected voltage by the DVR in order to improve the dynamic performance and rapidly restore the load voltage to the desired value. The principle of vector control is explained with the aid of Figure below. The grid voltage u₁(t), u₂(t), u₃(t) is sampled and converted to the dq-frame (synchronous reference frame). It is desired that the d, and the *q* components of the load voltage are set as 0 and 1 pu respectively. The injected voltage is calculated by subtracting the grid voltage from the desired load voltage in the dq-frame and it is converted back to the three-phase frame. The voltages Δu₁(t), Δu₂(t), Δu₃(t) are the voltage that the VSC should generate to restore the load voltage to the desired value.

This technique has a shorter response time compared with RMS calculation. However, only balanced voltage dips have been considered. To compensate for unbalanced dips, a separation technique for the positive and the negative sequence should be included in the control algorithm, or alternatively, a high switching frequency has to be used.

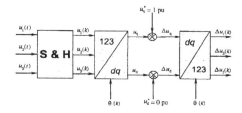

Fig. 15. *Principle of vector control.*

The DVC (Double Vector control) algorithm is implemented in the dq-frame and incorporates both current and voltage controllers with an inner current control loop and outer voltage control loop. However, only balanced voltage dips can be compensated. In case of unbalanced dips, the d- and q-components are not dc quantities and thus a conventional PI controller fails to track the reference signal properly. Because most faults are single- or double-phase, this algorithm uses a fast technique for separating positive and negative sequence components of the supply voltage, which are then controlled separately. Thus, two controllers have been implemented for the two sequences, each based on double vector control, i.e. constituted by a voltage-control loop and a current-control loop.

5. Simulation Results and Performance Evaluation

Case1: DVR performance results for symmetrical voltage dip from 320v to 260v at o.05sec to o.15sec

Here the dip in the source voltage is created during the interval 0.05 to 0.15sec. Now by the action of DVR it is injecting a voltage in series with the supply voltage in order to bring the voltage across the load to rated symmetrical voltage.

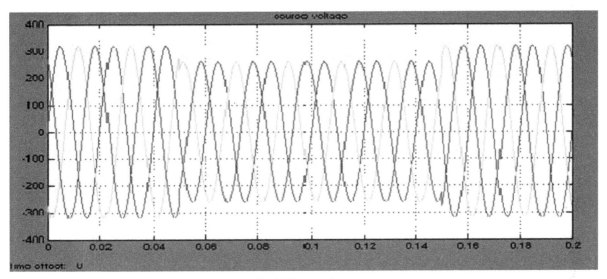

Fig. 16. *Source voltage*

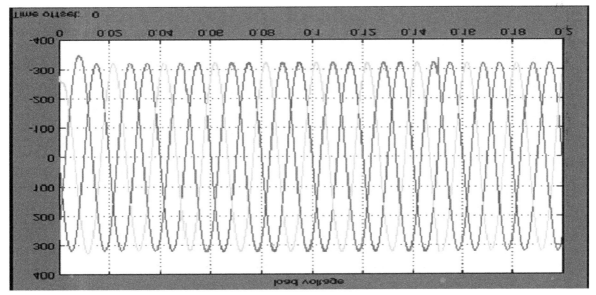

Fig. 17. *Load voltage*

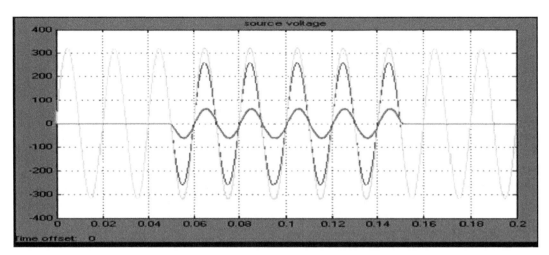

Fig. 18. *Phase 'a' source voltage, load voltage, injected voltage*

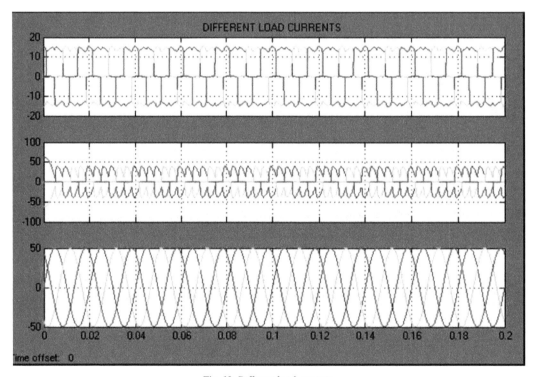

Fig. 19. *Different load currents*

6. Conclusions

This thesis has presented results from a DVR simulated at a low voltage level at a distribution end. The test voltage sags and swells were created using general sine wave generator blocks. The simulation results showed that the DVR could protect a sensitive load on a distribution network for three phase or two phase or single phase sags or swells. This model will yield a performance which simply does not depend on the load current, since this model has no impedance anywhere. But in practice the output voltage will get affected by load harmonics due to two reasons – the inverter output filter will call for harmonic drops when harmonic load currents flow through it and in the absence of feedback control, the system does

not correct anything to the right of the inverter. Secondly the finite bandwidth of inverter (due to a finite switching frequency) will make it fail in generating high frequency current produced at the source bus by high frequency component of load currents flowing in source impedance (which is taken as zero in this *simulink* model).

The control of DVR is not a complex problem; field experience justifies feed forward control. However providing a suitable DC side energy source to handle long periods of sag or swell throughout the day will be a problem. If it is a battery it requires a charger. Some researchers have proposed drawing charging power from the line using the same inverter during periods which sag or swell is little and can be handled by 90^0 voltage injection. But that makes the control pretty complex. If it is a AC-DC diode rectifier the DVR can handle only sags and not swells since during swells the inverter will absorb power (in the in phase injection strategy) and dump it on the DC side. So then it has to be a bilateral converter based AC-DC converter and then we get very close to what they call 'Unified Power Quality Conditioner'- then it is no more a DVR alone, but can easily become a UPQC.

Acknowledgements

I thank to our Institute Executive directors Mr. T Sai kumar & Mr. D Baba for providing creative environment for this work. Also I am very much thankful to our Institute principal Dr. K Ramesh for his kind permission and encouragement to write research paper. I would like to extend my heartfelt thanks to my colleagues. And finally I am very much obliged to my respected parents who inspiring me around the clock.

References

[1] "UNDERSTANDING POWER QUALITY PROBLEMS," voltage sags and interruptions, Math H.J.Bollen-IEEE PRESS.

[2] N.H.Woodly, L.Morgan, and A.sundram,"Experience with an inverter-based dynamic voltage restorer,"IEEETrans.Power delivery, vol.14, pp.1181-1186, July 1999.

[3] "Mini-DVR Dynamic Voltage Restorer With Active Harmonic Filter (Tests of Prototype)",2004 11th International Conference on Harmonics and Quality of Power

[4] "Study on the Dynamic Voltage Restorer Use in an Isolated Power Supply system" paper

[5] "Voltage Sag Compensation With Energy Optimized Dynamic Voltage Restorer"IEEE Transactions on power delivery,vol.18,No.3,July2003

[6] "Review of Dynamic Voltage Restorer for Power Quality Improvement." 30th Annual conference of the IEEE industrial electronics society,nov 2006

[7] "Control strategies for Dynamic Voltage Restorer compensating for voltage sags with phase jump" paper

[8] "Static Series Compensator for voltage dips mitigation" IEEE Bologna powertech conference, June 23-26,Bologna, Italy

[9] C. Meyer, C. Romaus and R. De Doncker, "Optimized Control Strategy for a MediumVoltage DVR", IEEE Trans. Power Electronics, Vol. 23(6), Nov. 2008, pp. 2746-2754.

[10] Q. Wang and S. Choi, "An Energy-Saving Series Compensation Strategy Subject to Injected Voltage and Input-Power Limits", IEEE Trans. Power Delivery, Vol. 23(2), Apr. 2008, pp. 1121-1131.

[11] C. Fitzer, M. Barnes and P. Green, "Voltage Sag Detection Technique for a Dynamic Voltage Restorer", IEEE Trans. Industry Applications, Vol. 40(1), Jan. 2004, pp. 203-212.

[12] B. Bae, J. Jeong, J. Lee and B. Han, "Novel Sag Detection Method for Line-Interactive Dynamic Voltage Restorer", IEEE Trans. Power Delivery, Vol. 25(2), Apr. 2010, pp. 1210-1211.

[13] J. Nielsen, M. Newman, H. Nielsen and F. Blaabjerg, "Control and Testing of a Dynamic Voltage Restorer (DVR) at Medium Voltage Level", IEEE Trans. Power Electronics, Vol. 19(3), May 2004, pp. 806-813.

[14] H. Kim and S. Sul, "Compensation Voltage Control in Dynamic Voltage Restorers by Use of Feed Forward and State Feedback Scheme", IEEE Trans. Power Electronics, Vol. 20(5), Sep. 2005, pp. 1169-1177.

[15] M. Marei, E. El-Saadany and M. Salama, "A New Approach to Control DVR Based on Symmetrical Components Estimation", IEEE Trans. Power Delivery, Vol. 22(4), Oct. 2007, pp. 2017-2024.

[16] M. Bongiorno, J. Svensson and A. Sannino, "An Advanced Cascade Controller for SeriesConnected VSC for Voltage Dip Mitigation", IEEE Trans. Industry Applications, Vol. 44(1), Jan. 2008, pp. 187-195.

[17] Abdelkhalek, A. Kechich, T. Benslimane, C. Benachaiba and M. Haidas, "More Stability and Robustness with the Multi-loop Control Solution for Dynamic Voltage Restorer", Serbian Journal of Electrical Engineering, Vol. 6(1), May 2009, pp. 75-88.

[18] H. P. Tiwari, S. K. Gupta, "Dynamic Voltage Restorer Based on Load Condition", International Journal of Innovation, Management and Technology, Vol. 1(1), April 2010, pp. 75-81.

[19] S. Ganesh, K. Reddy and B. Ram, "A Neuro Control Strategy for Cascaded Multilevel Inverter Based Dynamic Voltage Restorer", International Journal of Electrical and Power Engineering, Vol. 3(4), 2009, pp. 208-214.

[20] P. Margo, M. Heri, M. Ashari, M. Hendrik and T. Hiyama, "Compensation of Balanced and Unbalanced Voltage Sags using Dynamic Voltage Restorer Based on Fuzzy Polar Controller", International Journal of Applied Engineering Research, Vol. 3(3), 2008, pp. 879–890.

[21] H. Ezoji, A. Sheikholeslami, M. Rezanezhad and H. Livani, "A new control method for Dynamic Voltage Restorer with asymmetrical inverter legs based on fuzzy logic controller", Simulation Modelling Practice and Theory, Vol. 18, 2010, pp. 806–819.

[22] C. Zhan, V. Ramachandaramurthy, A. Arulampalam, C. Fitzer, S. Kromlidis, M. Barnes and N. Jenkins, "Dynamic Voltage Restorer Based on Voltage-Space-Vector PWM Control", IEEE Trans. Industry Applications, Vol. 37(6), Nov. 2001, pp. 1855-1863.

[23] H. Ding, S. Shuangyan, D. Xianzhong and J. Jun, "A Novel Dynamic Voltage Restorer and its Unbalanced Control Strategy Based on Space Vector PWM", Electric Power and Energy Systems, Vol. 24, 2002, pp. 693-699

14

Performance evaluation of cooperative ML-IDMA communications with DF protocol

Basma A. Mahmoud, Esam A. A. Hagras, Mohamed A. Abo El-Dhab

Department of Electronics & Communications, Arab Academy for Science, Technology & Maritime Transport, Cairo, Egypt

Email address:

basmaamahmoud@yahoo.com(B. A. Mahmoud), esamhagras_2006@yahoo.com(E. A. A. Hagras), mdahab@aast.edu(M. A. A. El-Dhab)

Abstract: Relaying and cooperative diversity allow multiple wireless radios to effectively share their antennas and create a virtual antenna array, thereby leveraging the spatial diversity benefits of multiple-input multiple-output antenna systems. In this paper, we consider the bit error rate performance analysis of a cooperative relay communication system on Multi-Layer-IDMA (ML-IDMA) using Maximal-Ratio-Combining (MRC) technique; we examine the effect of layers number on the performance and derive the average bit error probability of the Decode-and-Forward (DF) relay schemes by using the closed-form relay link Signal-to-Noise Ratio (SNR). Based on the analysis, we show that from the simulation results, degradation in the performance is observed when increasing number of layers but on the other hand we saved in the Band Width (BW).

Keywords: Cooperative Transmission, Multi Layers, Mrc, Df

1. Introduction

Multiple-Inputs Multiple-Output (MIMO) technology is widely used due to its ability to offer high diversity and multiplexing gain. The Impracticality of mounting multiple antennas on a mobile device favor other techniques to be used in wireless communication and these techniques are also designed to have less impact on the size and power consumption of the devices. Recently, it has been shown that, in a cooperative system, two or more users cooperate with each other to transmit information to the destination. The cooperative users can share each other's antennas to form a virtual multiple antenna system so that a single antenna device can also benefit from the spatial diversity provided by the cooperative users. In such a way, cooperative communication allows a source node with a single antenna to share the antennas of other nodes, resulting in a form of virtual MIMO system[1].

User cooperation systems that utilize different cooperative signaling methods are known to improve cellular system capacity and coverage. The relay node physical limitation and allowed signaling complexity are the two criteria that limit the used cooperation system. In[2, 3] several cooperative diversity protocols were developed and analyzed; Amplify-and-Forward (AF), Decode-and-Forward (DF), Detect-and-Forward (DtF), Estimate-and-Forward (EF) and Selective-DF (S-DF).

In AF protocol, relay nodes amplify the signals received from a source node and transmit the amplified version of the signals to a destination node. In EF, the relays forward an estimate of its received signals to the destination .For DtF, the relays detect the received signals and forward the detected symbols to the destination. For DF, the relay nodes decode the information received from the source and re-encode the signal before transmitting it to the destination. For S-DF, only those nodes that can correctly decode are selected to forward the signals to the destination[4].

In such a way, cooperative communication allows a source node with a single antenna to share the antennas of other nodes, resulting in a form of virtual multiple-input multiple-output (MIMO) system.

As a kind of non-orthogonal multiple access scheme, Interleave Division Multiple Access (IDMA) has been widely researched[5], which is regarded as a special form of Code Division Multiple Access (CDMA) by treating interleaving index sequences as multiple access codes. IDMA performance is better than the conventional CDMA regarding the power and bandwidth efficiency. IDMA has common advantages with CDMA, diversity against fading and mitigation of the user interference problem, are two important ones. A low-cost turbo-type Multi-User Detection (MUD) algorithm applicable to the system with large numbers of users, which is crucial for high-rate multiple access communication, is an important unique advantage of the IDMA that can add to its bandwidth efficiency as well as high transmission speed of data[6].

The principle of the IDMA systems is that the chip-level interleavers should be different for different users. In addition, a low-cost chip-by-chip iterative detection scheme can be utilized in the IDMA systems. Motivated by the concept of IDMA, Superposition Coded Modulation (SCM) partitions the data to multi layer, where each layer is treated by a user equivalently[7]. Multi-layer IDMA (ML-IDMA) is a special form of superposition coding scheme and it can be considered as a joint modulation/channel coding transmission scheme. Based on these backgrounds, according to the above observations, we propose a cooperative transmission scheme based on ML-IDMA.

In[6, 9-11], authors' have paid attention to the one layer IDMA cooperative system and without using any combiner technique at the destination. In this paper, we carry out the performance analysis of cooperative single user ML-IDMA scheme for equidistant relaying geometry with different number of layers and a Maximal-Ratio-Combining (MRC) technique; relay protocol that is used in this paper is Decode-and-Forward (DF). The rest of the paper is organized as follows: in Section 2, we discuss the ML-IDMA system with DF Protocol; the experimental result is presented in Section 3; while conclusions and future works are presented in Section 4.

2. ML-IDMA System with DF Protocol

Recently a wireless transmission is very seldom analogue and the relay has enough computing power, so DF is most often the preferred method to process the data in the relay. The received signal is first decoded and then re-encoded. So there is no amplified noise in the sent signal, as is the case using AF protocol.

In the case of IDMA with DF, the original data is extracted from the received signal by using iterative Multi-User Detection (MUD) at each relay. Decoded data is further encoded by the relay in the similar way as encoded by the source. Encoded signal is finally sent to the destination. In literature, DF is regenerative process, so it can eliminate the Source-Relay (S-R) noise from the inter-user signal. At low SNR if the decoding at the relay is unsuccessful then the performance of DF will be degraded[12].

We consider the cooperative system in Figs. (1-4). the system consists of a source (S), (M) relays, and a destination (D). The signal is transmitted through two phases. For the first phase the source broadcasts the signal to the relays, and the relays transmit their signals for the second phase to the destination.

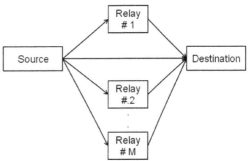

Figure 1. *System model for cooperative ML-IDMA relay system with multiple relays.*

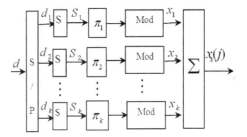

Figure 2. *ML-IDMA transmitter (Source)*

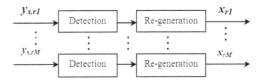

Figure 3. *Relays with Decode-and-forward protocol*

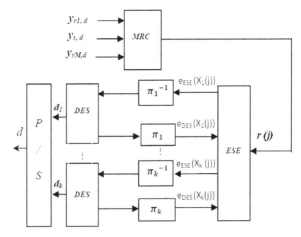

Figure 4. *ML-IDMA Receiver (Destination)*

In our design, the source/user generates an input data sequence $d = [d(1), d(2) ..., d(N)]$ which is converted by serial-to-parallel converter into K subsequences layers. Then the data, in each layer, is spread, interleaved and modulated, independently. Finally, all data in K layers are linearly superimposed to transmission then the source broadcasts the superimposed signal to the relays.

For K-layer, the data sequence $d_k = [d_k(1), d_k(2)..., d_k(I)]$ is first spread, generating a spread sequence $s_k = [s_k(1), s_k(2)...., s_k(J)]$ then the spread sequence s_k is interleaved by a distinct chip-level interleaver π_k to produce a permutated sequence s_k. After interleaving, the randomly sequence s_k is modulated to $x_k = [x_k(1), x_k(2) \ldots , x_k(J)]$ by BPSK. Where $I = N / K$, N is the user data length and I is the layer data length.

The transmitted signal x_s, after sum mission from K Layers, is then given by:

$$x_s(j) = \sum_{k=1}^{K} x_k(j) \quad j=0, 1\ldots, J \quad (1)$$

By assuming that each terminal has one antenna; each relay decode/detect the received signal and retransmit it after regeneration of the decoded/detected signal to the destination. Assume that the channel between the source and each relay is a quasi-static Rayleigh fading channel with Additive White Gaussian Noise (AWGN).

The received signal at the relay is given by:

$$y_{s,r} = \sqrt{P_t}\, h_{s,r}\, x_s + n_{s,r} \quad (2)$$

where, $h_{s,r}$ is the channel coefficient between the source and relay, and $n_{s,r}$ is a sample of an AWGN process with zero mean and variance σ2 per dimension between source and relay.

By normalizing the power to P_t=1, $y_{s,r}$ is given by:

$$y_{s,r} = h_{s,r}\, x_s + n_{s,r} \quad (3)$$

We will represent the signal at the relay after detection and regeneration as x_r. Where the detection algorithm that is done at the relay using DF protocol was the same as done at the destination (will be discussed later).

The received signal at the destination at the first phase is given by:

$$y_{s,d} = h_{s,d}\, x_s + n_{s,d} \quad (4)$$

where, $h_{s,d}$ is the channel coefficient between source and destination, and $n_{s,d}$ is a sample of an AWGN process with zero mean and variance σ2 per dimension between source and destination.

The relayed signal when received at destination is given by:

$$y_{r,d} = h_{r,d} x_r + n_{r,d} \quad (5)$$

where, $h_{r,d}$ is the channel coefficient between relay and destination, and $n_{r,d}$ is a sample of an AWGN process with zero mean and variance σ2 per dimension between relay and destination.

After receiving the relayed signal ($y_{r,d}$) and direct signal ($y_{s,d}$) from relay and source respectively, the total received signal ($r(j)$) at the destination for Maximum-Ratio-Combining (MRC) receiver is given by:

$$r = h_{s,d}^* y_{s,d} + \sum_{m=1}^{M}(h_{s,r_m}^* * h_{r_m,d}^*)\, y_{r_m,d} \quad (6)$$

where, M is the total number of relays.

Using a single-path / one relay system (M=1), we can rewrite Eq. (6) as:

$$r = (h_{s,r}^* * h_{r,d}^*)\, y_{r,d}\, 8 \quad (7a)$$

$$r = (h_{s,r}^* * h_{r,d}^*) h_{r,d} x_r + (h_{s,r}^* * h_{r,d}^*) n_{r,d} \quad (7b)$$

$$r(j) = h \sum_{k=1}^{K} x_k(j) + n(j) \quad (7c)$$

where, $h = (h_{s,r}^* * h_{r,d}^*) h_{r,d}$, $n = (h_{s,r}^* * h_{r,d}^*) n_{r,d}$ represents the composite noise.

Since each layer of ML-IDMA systems can be treated as one user of multi-user systems, the suboptimal iterative detection[13] and[14] can be applied in ML-IDMA systems. The receiver of ML-IDMA is illustrated in Fig. 4, which is composed by an Elementary Signal Estimator (ESE) and K de-spreaders (DESs). They are applied to solve inter-layer interference and the spreading constraint separately[15, 16]. The receiver performs the iterative processes to update the extrinsic information between ESE and DESs. After the last iteration, the DESs produce de-spreading signal.

For the detection of layer-k, we can rewrite Eq. (7c) as:

$$r(j) = h\, x_k(j) + \xi_k(j), \quad j=1, 2\ldots, J \quad (8)$$

where, $\xi_k(j)$, represents the interlayer interference with respect to layer-k.

Similar to[17], for simplicity, our derivation will be based on a single-path channel model with real channel coefficients and BPSK signaling. However, the principle here is not only limited to this case, and can be easy to apply into other cases such as multipath channels, higher order modulations and complex channels[13]-[14].

The ESE function is used to calculate the extrinsic log-likelihood Ratios (LLR) for estimating the transmitted signal. From the definition of the extrinsic LLR, the output of ESE function can be obtained by:

$$e_{ESE}(x_k(j)) = 2h \cdot \frac{r(j) - E(\xi_k(j))}{Var(\xi_k(j))} \quad (9)$$

where,

$$E(r(j)) = h \sum_k E(x_k(j)) \quad (10)$$

$$Var(r(j)) = |h|^2 \sum_k Var(x_k(j)) + \sigma^2 \quad (11)$$

$$E\left(\xi_k(j)\right) = E(r(j)) - h\, E(x_k(j)) \quad (12)$$

$$Var\left(\xi_k(j)\right) = Var(r(j)) - |h|^2 Var(x_k(j)) \quad (13)$$

The mean and variance of $x_k(j)$ can be calculated by the feedback from DESs, as follow:

$$E(x_k(j)) = \tanh\left[\frac{e_{DES}(x_k(j))}{2}\right] \quad (14)$$

$$Var(x_k(j)) = 1 - (E(x_k(j)))^2 \quad (15)$$

The mean and variance of the interlayer interference can be used to analyze and detect the signal of each layer. Then the updated extrinsic LLR from ESE function was proven to go through the layer-specific de-interleaver and gets into the DESs iteratively[18].

3. Simulation Results

In this paper, we carry out the performance analysis of cooperative single user ML-IDMA scheme for equidistant relaying geometry with different number of layers and a MRC technique. The applied relay protocol is Decode-and-Forward (DF), which is implemented for the system that uses Binary Phase Shift Keying (BPSK). MATLAB is used to simulate the obtained results. In this paper it is assumed that all stations are arranged at the edges of a square with a length of one. That means that all channels will have the same path loss and therefore the same average Signal-to-Noise Ratio (SNR).

To simulate the cooperative ML-IDMA scheme, we assume that the channel is Rayleigh fading channel, equidistant relaying geometry, BPSK signaling is used, frame length (N) = 512, Spreading Length (SP) =32, number of relays (R) =1..., 4, number of layers (K) =2, 4 and number of iteration (it) =3.

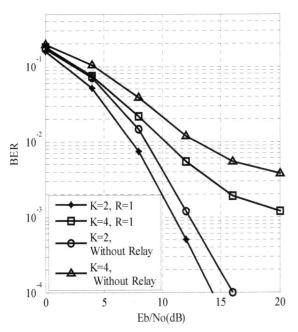

Fig 5. BER of cooperative ML-IDMA in Rayleigh channels with K=2, 4, N=512, it=3, SP=32 and R = 1

Fig. 5 shows that there is degradation in the BER performance by increasing the number of layers. In the case of without relay, K=2 and Eb/N0=16 dB, the BER performance was 10^{-4} while increasing K to 4, degraded BER performance to $5*10^{-3}$. The observed degradation in the BER performance by increasing the number of layers is due to increasing in the signal amplitude which cause signal distortion. For single relay (R=1), in the case of double layer (K=2), the improvement of the BER performance is 1.7 dB compared to without relay system. Also in the case of four layer (K=4) and Eb/N0 = 20 dB,

the BER improvement was found to be about $2.6 *10^{-3}$ when compared to without relay system as shown in Fig.5.

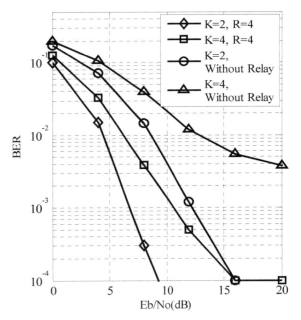

Fig 6. BER of cooperative ML-IDMA in Rayleigh channels with K=2, 4, N=512, it=3, SP=32 and R = 4

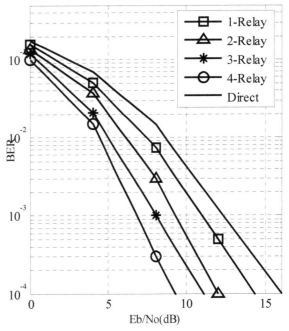

Fig 7. BER of cooperative ML-IDMA system in Rayleigh channels with R=1, 2, 3, 4, K=2, N=512, it=3 and SP=32

Fig.6 shows the improvement in BER performance by increasing the number of relays to R=4 at K=2 by about 8 dB when compared to without relay system. Fig. 7 show that R= 4 is superior to the case of R=0 by about 5 dB and is superior to the case of R=1 by about 6.7 dB at the BER of 10^{-4}. We can conclude that the co-operative environment, which added up two signals of different links, performs better than systems designed without relay environment. We can also see that when additional relay is deployed, the performance in the co-operative environment gets better. The analysis showed that the addition of different path powers in co-operative environment results in a lower Bit-Error-Rate (BER)

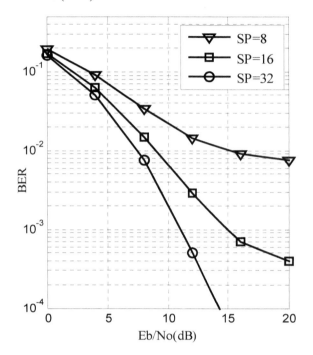

Fig 8. *BER of cooperative ML-IDMA system in Rayleigh channels with SP=8, 16, 32, R=1, K=2, N=512 and it=3*

Fig. 8 shows that increasing the Spreading Length (SP) can improve the performance significantly. By increasing the SP we were able to get larger spreading gain of the spread spectrum signal that improves the performance efficiency as the signal becomes larger.

4. Conclusion

In this paper, a cooperative communication scheme based on ML-IDMA technique is proposed for a network which has one source, one destination and multiple common relays. The proposed system is based on the chip-by-chip detection algorithm. Data reaches the destination through two different paths, i.e. a direct path from source to destination and a relayed path, where Decode-and-Forward DF relay protocol is applied to the data that finally reaches the MRC combiner at the destination.

The proposed system has been simulated to evaluate the performance of ML-IDMA technique with different numbers of layers with Maximal-Ratio-Combining (MRC) technique in the cooperative environment. The simulation results showed that by increasing the number of layers, degradation in the performance was obtained but on the other hand the Band Width (BW) was saved by (1/K).

The simulation also showed that the cooperative environment, which added up two signals of different links, performed better than the system designed without relays. When an additional relay was deployed, the performance in

the cooperative environment got better. This improvement means that the addition of different path powers in cooperative environment actually results in a lower Bit Error Rate (BER) in the E_b/N_0 vs. BER curve. We also observed that, the performance of cooperative environment with two or more relays is better in all investigated layers. Future work will be done to investigate the performance of coded ML-IDMA cooperative schemes, OFDM ML-IDMA, Coded OFDM ML-IDMA.

References

[1]   F.Lenkeit, D.Wübben,A.Dekorsy," An Improved Detection Scheme for Distributed IDM-STCs in Relay-Systems ",IEEE 76th Vehicular Technology Conference (VTC2012-Fall), Québec, Canada, 3. - 6. September 2012.

[2]   J. N. Laneman, and G. W. Wornell, "Distributed Space-Time-Coded Protocols for Exploiting Cooperative Diversity in Wireless Networks," IEEE Trans. Inf. Theory, vo.49, no.10, pp.2415-2425, Oct 2004.

[3]   G. Scutari, S. Barbarossa," Distributed space-time coding for regenerative relay networks," IEEE Trans. Wireless Commun., vol.4, no.5, pp.2387-2399, Sept. 2005.

[4]   Jinhong Yuan, Yonghui Li, and Li Chu, "Differential Modulation and Relay Selection WithDetect-and-Forward Cooperative Relaying," IEEE Trans. ON Vehicular Technology, vol. 59, no. 1, jan. 2010.

[5]   L. Ping, L. Liu, K. Wu, and W. K. Leung, "Interleave-division multipleaccess," IEEE Trans. Wireless Commun., vol. 5, no. 4, pp. 938-947, Apr. 2006.

[6]   Zhifeng Luo, Deniz Gurkan, Zhu Han, Albert Kai-sun Wong and Shuisheng Qiu," Cooperative Communication Based on IDMA," 5th International Conference on Wireless Communications, Networking and Mobile Computing, pp.1-4, Sep. 2009.

[7]   J. Tong, P. Li, Z. H. Zhang, and W. K. Bhargava, "Iterative Soft Compensation for OFDM Systems with Clipping and Superposition Coded Modulation," IEEE Trans. Commun., vol. 58, no. 10, pp. 2861–2870, Oct. 2010.

[8]   Zhifeng Luo, Deniz Gurkan, Zhu Han, Albert Kai-sun Wong and Shuisheng Qiu,"Cooperative Communication Based on IDMA," 5th International Conference on Wireless Communications, Networking and Mobile Computing, pp.1-4, Sep. 2009.

[9]   Xiaotian Zhou, Haixia Zhang and Dongfeng Yuan,"A multi-source cooperative scheme based on IDMA aided superposition modulation," International Conference on Information Theory and Information Security, pp.802-806, December 2010.

[10]   Weitkemper, P. , Wubben, D. and Kammeyer, K.-D.," Delay-diversity in multi-user relay systems with Interleave Division Multiple Access," IEEE International Symposium on Wireless Communication Systems 2008, pp.364-368, October 2008.

[11]   Chulhee Jang and Jae Hong Lee, "Novel IDM-cooperative diversity scheme and power allocation," in Proc. IEEE APWCS 2009, Seoul, Korea, pp.24-28, Aug. 2009.

[12] J. N. Laneman, D. N. C. Tse and G. W. Wornell, "Cooperative Diversity In Wireless Networks: Efficient Protocols And Outage Behavior", IEEE Trans. Info. Theory, Vol. 50, No. 12, pp. 3062-3080, November 2004.

[13] P. Li, L. H. Liu, K. Y. Wu, and W. K. Leung, "Interleave division multiple access," IEEE Trans. Wireless Commun., vol. 5, no. 4, pp. 938–947, Apr. 2006.

[14] P. Li, Q. H. Guo, and J. Tong, "The OFDM-IDMA approach to wireless communication systems," IEEE Wireless Commun. Mag., vol. 14, no. 3, pp. 18–24, Jun. 2007.

[15] Takyu O, Ohtsuki T, Nakagawa M. Companding system based on time clustering for reducing peak power of OFDM symbol in wireless communication. IEICE Transactions on Fundamentals of Electronics Communications and Computer Sciences 2006; E89- A(7): 1884–91.

[16] J. Akhtman, B. Z. Bobrovsky, and L. Hanzo, "Peak-to-average power ratio reduction for ODFM modems," in Proceedings of the 57th IEEE Semi-Annual Vehicular Technology Conference (VTC'03), vol. 2, Apr. 2003, pp. 1188–1192.

[17] P. Li, L. H. Liu, K. Y. Wu, and W. K. Leung, "A unified approach to multiuser detection and space-time coding with low complexity and nearly optimal performance," in Proc. 40th Allerton Conference, Allerton House, USA, pp. 170–179, Oct. 2002.

[18] Tao Peng, and Min Ye, "PAPR mitigation in superposition coded modulation systems using selective mapping" in Proceeding of the International Conference on Computer and Communication Technologies (ICCCT'2012), May 26-27, 2012, Phuket.

15

Working with TCP\IP based network monitoring system using Linux

Diponkar Paul[1, *], Shamsuddin Majamder[2]

[1]Department of Electrical and Electronic Engineering, Prime University, Dhaka, Bangladesh
[2]Department of Electrical and Electronic Engineering, World University of Bangladesh, Dhaka, Bangladesh

Email address:
dipo0001@ntu.edu.sg (D. Paul)

Abstract: Nagios is a stable, scalable and extensible enterprise-class network and system monitoring tool which allows administrators to monitor network and host resources such as HTTP, SMTP, POP3, disk usage and processor load. Originally Nagios was designed to run under Linux, but it can also be used on several UNIX o132perating systems. This chapter covers the installation and parts of the configuration of Nagios. The purpose of this paper is not only to introduce to everyone the concept of distributed monitoring with Nagios but capturing the beauty of it to improve the security of computer networks. Firstly, an introduction to Nagios will be discussed to provide readers a brief overview of what Nagios is. Next, it will discuss how distributed network monitoring is an essential part to information security. It will then proceed to introducing the requirements needed to build a distributed Nagios network monitoring environment and demonstrate how Nagios can be configured to construct a distributed monitoring environment that helps improve the state of security of distributed networks. In essence, companies hould be aware of the need for hiring specialized security analysts to perform round-the-clock systems monitoring to secure their resources.

Keywords: Nagois, CPAN, NET-SNMPD, CGI etc

1. Introduction

The document aims to assist users in installing, configuring, extending, troubleshooting and generally getting themost out of your Nagios system. It must be said that the documentation that ships with Nagios is excellent, and there are many of good tutorials that already exist. This book was written to complement the existing documentation that is already out there, and even go a litCtle further in some areas - especially for the beginners. For the true beginners in the audience, you'll find I have created a nice set of step by step instructions, walking you through the process of building a server from scratch, installing Nagios and configuring Nagios to monitor a typical network. Arguably the most useful source of Nagios information is the mailing list nagios-users. This list is frequented by expert Nagios developers and users alike and is *the* place to be if you're serious about Nagios. A great deal of information in this book has been obtained through the help of people on this list, I highly recommend subscribing and participating - to learn, and to help others learn. Almost every Nagios related question I've ever had has been answered and documented in the list archives. It's a good idea to check there before posting. There are several places that you can find information on Nagios. Here are some places you should bookmark. Nagios.org Documentation -- Nagios.org FAQ's --Nagios Mail Lists.The complexity of modern networks and systems is somewhat astounding, as any experienced System Administrator will tell you. Even seemingly small networks found in many Small/Medium Enterprises (SME's) can have extremely high levels of complexity in the systems they run. Nagios was designed as a rock solid framework for monitoring, scheduling and alerting. Nagios contains some very powerful features, harnessing them is not only a matter of understanding how Nagios works, but also how the system you're monitoring also works. This is an important realization. Nagios can't automatically teach you about complex systems, but it will be an valuable tool to help you in your journey. So what are the sorts of things Nagios can do? Nagios can do much more than this, but nevertheless here's a list of A more technically accurate and complete list of Nagios' features can be found in the official documentation (http://www.nagios.org/docs/). The Nagios package doesn't contain any checking tools (called plugins) at all. Does that statement sound crazy? Sure, but let me explain. Nagios focuses on doing what it does best - providing a robust, reliable and extensible framework for any type of check that a user can come up with. So how does Nagios perform it's checking? A huge number of plugins already exist that extend Nagios to perform every type of check maginable. And if there isn't an existing check that already exists, you're free to write your own. The nagios-plugins package is separately maintained and can be downloaded from various sources.We will cover the Nagios plugins later on. than Galstad is the creator of Nagios. Karl DeBisschop, Subhendu Ghosh, Ton Voon, and Stanley Hopcroft arethe main plugin developers. Many other people have contributed to the project over the years by submitting bug reports, atches, ideas, suggestions, add-ons, plugins, etc. A list of some of the contributors can be found at the Nagios website. As with anything, each tool has it's own set of strengths and weaknesses[1]. Some are the applications may seem similar, but are very different and range from full blown SNMP management solutions, to simple applications with not much flexibility. Big Brother, OpenNMS, OpenView and SysMon (there are dozens more) are often compared to Nagios, however they are quite different in many respects. In my travels as an IT professional, Nagios is the most commonly used monitoring tool by far. There are lots of specialist companies that offer monitoring as a service. A large number of these use Nagios.It's very useful to conceptually understand how Nagios works. The following is a very simplified view on how Nagios works.

Nagios runs on a server, usually as a daemon (or service). Nagios periodically run plugins residing (usually) on the same server, they contact (PING etc.) hosts and servers on your network or on the Internet. You can also have information sent to Nagios. You then view the status information using the web interface. You can also receive email or SMS notifications if something happens. Event Handlers can also be configured to "act" if something happens.For the cost of hardware these days, I usually recommend running a dedicated server. One complaint I often here from users (usually new to Linux/Unix or Open Source), is that Nagios is difficult to configure. The Nagios Book aims to dispel this myth, showing you how to build a Nagios server from scratch monitoring several hosts and sending alerts in a matter of hours. Note that it's not a competition when it comes to setting up any server, however I've heard reports of people complaining of taking days and even weeks to set up a Nagios server. This is simply not true, as you will find out! Please note that this document does not explain the how's and why's of how to build a server, that is beyond the scope of this material. My goal is to try and get users up and running quickly so they can experience the power of Nagios. Having said that, I have created some basic steps to build a FreeBSD server from scratch. These instructions were written and tested using FreeBSD 6.0, however they should also work just as well on earlier versions. If you're using Linux, some steps may be slightly different. I hope to add steps for each popular distro very soon. I plan to add some other distro specific information here eventually. If you have installation notes for another platform, 'We will glaadly add it.

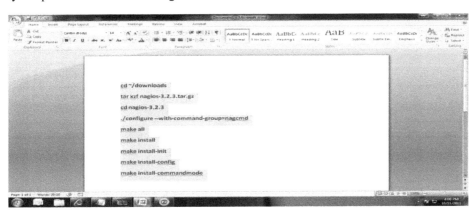

Figure 1: Nagios install step by step command line.

The simplest method of installation is for you to install the Nagios packages that are supplied with the distribution you are using. However, Nagios 2.0 is relatively new, so you may have to make do with an older Nagios version using this method. Configuring this is quite different from the version 2.0 described here, which is why it is recommended that you take things into your own hands and compile Nagios yourself if the distributor does not provide any Nagios 2.0 packages. If you are compiling Nagios yourself, you also have an

influence on directory structures and several other parameters. A Nagios system compiled in this way also provides an almost complete main configuration file, in which, initially, nothing has to be changed. But it should be mentioned here that compiling Nagios yourself might involve a laborious search for the necessary development packages, depending on what is already installed on the computer. OK, so you know all about what Nagios can do, it's time to get on with it and choose an installation method[2].

You have two main choices when installing Nagios: For compiling Nagios itself you require gcc, make, autoconf and automake. Required libraries are libgd1 and openssl2. The development packages for these must also be installed (depending on the distribution, with either the ending -dev or -devel): libssl-dev, libgd-dev, libc6-dev. For the plugins it is recommended that you also install the following packages at the same time: ntpdate,3 snmp,4 smbclient,5 libldap2, and libldap2-dev,6 as well as the client and developer packages for the database to be used (e.g., postgresqlclient and postgresql-dev). The three commands unpack the source code into the directory created for this purpose, /usr/local/src. When this is done, a subdirectory with the name nagios-2.0b3 is also created. Before the actual compilation and installation, the groups required for operation, namely nagios and nagcmd, are set up with groupadd, and the user nagios, who is assigned to these groups and with whose permissions the Nagios server runs is set up with useradd: linux:~ # groupadd -g 9000 nagios linux:~ # groupadd -g 9001 nagcmd linux:~ # useradd -u 9000 -g nagios -G nagcmd -d /usr/local/nagios \ -c "Nagios Admin" nagios Instead of the user (9000) and group IDs (9000 or 9001) used here, any other (available) ID may be used. The primary group nagios of the user nagios should remain reserved exclusively for this user.1 http://www.boutell.com/gd/2 http://www.openssl.org/ Depending on the distribution, the required RPM and Debian packages are sometimes named differently. Here you need to refer to the search help in the corresponding distribution. For Debian, the homepage will be of help. If a configure instruction complains, for example, of a missing gd.h file,

you can search specifically at http://www.debian.org/distrib/packages for the contents of packages. For the configure command, parameters are specified that differ from the standard; The values chosen here ensure that the installation routine selects the directories used here in the book and that all parameters are correctly set when the main configuration file is generated. This considerably simplifies the fine-tuning of the configuration. If --prefix is not specified, Nagios installs itself in the directory /usr/local/nagios. We recommend that you stick to this directory.The system normally stores its configuration files in the directory etc beneath its root directory. In general it is better to store these in the /etc hierarchy, however. Here we use /etc/nagios.9 Variable data such as the log file and the status file are by default stored by Nagios in the directory /usr/local/nagios/var. This is in the /usr hierarchy, which should only contain programs and other read-only files, not writable ones. In order to ensure that this is the case, we use /var/nagios.10 Irrespective of these changes, in most cases configure does not run through faultlessly the very first time, since one package or another is missing. For required libraries such as libgd, Nagios almost always demands the relevant developer package with the header files (here, libgd-dev or libgd-devel). Depending on the distribution, their names end in -devel or -dev. After all the tests have been run through, configure presents a summary of all the important configuration parameters:

2. Installing and Testing Plugins

Figure 2: Nagios Plugins install step by step command line.

What is now still missing are the plugins. They must be downloaded separately from http://www.nagios.org/ and installed. As independent programs, they are subject to a different versioning system than Nagios. The current version at the time of going to press was version 1.4, but you can, for example, also use plugins from version 1.3.1 if you don't mind doing without the most recent features. Although the plugins are distributed in a common source distribution, they are independent of one another, so that you can replace one version of an individual plugin with another one at any time, or with one you have written yourself. Installation The installation of the plugin sources takes place, like the Nagios ones, in the directoryIf you are not using Radius, you need have no qualms in ignoring the corresponding error messages.

Otherwise you should install the missing packages and repeat the configure procedure. The quite frequently required SNMP functionality is missing a Perl module in the example. This is installed either in the form of the distribution package or online via the CPAN archive:13 linux:~ # perl -MCPAN -e install Net::SNMP If we are running the CPAN procedure for the first time, it will guide you interactively through a self-explanatory setup, and you can answer nearly all of the questions with the default option. Running make in the directory nagios-plugins-1.4 will compile all plugins. Afterwards you have the opportunity to perform tests, with make check. Because these have not been particularly carefully programmed, you will often see many error messages that have more to do with the test itself than with the plugin. if you still want to try it, then the Cache Perl module must also be installed. Irrespective of make check, the most important plugins should be tested manually anyway after the installation. make install finally anchors the plugins in the subdirectory libexec (which in our case is /usr/local/nagios/libexec), but not all of them: the source directory contrib. contains a number of plugins that make install does not install automatically. Most plugins in this directory are shell or Perl scripts. Where needed, these are simply copied to the plugin directory /usr/local/nagios/libexec. The few C programs there are must first be compiled, which in some cases may be no laughing matter, since a corresponding makefile, and often even a description of the required libraries, is missing. If a simple make is not sufficient, as in the case of linux:nagios-plugins-1.4/contrib # make check_cluster214 cc check_cluster2.c -o check_cluster2 then it is best to look for help in the mailing list nagiosplug-help.15 The compiled program must also be copied to the plugin directory. The Comprehensive Perl Archive Network at http://www.cpan.org/.With check_cluster, hosts and services of a cluster can be monitored. Here you usually want to be notified if all nodes or redundant services provided fail at the same time. If one specific service fails on the other hand, this is not critical, as long as other hosts in the cluster provide this service.

http://lists.sourceforge.net/lists/listinfo/nagiosplug-help

Because plugins are independent programs, they can already be used manually for test purposes right now—before the installation of Nagios has been completed. In any case you should check the check_icmp plugin, which plays an essential role: it checks whether another computer can be reached via ping and is the only plugin to be used both as a service check and a host check. If it is not working correctly, Nagios will also not work correctly, since the system cannot perform any service checks as long as it categorizes a host as "down". describes check_icmp in, which is why there is only short introduction here describing its manual use. In order for the plugin to function correctly it must, like the /bin/ping program, be run as the user root. This is done by providing it In order for the Web front end of Nagios to function, the Web server must know the CGI directory and the basis Web directory. The following description, with a slight deviation, applies to both Apache 1.3 and Apache 2.0. 1.3.1 Setting Up Apache As long as you have not added a different address for the front end, through the configure script with -with-cgiurl, it can be addressed under /nagios/cgi-bin. Since the actual CGI scripts are located in the directory /usr/local/nagios/sbin, a corresponding script alias is set in the Apache configuration: Here the directives Order and Allow also allow access only from the specified network. It is recommended that you write the above details in your own configuration file, called nagios.conf, so that this configuration is not lost during an Apache update, and place it in the Apache directory for individual configurations. This is usually to be found under /etc/apache/conf.d, but depending on the distribution and the

Apache ersion, this could also be under /etc/httpd/conf.d or /etc/apache2/conf.d. In any case the Apache configuration file must integrate this directory with the directive Include. More recent SuSE distributions only accept files in the subdirectory conf.d that end in .conf. In the state in which it is delivered, Nagios allows only authenticated users access to the CGI directory. This means that users not "logged in" have no way to see anything other than the home page and the documentation. They are blocked off from access to other functions. There is a good reason for this: apart from status queries and other display functions, Nagios has the ability to send commands via the Web interface. The interface for external commands is used for this purpose (Section 13.1, page 240). If this is active, checks can be switched on and off via the Web browser, for example, and Nagios can even be restarted. Only authorized users should be in a position to do this. The easiest way to implement a corresponding authentication is via a .htaccess file in the CGI directory /usr/local/nagios/sbin.17 The document directory, on the other hand, requires no special protection. In addition, the parameter use_authentication in the CGI configuration file cgi.cfg18 of Nagios must be set to 1: The access rule described here, via .htaccess in the CGI directory, adheres to the official Nagios documentation. Those more familiar with Apache will have other configuration possibilities available, Name is just a comment that the browser displays if the Web server requests authentication. AuthType Basic stands for simple authentication, in which the password is transmitted without encryption, as long as no SSL connection is used. It is best to save the password file—here htpasswd—in the Nagios configuration directory /etc/nagios. The final parameter, require valid-user, means that all authenticated users have access (there are no restrictions for specific groups; only the user-password pair must be valid). In combination with its own modules and those of third parties, Apache allows a series of other authentication methods. These include authentication via an LDAP directory, via Pluggable Authentication Modules (PAM),19 or using SMB via a Windows server. Here we refer you to the relevant literature and the highly detailed documentation on the Apache home page at http://httpd.apache.org/. The (basically freely selectable) name of the password file will be specified here so that it displays what type of password file is involved. It is generated with the htpasswd2 program included in Apache (in Apache 1.3 the program is called htpasswd)Even though configuration of the Web interface is now finished, at the moment only the documentation is properly displayed: Nagios itself must first be correspondingly adjusted—as described in detail in the following chapter– before it can make usable monitoring data available in this way.Although the Nagios configuration can become quite large, you only need to handle a small part of this to get a system up and running. Luckily many parameters in Nagios are already set to sensible default settings. So this chapter will be primarily concerned with the most basic and frequently used parameters, which is quite sufficient for an initial configuration. Further details on the configuration are provided by the chapters on individual Nagios features: in Chapter 6 about network plugins (page 85) there are many examples on the configuration of services. All parameters of the Nagios messaging system are explained in detail in Chapter 12, page 215, and the parameters for controlling the Web interface are described in Chapter 16 from page 273. In addition to this, Nagios includes its own extensive documentation, once it is installed, in the directory /usr/local/nagios/share/docs, which can also be reached from the Web interface. This can always be recommended as a useful source for further information, which is why each of the sections below refer to the corresponding location in the original documentation. The installation routine in make install-config stores examples of individual configuration files in the directory /etc/nagios. They all end in -sample, so that a possible update will not overwrite the files needed for productive operation. All subsequent work should be carried out as the user nagios. If you are editing files as the superuser, you must ensure yourself that the contents of directory /etc/nagios afterwards belong to the user nagios again. With the exception of the file resource.cfg—this may contain passwords, which is why only the owner nagios should have the read permission set—all other files may be readable for all.The central configuration takes place in nagios.cfg. Instead of storing all configuration options there, it makes links to other configuration files (with the exception of the CGI configuration). The easiest method is first to copy the example file: nagios@linux:/etc/nagios$ cp nagios.cfg-sample nagios.cfg Those who compile and install Nagios themselves have the advantage that at first they do not even need to adjust nagios.cfg, since all paths are already correctly set.1 And that's as much as you need to do. Nevertheless one small modification is recommended, which helps to maintain a clear picture and considerably simplifies configuration where larger networks are involved. The parameter concerned is cfg_file, which integrates files with object definitions). The file nagios.cfg-sample, included in the package, As an alternative to cfg_file, you can also use the parameter cfg_dir: this requests you to specify the name of a directory from which Nagios should integrate all configuration files ending in .cfg (files with other extensions are simply ignored). This also works recursively; Nagios thus evaluates all *.cfg files from all subdirectories. With the parameter cfg_dir you therefore only need to specify a signal directory, instead of calling all configuration files, with cfg_file, individually. The only restriction: these must be configuration files that describe objects. The configuration files cgi.cfg and resource.cfg are excluded from this, which is why, like the main configuration file nagios.cfg, they remain in the main directory /etc/nagios. For the object-specific configuration, it is best to create a directory called /etc/ nagios/mysite, then remove all cfg_file directives in nagios.cfg (or comment them out with a # at the beginning of the line) and replace them with the following:The main directory /etc/nagios contains only three configuration files and the password file for protected Web access. For the sake of clarity, the configuration examples *-sample should be

moved to the directory sample. In this doc we will include all objects of a type in a file of its own, that is, all host definitions in the file hosts.cfg, all services in services.cfg, and so on. But you could just as well save each of the host definitions in a separate file for each host and use a directory structure to reflect this:http://mama.indstate.edu/users/ice/tree/Extended Service Information, like Extended Host Information. Not all object types are absolutely essential; especially at the beginning, you can easily do without the *dependency, *escalation, and *extinfo objects, as well as the servicegroup. looks at escalation and dependencies in detail. The extended information objects are used to provide a "more colorful" graphical representation, but they are not at all necessary for running Nagios. We refer here to the original documentation.3 Notes on the object examples below Although the following chapters describe individual object types in detail, only the mandatory parameters are described there and those that are absolutely essential for meaningful operation. Mandatory parameters here are always printed in bold type. The first (comment) line in each example lists the file in which the recorded object definition is to be stored. When you first start using Nagios, it is recommended that you restrict yourself to a minimal configuration with only one or two objects per object type, in order to keep potential sources of error to a minimum and to obtain a running system as quickly as possible. Afterwards extensions can be implemented very simply and quickly, especially if you take on board the tips mentioned in Section 2.11 on templates Time details in general refer to time units. A time unit consists of 60 seconds by default. It can be set to a different value in the configuration file nagios.cfg, using the parameter interval_length. You should really change this parameter only if you know exactly what you are doing.In order for the Web front end to work correctly, Nagios needs the configuration file cgi.cfg. The example included, called cgi.cfg-sample, can initially be taken over one-to-one, since the paths contained in it were set correctly during installation: nagios@linux:/etc/nagios$ cp sample/cgi.cfg-sample ./cgi.cfg Important: the file cgi.cfg should be located in the same directory as nagios.cfg, because the CGI programs have been compiled in this path permanently. If cgi.cfg is located in a different directory, the Web server must also be given an environment variable with the correct path, called NAGIOS_CGI_CONFIG. How this is set in the case of Apache is described in the corresponding online documentation at http://httpd.apache.org/docs-2.0/env.html. Out of the box, only a few parameters are enabled in the CGI configuration file. pre-flight check Although warnings displayed here can in principle be ignored, this is not always what the inventor had in mind: perhaps you made a mistake in the configuration, and Nagios is ignoring a specific object, which you would actually like to use. The first warning in the example refers to a host called linux02, which has not been allocated any services. Since Nagios works primarily with service checks, and uses host checks only if it needs them, a computer should basically always be allocated at least one service. Nagios issues a warning, as here, if no service at all has been defined for a particular host.It is also recommended, however, to always define a "PING" service for every host, although this is not absolutely essential. Even if the same plugin, check_icmp, is used here as with the host check, this is not the same thing: the host check is satisfied with a single response packet, after all, it only wants to find out if the host "is alive". As a service check, check_icmp registers packet run times and loss rates, which can be used to draw conclusions, if necessary, concerning existing problems with a network card. The second warning refers to a contact named wob, who, although defined, is not used, because he does not belong to any contact group. In contrast to warnings, genuine errors must be eliminated, because Nagios will usually not start if the parser finds an error, as in the

following example:Here the configuration mistakenly contains a host called linux03, for which there is no definition. If you read through the error message carefully, you will quickly realize that the error can be found in the file /etc/nagios/mysite/services.cfg. In the definition of independencies (host and service dependencies, see Section 12.6 page 234) there is a fundamental risk that circular dependencies could be specified by mistake. Because Nagios cannot automatically resolve such dependencies, this is also checked before the start, and if necessary, an error is displayed. When using the parents parameter, it is also possible that two hosts may inadvertently serve mutually as "parents"; Nagios also test this.

3. Getting Monitoring Started

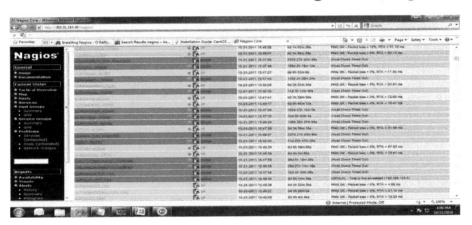

Figure 3: Host and service status page.

is causes Nagios to reread the configuration, end tests for hosts and services that no longer exist, and integrate new computers and services into the test. However, with each reload there is a renewed scheduling of checks, meaning that Nagios plans to carry out all tests afresh. To prevent all tests from being started simultaneously at bootup, Nagios performs a so-called spreading. Here the server spreads the start times of the tests over a configurable period.1 For a large number of services, it can therefore take a while before Nagios continues the test for a specific service. For this reason you should never run reloads at short intervals: in the worst case, Nagios will not manage to perform some checks in the intervening period and will perform them only some time after the most recent reload. Before being reloaded, the configuration is tested to eliminate any existing errors.With the Simple Network Management Protocol, SNMP, local resources can also Be queried over the network (see also Client 4 in Figure 5.1, page 80). If an SNMP daemon is installed (NET-SNMPD is very extensively used, and is described in Section 11.2.2 from page 187), Nagios can use it to query local resources such as processes, hard drive and interface load. The advantage of SNMP lies in the fact that it is widely used: there are corresponding services for both UNIX and Windows systems, and almost all modern network components such as routers and switches can be queried via

SNMP. Even uninterruptable power supplies (USPs) and other equipment sometimes have a network connection and can provide current status information via SNMP. Apart from the standard plugin check_snmp, a generic SNMP plugin, there are various specialized plugins that concentrate on specific SNMP queries but are sometimes more simple to use. So check_ifstatus and check_ifoperstatus[5], for example, focus precisely on the status of network interfaces. If you are grappling with SNMP for the first time, you will soon come to realize that the term "readable for human beings" did not seem to be high up on the list of priorities when the protocol was defined. SNMP queries are optimized for machine processing, such as for a network monitoring tool. If you use the tool available from the vendor for its network components, SNMP will basically remain hidden to the user. But to use it with Nagios, you have to get your hands dirty and get involved with the protocol and its underlying syntax. It takes some getting used to, but it's not really as difficult as it seems at first sight.

4. Plugins for Network Services

Every plugin that is used for host and service checks is a separate and independent program that can also be used independently of Nagios. The other way round, it is not so easy: in order for Nagios to use an external program, it must stick to certain rules. The most important of these concerns the return status that is returned by the program. Using this, A plugin therefore does not distinguish by using the pattern "OK—Not OK", but is more differentiated. In order for it to be able to categorize a status as WARNING, it requires details of up to what measured value a certain event is regarded as OK, when it is seen as a WARNING, and when it is CRITICAL. An example: apart from the response time, a ping also returns the rate of packet loss. For a slow network connection (ISDN, DSL), a response time of 1000 milliseconds could be seen as a warning limit and 5000 milliseconds as critical, because that would mean that interactive working is no longer possible. If there is a high load on the network connection, occasional packet loss could also occur,1 so that 20 percent packet loss can be specified as a warning limit, 60 percent as the critical limit The classic reachability test in UNIX systems has always been a ping, which sends an "ICMP echo request" packet and waits for an "ICMP echo response" packet. The Nagios plugin package includes two programs that carry out this ping check: check_icmp and check_ping. Even though check_ping is used in the standard configuration, you should replace it with the more efficient check_icmp, which has been included since plugin version 1.4. Whereas check_ping calls the UNIX program /bin/ping, which is why there are always compatibility problems [4] with the existing ping version, check_icmp sends ICMP without any external help programs. check_icmp basically works more efficiently, since it does not wait for one second between individual packets, as ping Four pseudo plugins are available for testing the POP and IMAP protocols: check_ pop, check_spop, check_imap, and check_simap. They are called pseudo plugins because they are just symbolic links to the plugin check_tcp. By means of the name with which the plugin is called, this determines its intended use and correspondingly sets the required parameters, such as the standard port, whether something should be sent to the server, the expected response and how the connection should be terminated. The options are the same for all plugins, which is why we shall introduce them all together:Instead of ip, the host name or IP address of the target computer is given. For systems with several virtual environments, you will land in the default environment, and for most Web hosting providers you will then receive an error message: nagios@linux:nagios/libexec$ ./check_http -I www.swobspace.de HTTP WARNING: HTTP/1.1 404 Not Found -u url or path / --url=url or path The argument is the URL to be sent to the Web server. If the design document lies on the server to be tested, it is sufficient to enter the directory path, starting from the document root of the server: nagios@linux:nagios/libexec$ ./check_http -H linux.swobspace.net\ -u /mailinglisten/index.html HTTP OK HTTP/1.1 200 OK - 5858 bytes in 3.461 seconds If this option is not specified, the plugin asks for the document root /. -p port / --port=port This is an alternative port specification for HTTP.It is not so simple to monitor UDP ports, since there is no standard connection setup, such as the

three-way-handshake for TCP, in the course of which a connection is opened, but data is not yet transferred. For a stateless protocol such as UDP there is no regulated sequence for sent and received packets. Nagios provides three plugins for monitoring databases: check_pgsql for PostgreSQL, check_mysql for MySQL, and check_oracle for Oracle. The last will not be covered in this book.13 They all have in common the fact that they can be used both locally and over the network. The latter has the advantage that the plugin in question does not have to be installed on the database server. The disadvantage is that you have to get more deeply involved with the subject of authentication, because configuring a secure local access to the database is somewhat more simple. For less critical systems, network access by the plugin can be done without a password. To do this, the user nagios is set up with its own database in the database management system to be tested, which does not contain any (important) data.There are two possibilities for monitoring uninterruptible power supplies (UPS): the Network UPS Tools support nearly all standard devices. The apcupsd daemon is specifically tailored to UPS's from the company APC The following rule generally applies: no plugin directly accesses the UPS interface. Rather they rely on a corresponding daemon that monitors the UPS and provides status information. This daemon primarily serves the purpose of shutting down the connected servers in time in case of a power failure. But it also always provides status information, which plugins can query and which can be processed by Nagios. Both the solution with the Network UPS Tools and that with apcupsd are fundamentally network-capable, that is, the daemon is always queried via TCP/IP (through a proprietary protocol, or alternatively SNMP). But you should be aware here that a power failure may affect the transmission path, so that the corresponding information might no longer even reach Nagios. Monitoring via the network therefore makes sense only if the entire network path is safeguarded properly against power failure. In the ideal scenario, the UPS is connected directly to the Nagios server. Calling the check_ups plugin is no different in this case from that for the network configuration, since even for local use it communicates via TCP/IP—but in this case, with the host localhost). The Network UPS Tools The Network UPS Tools is a manufacturer-independent package containing tools for monitoring uninterruptible power supplies. Different specific drivers take care of hardware access, so that new power supplies can be easily supported, provided their protocols are known. The remaining functionality is also spread across various programs: while the daemon upsd provides information, the program upsmon shuts down the computers supplied by the UPS in a controlled manner. It takes care both of machines connected via serial interface to the UPS and, in client/server mode, of computers supplied via the network. The homepage http://www.networkupstools.org/ lists the currently supported models and provides further information on the topic of UPS. Standard distributions already contain the software, but not always with package names that are very obvious: in SuSE and Debian they are known by the name of nut. To query the information provided by the daemon upsd, there is the check_ups plugin from the Nagios Plugin package. It queries the status of the UPS through the network UPS Tools' own network protocol. A subproject also allows it to query the power supplies via SNMP.19 However, further development on it is not taking place at the present time.The following example tests the above defined local UPS with the name upsfw. The -T switch should ensure that the output of the temperature is given in degrees Celsius, which only partially works here: the text displayed by Nagios before the pipe sign | contains the correct details, but in the performance data after the |, the plugin version 1.4 still shows the information in degrees Fahrenheit.The check_procs plugin monitors processes according to various criteria. Usually it is used to monitor the running processes of just one single program. Here the upper and lower limits can also be specified. nmbd, for example, the name service of Samba, always runs as a daemon with two processes. A larger number of nmbd entries in the process table is always a sure sign of a problem; it is commonly encountered, especially in older Samba versions.Services such as [7] Nagios itself should only have one main process. This can be seen by the fact that its parent process has the process ID 1, marking it is a child of the init process. It was often the case, in the development phase of Nagios 2.0, that several such processes were active in parallel after a failed restart or reload, which led to undesirable side effects. You can test to see whether there really is just one single If you'd like to build your own SMS gateway using Gnokii, I wrote a "SMS Gateway How To" a few years ago. You can read the original article here (http://www.chrisburgess.com.au/sms-gateway-how-to/), or read on for the simplified steps. I am assuming you are using the same server to run both Nagios and Gnokii. Installing and Configuring Gnokii for SMS Alerts cd /usr/ports/comms/gnokii make install clea

If we'd prefer another messaging type, the Nagios mail list archives mention all sorts of methods people use.

Optimizing and Enhancing Nagios

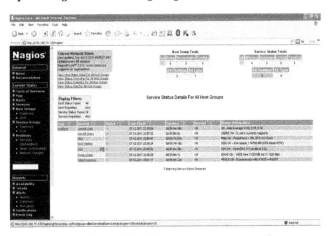

Figure 5: Specific service status page.

So far, we've modified the sample file called minimal.cfg (originally minimal.sample). Using the minimal sample is ideal for getting started, and by all means, you're free to use run Nagios in that configuration if you wish. At this stage however, it's worth pointing out that you can split your configurations into separate config files to help make things more manageable. Straight from the documentation, we can see how this works: NagiosQL is a web front end for simplifying the set up of the configuration files for Nagios[3]. In this guide I will only be installing NagiosQL. If we also need to install Nagios, an excellent guide to use for installing from source can be found at http://nagios.sourceforge.net/docs/3_0/quickstart-fedora.ht ml. Generally we prefer to install using an PM package from a Yum repository, [6] but I have not yet found a repository that contains the Nagios 3 package for entOS/Red Hat. First install the MySQL and PHP packages needed to support NagiosQL. I am reinstalling, so hopefully the packages I've listed will pick up everything that is required[8]:

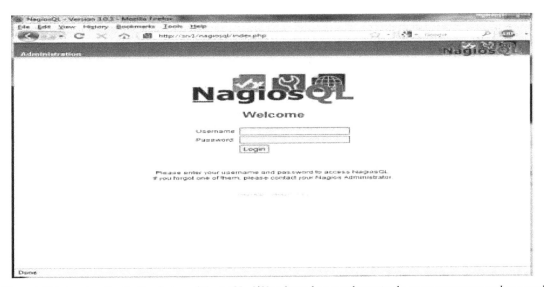

Figure 4: This is web authentication feature of NagiosQL . If User login this page then write the correct user name and password.

Host and Service Add at Nagiosql:

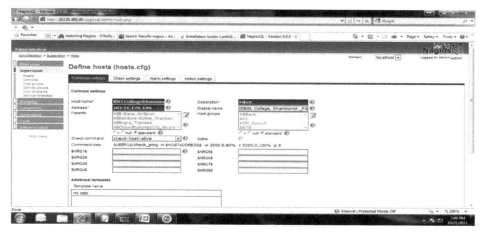

Figure 6: *Admin user add the host and service with required field like as IP Address, time, hostgroup, email.*

5. Conclusion

All in all, it is the people behind the scenes that play a crucial role in managing security in today's networks. Nagios is just a network monitoring tool. And although it is useful to further enhance the confidentality, integrity and availability of resources, it really depends on how intelligently the plugins work and how security analysts configure them to work on crucial company's servers. "Monitoring provides immediate security in a way that just doing a vulnerability assessment or dropping a firewall into a network can never provide. Monitoring provides dynamic security in a way that yet another security product can never provide. And, as security products are added into a network - firewalls, IDSs, specialized security devices - monitoring only gets better"35. This is because faster detection and response can be achieved through active network monitoring of events from a number of security components. It is always crucial for security [9] analysts to maintain monitoring 24x7 with an 'active' mind. In other words, they must work intelligently. For instance, if there are no new alerts generated for some time, security analysts should always make it a habit to check around to make sure things are still moving fine. With GNU license open source software like Nagios,[6] it is also advisable for one to subscribe to relevant mailing lists to keep in touch with new bug fixes and patch to the latest version of software applications used. Currently, Nagios is at version 1.1 and is expected to roll into 2.0 by summer of 2003 with further enhancements including passive host checks, which enable the central monitoring server to be able to receive not only remote services results but host status as well.For more interests in the next version, one can view http://www.Nagios.org/upcoming.php. "It takes constant monitoring. It's not one tool over another; it's the mind-set of the staff who review our systems, read information, put in proper patches and do proper testing.

References

[1]   Lamsal, P. "Management of the Next Generation IP Core Network." 16th April 1999. URL: http://www.tml.hut.fi/Opinnot/Tik-110.551/1999/papers/12ManagementOfIPngCore/ipcore.html

[2]   Winkler, Ira. "Ounce of Prevention." November 1999. URL: http://www.infosecuritymag.com/articles/1999/winkler.shtml

[3]   Walker, L. "The View From Symantec's Security Central." 9th January 2003. URL: http://www.washingtonpost.com/wp-dyn/articles/A28625-2003Jan8.html

[4]   Messmer, E. and Pappalardo, D. "A Year After Meltdown: No Silver Bullet for DoS." 2nd May 2001. URL: http://www.nwfusion.com/news/2001/0205ddos.html FA27 2F94 998D FDB5 DE3D F8B5 06E4 A169 4E46

[5]   CERT/CC "CERT Advisory CA-2002-03 Multiple Vulnerabilities in Implementations of the Simple Network Management Protocol (SNMP)." 4th August 2003. URL: http://www.cert.org/advisories/CA-2002-03.html

[6]   Kamthan, P. "CGI Security: Better Safe Than Sorry." 19th September 1999. URL: http://tech.irt.org/articles/js184/[28] - [30] Refer to [15].

[7]   Polombo, D. "Prelude HOWTO." 16th September 2002. URL: http://www.prelude-ids.org/article.php3?id_article=6[32] Refer to [11].

[8]   Habib, A., Hefeeda, M. M. and Bhargava, B. K. "Detecting Service Violations and DoS Attacks." 2002. URL: http://216.239.39.104/search?q=cache:m4uV_rBx9hIJ:www.isoc.org/isoc/conferences/ndss/03/proceedings/papers/12.pdf+detecting+service+violations+and+dos+attacks&hl=en&ie=UTF-8

16

Authority system to prevent privacy protection in Peer-to-Peer network system

S. Uvaraj[1], N. Kannaiya Raja[2]

[1]Arulmigu Meenakshi Amman College of Engineering, Kanchipuram
[2]Defence Engineering College, Ethiopia

Email address:
ujrj@rediffmail.com(S. Uvaraj), kannaiyaraju123@gmail.com(N. K. Raja)

Abstract: Collusive piracy is the main source of intellectual property violations within the boundary of a P2P network. Paid clients (colluders) may illegally share copyrighted content files with unpaid clients (pirates). Such online piracy has hindered the use of open P2P networks for commercial content delivery. We proposed a proactive content poisoning scheme to stop colluders and pirates from alleged copyright infringements in P2P file sharing. The basic idea is to detect pirates timely with identity-based signatures and time-stamped tokens. The scheme stops collusive piracy without hurting legitimate P2P clients by targeting poisoning on detected violators, exclusively. We developed a new peer authorization protocol (PAP) to distinguish pirates from legitimate clients. Detected pirates will receive poisoned chunks in their repeated attempts. Pirates are thus severely penalized with no chance to download successfully in tolerable time. Based on simulation results, we find 99.9 percent prevention rate in Gnutella, KaZaA, and Freenet. We achieved 85-98 percent prevention rate on eMule, eDonkey, Morpheus, etc. The scheme is shown less effective in protecting some poison-resilient networks like BitTorrent and Azureus. Our work opens up the low-cost P2P technology for copyrighted content delivery. The advantage lies mainly in minimum delivery cost, higher content availability, and copyright compliance in exploring P2P network resources.

Keywords: Peer-to-Peer, Content Delivery Network, Reputation System, Colluder, Content Poisoning and Network Security

1. Introduction

PEER-TO-PEER (P2P) networks are most cost-effective in delivering large files to massive number of users [3]. Unfortunately, today's P2P networks are grossly abused by illegal distributions of music, games, video streams, and popular software. These abuses have not only resulted in heavy financial loss in media and content industry, but also hindered the legal commercial use of P2P technology. The main sources of illegal file sharing are peers who ignore copyright laws and collude with pirates. To solve this peer collusion problem, we propose a copyright-compliant system for legalized P2P content delivery. Our goal is to stop collusive piracy within the boundary of a P2P content delivery network. In particular, our scheme appeals to protecting large-scale perishable contents that diminish in value as time elapses. Traditional content delivery networks (CDNs) use a large number of surrogate content servers over many globally distributed WANs. The content distributors need to replicate or cache contents on many servers. The bandwidth demand and resources needed to maintain these CDNs are very expensive. A P2P content network significantly reduces the distribution cost. P2P networks improve the content availability, as any peer can serve as a content provider. P2P networks are inherently scalable, because more providers lead to faster content delivery.

We use identity-based signatures (IBS) [4] to secure file indexes. IBS offers similar level of security as PKI-based signatures with much less overhead. We apply discriminatory content poisoning against pirates. We focus on protection of decentralized P2P content networks. Protecting centralized P2P networks like Napster is much simpler than the scheme we proposed because of centralized indexing.

1.1. Types of Clients

- Honest clients
- Colluders clients
- Pirates clients

Honest or legitimate clients are those that comply with the copyright law not to share contents freely. Pirates are peers attempting to download some content files without paying or authorization. The colluders are those paid clients who share the contents with pirates. Pirates and colluders coexist with the law-abiding clients. Content poisoning is implemented by deliberate falsification of the file requested by pirate.

2. Our Approach and Contributions

Content poisoning is often treated as a security threat to P2P networks. To our best knowledge, using selective content poisoning to prevent collusive piracy has not been explored in the past. We offer the very first proactive poisoning approach to curtailing copyright violation in P2P networks. We make the following specific contributions towards P2P content delivery.

2.1. Distributed Detection of Colluders and Pirates

We develop a protocol that identifies a peer with its endpoint address. File index format is changed to incorporate a digital signature based on this identity. A peer authentication protocol is developed to establish the legitimacy of a peer when it downloads and uploads the file. Using IBS, our system enables each peer to identify unauthorized peers or pirates without the need for communication with a central authority.

2.2. Proactive Content Poisoning of Detected Pirates

Our protocol requires sending poisoned chunks to any detected pirate requesting a protected file. If all clients simply deny download request without poisoning, the pirates can still accumulate clean chunks from colluders that are willing to share. With poisoning, pirates are forced to discard even clean chunks received.

2.3. Containment of Peer Collusion to Stage Piracy

We recognize that peer collusion is inevitable: a paid customer may intentionally collude with pirates; a pirate may also hack into client hosts and turn them into unwilling colluders. Our system is designed so that even with large number of colluders, a pirate will still suffer from intolerably long download time. We also present a random collusion detection mechanism to further enhance our system.

2.4. Trusted P2P Platform for Copyrighted

Content Delivery Hardware investment for P2P content delivery is much lower than that required in any existing CDNs. Our system only uses a few distribution agents to serve large number of clients. The system is highly scalable, robust to peer and link failures, and easily deployed. All claimed advantages are backed by performance analysis and simulation results.

3. Copyright-Protected P2P Networks

This section specifies the system architecture, client joining process, pirate poisoning mechanism, and colluder detection that we built in the newly proposed copyright-protection scheme.

3.1. Trusted P2P Network Architecture

A protected P2P content delivery network, consisting of paid clients, colluders, pirates, and distribution agents. The design goal is to prevent pirates from downloading copyrighted files from colluders. Proactive poisoning is applied to pirates only without hurting paid clients. Only a handful of agents are used to handle the bootstrap and distribution of requested digital contents.

To join the system, clients submit the requests to a transaction server which handles purchasing and billing matters.

A private key generator (PKG) is installed to generate private keys with identity-based signatures (IBS) for securing communication among the peers. The PKG has a similar role of a certificate authority (CA) in PKI services. The difference lies in that CA generates public keys distributed in IEEE 509 certificates, while PKG takes much lower overhead to generate private keys, which are used by local hosts.

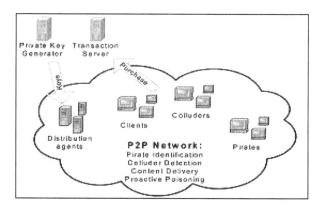

Fig 1. *A secured P2P platform for copyright-protected content delivery*

This open network is accessed by a large number of paid clients, some colluders or pirates, and a few distribution agents. The system design prevents pirates from downloading copyrighted files from colluders. The transaction server and PKG are only used initially when peers are joining the P2P network. With IBS, the communication between peers does not require explicit public key, because the identity of each party is used as the public key. In our system, file distribution and copyright protection are completely distributed. Based on past experience, the number of peers sharing or requesting the same file at any point of time is around hundreds. Depending on the variation of the swamp size, only a handful of distribution agents is needed. For example, it is sufficient to use 10 PC-based distribution agents to handle a swamp size of 2,000 peers.

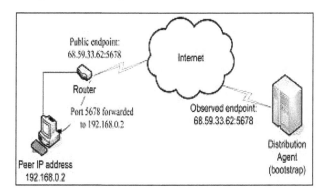

Fig 2. The bootstrap agent observes end-point address p= 68:59:33:62: 5678 in a trust-enhanced P2P network

Fig. 2 depicts an example: A peer has an IP address 192.168.0.2 leased from its local router. It is listening to port 5,678 forwarded by the router. When communicating with the bootstrap agent, the peer announces its listening port number. The bootstrap agent calls an Observe () subroutine, which verifies that the same peer is indeed reachable via the claimed port, although its public IP address is actually 68.59.33.62. Hence, the peer is identified by 68.59.33.62:5678. The detail of Observe () is as follows: when a peer sends message to its bootstrap agent through outgoing port, agent attaches a random number (nonce) in the reply. The agent then sends a message to the advertised listening port 68.59.33.62:5678, asking the peer to send back the nonce. If the peer replies correctly, then its endpoint is verified. The endpoint address is used as peer's public key. There is no need to encrypt the file body. This reduces the system overhead. Enabling peers behind NAT without a static listening port requires a hole-punching mechanism. The system uses the bootstrap agent to forward the incoming requests. The identities of all agents, except the bootstrap agent, are hidden from clients. This stops a malicious node to blacklist or attack the distribution agents.

4. System Implementation

In a P2P content distribution network, only the content owner can verify the user ID/password pair; peers cannot check each other's identity. Revealing a user's identity to other peers violates his or her privacy. To solve this problem, we developed a PAP protocol. First, we apply IBS to secure file indexing. Then we outline the procedure to generate tokens. Finally, we specify the PAP protocol that authorizes file access to download by peers. Secure File Indexing In a P2P file-sharing network, a file index is used to map a file ID to a peer endpoint address. When a peer requests to download a file, it first queries the indexes that

match a given file ID. Then the requester downloads from selected peers pointed by the indexes. To detect pirates from paid clients, we propose to modify file index to include three interlocking components: an authorization token, a timestamp, and a peer signature. Each legitimate client has a valid token assigned by its bootstrap agent. The timestamp indicates the time when a token expires. Thus, the peer needs to refresh the token periodically.

4.1. Protection in Peer Joining Process

For a peer to join the network it first logs in to a transaction server to purchase the content after transaction, the client receives a digital receipt containing the content title, client ID, etc. This receipt is encrypted such that only content owner and distribution agent can decrypt. The client receives the address of the bootstrap agent as its point of contact. The joining client authenticates with the bootstrap agent using the digital receipt. The session key assigned by the transaction server secures their communication. Since the bootstrap agent is set up by the content owner, it decrypts the receipt and authenticates its identity. The bootstrap agent requests a private key from PKG and constructs an authorization token, accordingly. Let k be the private key of content owner and id be the identity of the content owner. We use Ek(msg) to denote the encryption of message with key k. The Sk(msg) denotes a digital signature of plaintext msg with key k. The client is identified by user ID and the file by file ID. Each legitimate peer has a valid token. The token is only valid for a short time so that a peer needs to refresh the token periodically. To ensure that peers do not share the content with pirates, the trusted P2P network modifies the file-index format to include a token and IBS peer signature. Peers use this secured file index in inquiries and download requests. Seven messages are specified below to protect the peer joining process:

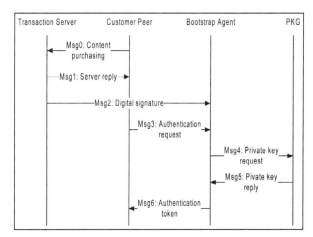

Fig 3. The protected peer joining process for copyrighted P2P content delivery. Seven messages are used to secure the communicationsamong four parties involved

The protected peer joining process for copyrighted P2P content delivery. Seven messages are used to secure the communications among four parties involved.

Seven Messages Are Specified Below To Protect The Peer Joining Process:
Msg0: Content purchase request;
Msg1: BootstrapAgentAddress, Ek (digital_receipt, Bootstrap-Agent_session_key);
Msg2: Adding digital signature Ek (digital_receipt);
Msg3: Authentication request with userID, fileID, Ek (digital_receipt);
Msg4: Private key request with privateKeyRequest (observed peer address);
Msg5: PKG replies with privateKey;
Msg6: Assign the authentication token to the client.

4.2. Pirate Identification

In a P2P content distribution network, only the content owner can verify the user ID/password pair; peers cannot check each other's identity. Revealing a user's identity to other peers violates his or her privacy. To solve this problem, we are using a PAP protocol. First, we apply IBS to secure file indexing. Then we outline the procedure to generate tokens. Finally, we specify the PAP protocol that authorizes file access to download by peers.

4.2.1. Secure File Indexing

In a P2P network, a file index is used to map a file ID to a peer endpoint address. When a peer requests to download a file, it first queries the indexes that match a given file ID. Then the requester downloads from selected peers pointed by the indexes. To detect pirates from paid clients, we propose to modify file index to include three interlocking components: an authorization token, a timestamp, and a peer signature. Each legitimate client has a valid token assigned by its bootstrap agent. The timestamp indicates the time when a token expires. Thus, the peer needs to refresh the token periodically. This short-lived token is designed for protecting copyright against colluders. The cost at each distribution agent to refresh the client tokens is rather limited, as shown via experiments. The peer signature is signed with the private key generated by PKG. This signature proves the authenticity of a peer. Download requests make explicit references to file indexes. The combined effects of the three extra fields ensure that all references to the file indexes are secured. Peers identify the pirates by checking the validity of the token and the signature in a file index. These features secure the P2P network operations to safeguard the sharing of clean contents among the paid clients.

4.2.2. Token Generation

First, both the transaction server and the PKG are fully trusted. Their public keys are known to all peers. The PAP protocol consists of two integral parts: token generation and authorization verification. When a peer joins the P2P network, it first sends authorization request to the bootstrap agent. All messages between a peer and its bootstrap agent are encrypted using the session key assigned by the

transaction server at purchase time.

The authorization token is generated by Algorithm 1 specified below. A token is a digital signature of a three-tuple: {peer endpoint, file ID, timestamp} signed by the private key of the content owner. Since bootstrap agent has a copy of the digital receipt sent by transaction server, verifying the receipt is thus done locally. The Decrypt (Receipt) function decrypts the digital receipt to identify the file L. The Observe (requestor) returns with the endpoint address p. The Owner Sign (λ; p; ts) function returns with a token. Upon receiving a private key, the bootstrap agent digitally signs the file ID, endpoint address, and timestamp to create the token. The reply message contains a four-tuple: {endpoint address, peer private key, timestamp, token}. The reply message from bootstrap agent is encrypted using the assigned session key.

Algorithm 1. Token Generation
Input: Digital Receipt
Output: Encrypted authorization token T
Procedures :
Step 01: if Receipt is invalid,
Step 02: deny the request;
Step 03: else
Step 04: λ = Decrypt(Receipt);
// λ is file identifier decrypted from receipt
Step 05: p = Observe(requestor);
// p is endpoint address as peer identity
Step 06: k = PrivateKeyRequest (p);
// Request a private key for user at p
Step 07: Token T = OwnerSign(f; p; ts)
// Sign the token T to access file f
Step 08: Reply = {k; p; ts;T} // Reply with key, endpoint address, timestamp, and the token
Step 09: SendtoRequestor {Encrypt(Reply)}
// Encrypt reply with the session key
Step 10: end if

4.3. Proactive Poisoning

The PAP protocol is formally specified below. A client must verify the download privilege of a requesting peer before clean file chunks are shared with the requestor. If the requestor fails to present proper credentials, the client must send poisoned chunks. In PAP, a download request applies a token T, file index ø, timestamp ts, and the peer signature S. If any of the fields are missing, the download is stopped. A download client must have a valid token T and signature S. Two pieces of critical information are needed: public key K of PKG and the peer endpoint address p. Algorithm 2 verifies both token T and signature S. File index ø(λ,p) contains the peer endpoint address p and the file ID λ. Token T also contains the file index information and ts indicating the expiration time of the token. The Parse (input) extracts timestamp ts, token T, signature S, and index ø from a download request. The function Match (T; ts, K) checks the token T against public key K. Similarly, Match (S; p) grants access if S matches with p.

Algorithm 2. Peer Authorization Protocol

Input: T = token, ts = timestamp, S = peer signature, and

$ø(\lambda,p)$ = file index for file λ at endpoint p

Output: Peer authorization status

True: authorization granted

False: authorization denied

Procedures :

Step 01: Parse (input) = {T; ts;S; $ø(\lambda,p)$}
// Check all credentials from a input request
Step 02: p = Observe(requestor);
// detect peer endpoint address p //
Step 03: if {Match (S; p) fails},
//Fake endpoint address p detected //return false;
Step 04: endif
Step 05: if {Match(T; ts;K) fails},
return false;
// Invalid or expired token detected //
Step 06: endif
Step 07: return true;

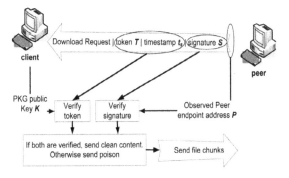

Fig 4. *The PAP enables instant detection of a pirate upon submitting an illegal download request.*

4.4. Data Flow

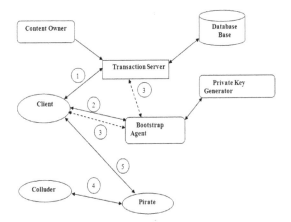

Fig 5. *Dataflow diagram*

There are four types of peers coexist in the P2P network: clients (honest or legitimate peers), colluders (paid peers sharing contents with others without authorization),

distribution agents (trusted peers operated by content owners for file distribution), and pirates (unpaid clients downloading content files illegally).

1. Client joining process
2. Get token from Bootstrap Agent
3. Download process
4. Pirate receive clean file chunks
5. Pirate receives poisoned chunks.

Clients are those that comply with the copyright law not to share contents freely. Pirates are peers attempting to download some content files without paying or authorization. The colluders are those paid clients who share the contents with pirates.

A private key generator (PKG) is installed to generate private keys with IBS for securing communication among the peers. The PKG has a similar role of a certificate authority (CA) in PKI services. The difference lies in the fact that CA generates the public/private key pairs, while PKG only generates the private key.

5. Proactive Content Poisoning

In this approach, modified file index format enables pirate detection. PAP authorizes legitimate download privileges to clients. Content distributor applies content poisoning to disrupt illegal file distribution to unpaid clients. The system is enhanced by randomized collusion detection. In our system, a content file must be downloaded fully to be useful. Such a restraint is easily achievable by compressing and encrypting the file with a trivial password that is known to every peer. This encryption does not offer any protection of the content, except to package the entire file for distribution.

Fig. 4 illustrates the proactive content poisoning mechanisms built in our enhanced P2P system. If a pirate sends download request to a distribution agent or a client, then by protocol definition, it will receive poisoned file chunks. If the download request was sent to a colluder, then it will receive clean file chunks. If a pirate shares the file chunks with another pirate, then it could potentially spread the poison. Therefore, it is critical to send poisoned chunks to pirates, not simply denying their requests. Otherwise, even if all clients deny pirate's requests, the pirate still can assemble a clean copy from those colluders who have responded with clean chunks. With poisoning, we exploit the limited poison detection capability of P2P networks and force a pirate to discard the clean chunks downloaded with the poisoned chunks. The rationale behind such poisoning is that if a pirate keeps downloading corrupted file, the pirates will eventually give up the attempt out of frustration.

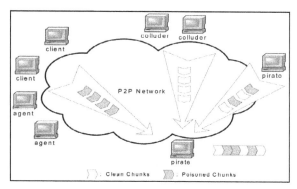

Fig 6. *Proactive poisoning mechanisms built in trusted P2P network,where clean chunks (white) and poisoned chunks (shaded) are mixed infile streams received by pirates, but legitimate clients receive only clean Chunks.*

6. Conclusion and Future Enhancement

Here we conclude the Authority System to Prevent Privacy Protection in P2P Network system to stop paid peers and unpaid peers from suspected copyright male faction in file sharing network. And also when a pirate is detected, the distributed agent sends the falsified chunk file to the particular pirate client with proper counteractive actions. Combining DRM and reputation system to protect P2P content delivery networks will lead to a total solution of the online piracy problem. There are many other forms of online or offline piracy that are beyond the scope of this study. For example, our protection scheme does not work on a private or enclosed network formed by pirate hosts exclusively. We did not solve the randomized piracy problems using email attachments, FTP download directly between colluders, or replicated CDs or DVDs. In future we can focus on prototyping and benchmark experiments which are needed in Real-Life Open P2P Networks Here we can only prove the protection concept, lacking of sustained accuracy. Proactive chunk poisoning can be made selectively to reduce the processing overhead. However, further studies are needed to upgrade the performance of the copyright-protected system in real-life P2P benchmark applications.

References

[1]   N. Anderson(Sept. 2007), "Peer-to-Peer Poisoners: A Tour of Media- Defender," Ars Technica.

[2]   S. Androutsellis-Theotokis and D. Spinellis (2004), "A Survey of Peerto- Peer Content Distribution Technologies," ACM Computing Surveys, vol. 36, pp. 335-371.

[3]   S. Chen and X.D. Zhang(May 2006), "Design and Evaluation of a Scalable and Reliable P2P Assisted Proxy for On-Demand Streaming Media Delivery," IEEE Trans. Knowledge and Data Eng., vol. 18, no. 5, pp. 669-682.

[4]   N. Christin, A.S. Weigend, and J. Chuang(2005), "Content Availability,Pollution and Poisoning in File-Sharing P2P Networks," Proc. ACMConf. e-Commerce, pp. 68-77.

[5]   E. Damiani, D.C. di Vimercati, S. Paraboschi, P. Samarati, and F. Violante(2002), "A Reputation-Based Approach for Choosing Reliable Resources in Peer-to-Peer Networks," Proc. ACM Conf. Computer and Comm. Security (CCS '02), pp. 207-216.

[6]   M. Fetscherin and M. Schmid(2003), "Comparing the Usage of Digital Rights Management Systems in the Music, Film, and Print Industry,"Proc. Conf. e-Commerce.

[7]   B. Gedik and L. Liu (June 2005), "A Scalable P2P Architecture for Distributed Information Monitoring Applications," IEEE Trans. Computers, vol. 56, no. 6, pp. 767-782.

[8]   [T. Kalker, D.H.J. Epema, P.H. Hartel, R.L. Lagendijk (June 2004), and M. Van Steen, "Music2share—Copyright-Compliant Music Sharing in P2P Systems," Proc. IEEE, vol. 92, no. 6, pp. 961-970.

[9]   B. Krishnamurthy, C. Wills, and Y. Zhang (Nov. 2001), "On the Use and Performance of Content Distribution Networks," Proc. Special Interest Group on Data Comm. on Internet Measurement Workshop (SIGCOMM).

17

Analysis of effectiveness of communication overheads in the parallel computing system using stochastic colored petri nets

Nguyen Minh Quy[1], Huynh Quyet Thang[2], Ho Khanh Lam[1]

[1]Faculty of Information Technology, Hung Yen University of Technology and Education, Hung Yen, Vietnam
[2]School of Information Communication and Technology, Hanoi University of Science and Technology, Ha Noi, Vietnam

Email address:

quyutehy@gmail.com (N. M. Quy), thanghq@soict.hust.edu.vn (H. Q. Thang), lamhokhanh@gmail.com (H. K. Lam)

Abstract: In architectures of parallel computing system, which has a large number of processing nodes, communication overhead is an important metric to evaluate and minimize by improving computation speedup solutions. In this paper, we propose using Stochastic Colored Petri Net to give models of parallel computing multi-processing systems for analyzing and evaluating effectiveness of communication overheads to system performance.

Keywords: SCPN, Parallel Computing System, Communication Overhead, Interconnecting Network

1. Introduction

We proposed the extension of the speedup formula of the Amdahl's law by including the communication overhead, and using the Closed Product Form Queueing Network (CPFQN) to analyze and evaluate effectiveness of the communication overhead (with changes of the data size, number of processing nodes and interconnecting network structures of processing nodes) on the speedup and performance of parallel computing multi-processing systems [1]. In this paper, we analyze and evaluate effectiveness of the communication overhead to performance of parallel computing system; another method is presented using Stochastic Colored Petri Nets (SCPN), in which transition times have an exponential distribution.

The Colored Petri Net (CPN) is a graphical oriented language for design, specification, simulation and verification of systems. It is in particular well-suited for systems in which synchronization, communication and resource sharing are important [2]. The CPN is a combination of Petri nets and programming language in which control structures, synchronization, communication and resource sharing are described by CPN. Data and data processing are described by functional programming language. Each CPN could be transformed to an equivalent PN and vice versa, i.e. if CPN has scalar data type (e.g.

integers, strings and real numbers) then equivalent PN could be unlimited.

A CPN is defined as follow [3]:

$$CPN = (\Sigma, P, T, A, N, C, G, E, I)$$

Where:

Σ - Finite set of non-empty types, also called finite and non-empty colour sets.

P - Finite set of places.

T - Finite set of transitions.

A - Set of arcs such that $P \cap T = P \cap A = T \cap A = \phi$.

N- Node function, defined from a into "colour on arcs" $P x T \cup T x P$, is written by $N : A \rightarrow P x T \cup T x P$.

$C : P \rightarrow \Sigma$ - colour function assigned to each place $p \in P$ one colour set $C(p) \in \Sigma$.

$G : T \rightarrow EXP$ - Guard function assigned to each transition $t \in T$ one guard expression as "Boolean function of probability" and is written as:

$\forall t \in T : [Type(g(t)) = Boolean \wedge Type(Var(g(t))) \subseteq \Sigma]$ where $Type(Vars)$ is set of types $\{Type(v) | v \in Vars\}$, $Vars$ is set of variations, $Vars(G(t))$ are variations in $g(t)$.

$E : F \rightarrow EXP$- An arc expression function assigned to each arc $a \in A$ one expression including multi-set

$C(p)_{MS}$:

$$\forall a \in A : \left[Type(E(a)) = C(p(a))_{MS} \wedge Type(var(E(a))) \subseteq \Sigma \right]$$

where, $p(a)$ is the place of $N(a)$ and $C(p)_{MS}$ denotes the set of all multi-sets over C

$I : P \rightarrow EXP$ – An initialization function assigned to each place $p \in P$ one expression including multi-set $C(p)_{MS}$:

$$\forall p \in P : \left[Type(I(p)) = C(p)_{MS} \wedge Var(I(p)) = \phi \right]$$

Graphical representation of CPN consists of ellipses and circles denoted places which describe states of the system (buffers). The rectangles designate transitions which describe actions (processes). The arrows designate arcs. Expressions of arcs describe variable status of CPN when transitions occur. Each place contains a set of marker called tokens. In the CPN, each of these tokens carries a data value, which belongs to a given type and could be different from other one, while all tokens are the same in common Petri nets. CPNs create condensation network models by using definitions of colours. In the SCPN, there are immediate transitions and transitions associated with delay times of firing, that are called timed transitions.

Nowadays, there are some researches that used CPN to model, analyze and evaluate many systems in different fields, especially on information technology and communication. Recently, CPN was used to model parallel systems in [4], [5] and [6]. However, communication overhead problem of modern parallel computing systems is slightly assessed. Kotov et al. [7] presented a mesh network structure and used Design/CPN tool to analyze and evaluate. In this study, we present the cluster structure of processing nodes to analyze and evaluate since the cluster structures are typical in architectures of parallel computing systems.

2. SCPN Model of Parallel Computing Multi-Processing System

There are some versions for modeling tool of CPN such as CPN tools (Aarhus University, Denmark) and TimeNet (Armin Zimmermann and Knoke, Technical University of Berlin, Germany). We used TimeNet 4.0.2 tool to model the parallel computing multi-processing system. We chose the cluster interconnection parallel system with commonly used configurations 2Dtorus, in which each processing node connects to four adjacent nodes (degree of node is four). The present model consists of five processing nodes: node P0 connects to four adjacent nodes P1, P2, P3 and P4. The structure and technology of each node Pi (i=0,1,2,3 and 4) are identical as shown in Figure 1.

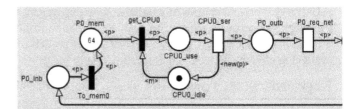

Figure 1. *SCPN of processor*

Table 1. *List of places in processor*

Name	Token type	Token list	Queue	Meaning
P0_mem0	int	8, 16, 32, 64	Random	Processors: number of initial tokens representing number of data packet with content =1 for P0 (expression on output arc p == 1)
CPU0_use	int		Random	CPU processes data packet
CPU0_idle	int	1	Random	Idle CPU, initial state is established by one token
P0_outb	int		Random	Output buffer of processor, initial state is empty
P0_inb	int		Random	Input bufer of processor, initial state is empty

Table 2. *Token list of processor has identical content for all data packets (P0: content = 1, P1: content =2, P2: content =3, P3: content = 4, and P4: content = 5).*

Tokenlist	
Number	Content
1	1.0
2	1.0
3	1.0
4	1.0
5	1.0

Tokenlist	
Number	Content
6	1.0
7	1.0
8	1.0
9	1.0
10	1.0
11	1.0
12	1.0
13	1.0

Table 3. *Transitions having firing delay of processor*

Name	timeFunction	localGuard	Meaning
CPU0_ser	EXP (CPU_service_time)	P	Time for completely processing data packet of CPU (CPU_service_time)
P0_req_net	EXP (net_access_time)	P	Time for assessing communication request from processor to interconnect network

Table 4. *Immediate transitions of processor (time delay = 0)*

Name	Priority	weight	localGuard	Meaning
get_CPU0	1	1.0E0	p	Immediately fires as having data packet in memory
To_mem0	1	1.0E0	p	Immediately fires as communication from other processor across interconnect network

Model of cluster interconnect network is presented in *Figure 2.*

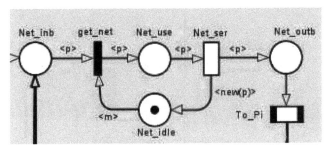

(a) SCPN of interconnect

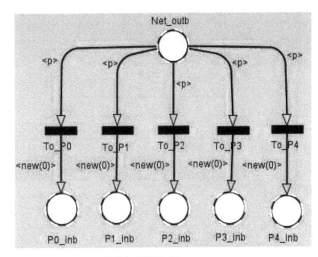

(b) Sub-SCPN of Interconnect

Figure 2. *SCPN of interconnecting network*

Table 5. *List of places of processor*

Name	Token type	Tokenlist	Queue	Meaning
Net_inb	int		Random	Input buffer of interconnect
Net_use	int		Random	Interconnect processes data packet
Net_idle	int	1	Random	Idle interconnect, initial state is established by one token
Net_outb	int		Random	Output buffer of Interconnectr, initial state is empty

Table 6. *Transitions having firing delay*

Name	timeFunction	localGuard	Meaning
Net_ser	EXP(net_server_time)	p	Communication time across interconnect (net_server_time)

Table 7. *Immediate transition of interconnect (time delay = 0)*

Name	Priority	weight	localGuard	Meaning
get_net	1	1.0E0	p	fires as having request of communication interconnect
To_P0	1	1.0E0	p	fires as having token at output buffer of interconnect: transition of data packet to processors

In the parallel computation, a processor node locally implements some tasks such as computation and sends of computing results to other processors that could be adjacent or remote depending on parallel algorithm of applications.

Therefore, the data packets from a processor will be sent to other processors without return itself. The communication overhead could be smallest if sending to adjacent processors or average value based on H- average routing distance (average hop count) of each configuration. For example, with 2Dtorus H = √n/2, *n* is node number of processors. Node P0 only transfers data packet across network to P1, P2, P3 and P4. Figure 3 presents SCPN of parallel computing system with 2Dtorus cluster interconnect.

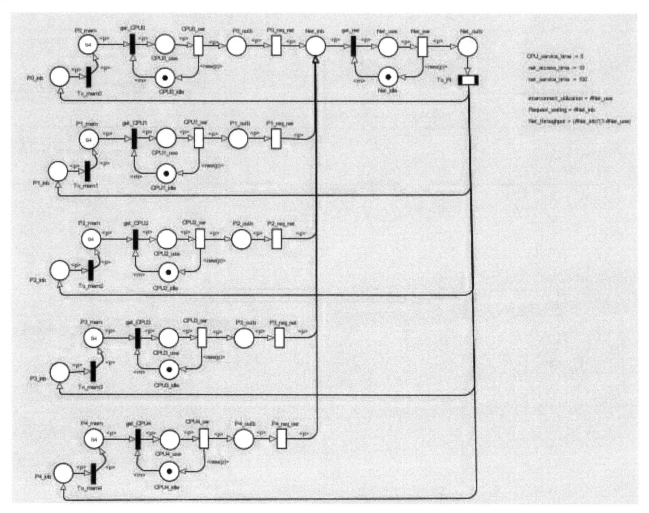

Figure 3. SCPN of the cluster 2Dtorus interconnection parallel computing system

In this model, initial states of places P0_mem (P1_mem, P2_mem, P3_mem and P4_mem) of respective processor nodes (P0, P1, P2, P3 and P4) have been established with number of tokens 8, 16, 32 and 64 (preset numbers of packets) for simulation scenarios. The firing of immediate transitions get_CPU0 (get_CPU1, get_CPU2, get_CPU3 and get_CPU4) presents that the data packets of an application are sent to the CPU of the processor node for processing. Places CPU0_use (CPU1_use, CPU2_use, CPU3_use and CPU4_use) have tokens present processing states of respective CPUs. Processing delay time, CPU_service_time, of processors is exponential distributed and is defined by value EXP (CPU_service_time). The data packets, which need to send to processors across interconnect, are transferred to output buffer by the processor which present by places P0_outb (P1_outb, P2_outb, P3_outb and P4_outb). From output buffers, processors sent to interconnect and are defined by interconnect access delay, net_access_time, that is exponential distribution, defined by value EXP(net_access_time). The communication requests from processors are input to input buffer of interconnect represented as place Net_inb. Normally, interconnect is connected to buffers by switches or routers to process information routing. When Net_inb has communication request, get_net transition is activated and interconnect communicates data packets (information routing, transition of data packets). Status of communicating execution of interconnect is represented by Net_use position. Net_ser transition has communication delay (or activate delay), net_service_time, that is an exponential distribution, EXP(net_service_time). After communication delay across interconnect, the data packets are transferred to output buffer of interconnect and To_P0 transitions (To_P1, To_P2, To_P3 and To_P4) (in Figure 2b) are immediately activated to transfer the data packets to input buffers of target processor nodes, i.e. P0_inb positions (P1_inb, P2_inb, P3_inb and P4_inb), respectively. At processor nodes, occurrence of token at P0_inb position (P1_inb, P2_inb, P3_inb and P4_inb) makes To_mem0 transitions

(To_mem1, To_mem2, To_mem3 and To_mem4) immediately activated. The data packets are transferred to respective memories. Communication procedure among processor nodes in the parallel system finishes with a data packet. In the present study, data packet size is not important since the number of data packet transferred at a setting period is effective to the communication overhead. Moreover, the number of processing node, n, interconnect configuration and data packet size are effective to parameters: net_access_time, net_service_time. These parameters will be changed in simulation scenarios.

The functions, <p>, at arcs ensure that tokens are transferred to respective tokens in processor positions with the same content. The functions, <m>, at arc ensure new free status for intranet and interconnect networks. The functions, new (0), determine completely communicating implementation of requests from processors. Performance parameters of this model are determined and presented in Table 8.

Table 8. Performance parameters

Parameters	Value	Meaning
Interconnect_utilization	#Net_use	Utilization of interconnect network is number of token (number of requests of served processors) at Net_use position. Utilization of interconnect is high then communication overhead across interconnect is too high and throughput is decreased.
Request_waiting	#Net_inb	Number of token having at Net_inb position presents communication requests (number of data packets) from processor waiting communication acquisition of Interconnect.
Net_throughput	(#Net_inb)*(1-#Net_use)	Throughput of interconnect

3. Results and Discussion

For simulation, there are two following scenarios:

Scenario 1: set CPU_service_time:=5, net_access_time:=10, net_service_time:=100 and processors have 8, 16, 32 and 64 data packets (number of tokens). Determine performance parameters at maximum, mean and minimum values for Interconnect_utilization, Request_waiting and Net_throughput. Results simulated on TimeNet4.0.2 are represented in Figs. 4, 6, 8 and 10 with period of second. The ordinate axis presents performance parameters and the horizontal axis presents modeling time in second.

Scenario 2: set CPU_service_time:=5, net_access_time:=10, net_service_time:=300 and processors have 8, 16, 32 and 64 data packets (number of tokens). Determine performance parameter at maximum, mean and minimum values for Interconnect_utilization, Request_waiting and Net_throughput. Results simulated on TimeNet4.0.2 are represented in Figs. 5, 7, 9 and 11 with period of second. The ordinate axis presents performance parameters and the horizontal axis presents modeling time in second.

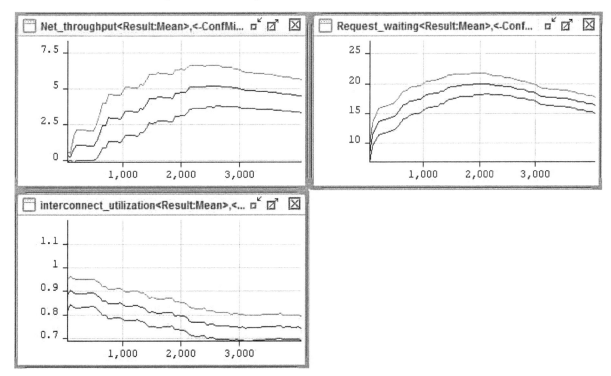

Figure 4. *Scenario 1 with 8 data packets, CPU_service_time=5, net_access_time=10 and*

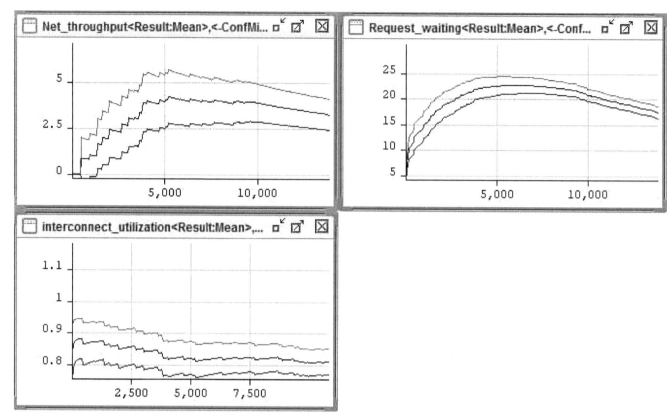

Figure 5. Scenario 2 with 8 data packets, CPU_service_time=5, net_access_time=10 and net_service_time=300.

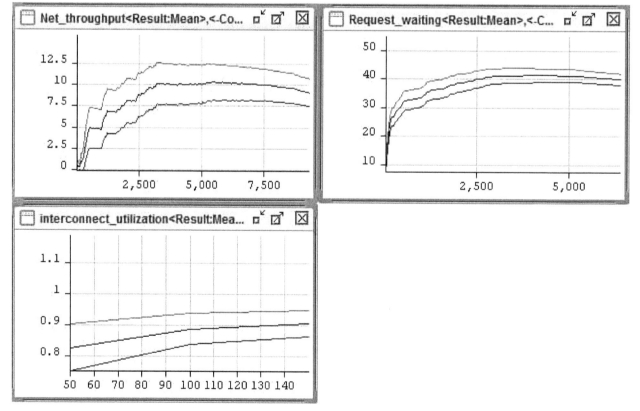

Figure 6. Scenario 1 with 16 data packets, CPU_service_time=5 and net_access_time=10, net_service_time=100.

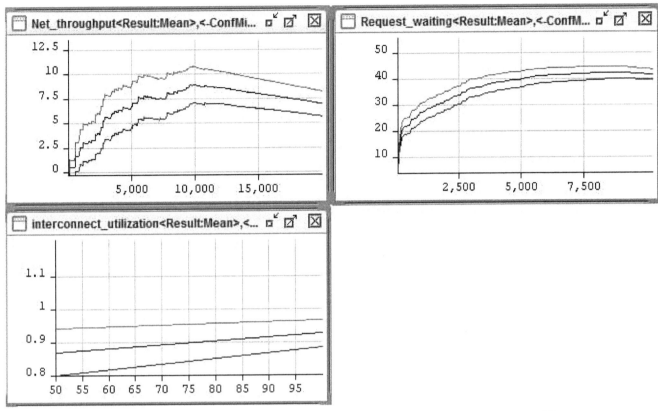

Figure 7. Scenario 2 with 16 data packets, CPU_service_time=5, net_access_time=10 and net_service_time=300.

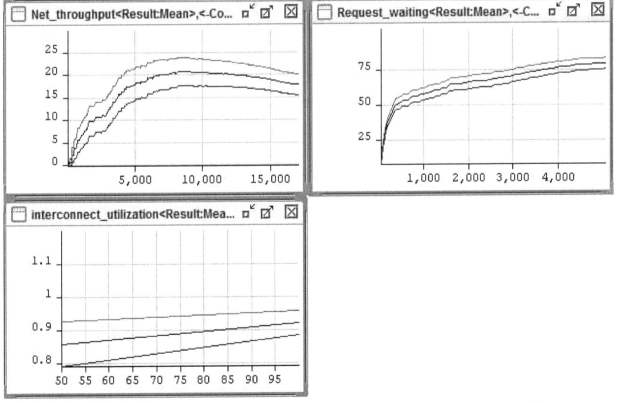

Figure 8. Scenario 1 with 32 data packets, CPU_service_time=5, net_access_time=10 and net_service_time=100.

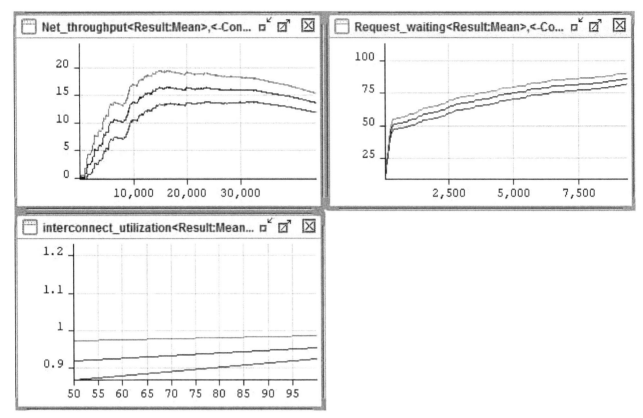

Figure 9. Scenario 2 with 32 data packets, CPU_service_time=5, net_access_time=10 and net_service_time=300.

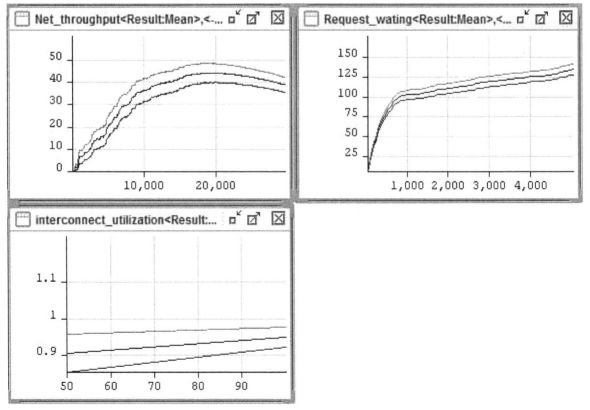

Figure 10. Scenario 1 with 64 data packets, CPU_service_time=5, net_access_time=10 and net_service_time=100.

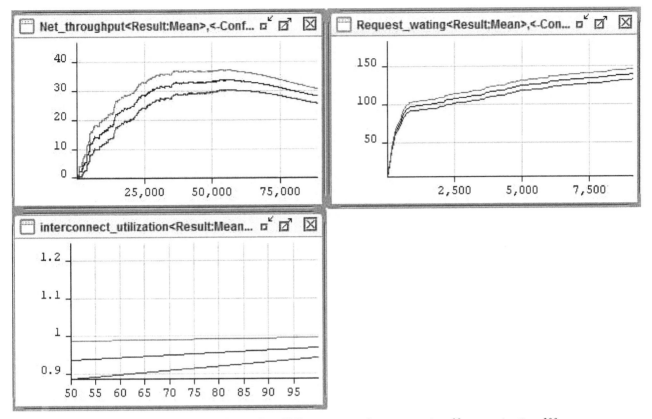

Figure 11. Scenario 2 with 64 data packets, CPU_service_time=5, net_access_time=10, net_service_time=300.

From simulation results of SCPN, with a large number of data packets, the utilization is large and numbers of data packets that must be wait are too large. The throughput is decreased as keeping mean values of CPU_service_time, net_access_time and net_service_time. When the communication delay time of interconnect (net_service_time) increases, performance of interconnect is decreased. By using SCPN, models of SCPN for any network interconnect structures could be established and their performance parameters could be also assessed. In the present study, a model of SCPN for 2Dtorus structure is presented to illustrate SCPN using in assessing performance of multi-processing parallel computing systems.

Our purpose is putting out the method using SCPN for assessing performance of multi-processing parallel computing systems depended on communication overhead of interconnect network of processing nodes, in which topologies of interconnect network and number of data packets transmitted between processing nodes are considered.

4. Conclusions

In the present study, we have successfully given models of parallel computing multi-processing systems for analyzing and evaluate effectiveness of communication overheads to system performance using Stochastic Colored Petri Net. The cluster structure of processing nodes have been also analysed and evaluated. A model of SCPN for 2Dtorus structure was used to illustrate SCPN using in

assessing performance of multi-processing parallel computing systems. With a large number of data packets, the utilization is large and numbers of data packets that must be waiting are too large. The throughput is decreased as keeping mean values of CPU_service_time, net_access_time and net_service_time. When the communication delay time of interconnect increases, performance of interconnect is decreased. We could have a more complicated SCPN for a multi-processing parallel computing system if the MPI protocol commonly used in parallel architectures is considered, and select some types of applications for simulation as matrix multiplication, image processing.

References

[1] Nguyen Minh Quy, Ho Khanh Lam, Huynh Quyet Thang, "Analysis of Effectiveness of Communication Overheads in the Parallel Computing System Using the Closed Product Form Queuing Network", RIVF-2013: The 10th IEEE RIVF International Conference on Computing and Communication Technologies, Hanoi, Vietnam, 10-13 November 2013, pp. 131-134.

[2] K. Jensen, "An Introduction to the Theoretical Aspects of Coloured Petri Nets". Lecture Notes in Computer Science vol. 803, Springer-Verlag 1994,230-272.

[3] K. Jensen,"Coloured Petri Nets. Basic Concepts, Analysis Methods and Practical Use". Monographs in Theoretical Computer Science, Springer-Verlag, 2nd corrected printing 1997, ISBN: 3-540-58276-2.

[4] Homayun Motameni, Zohre Ramezani and Zahra Usefi, "Modeling and Simulation of Parallelism by Colored Petri Nets". World Applied Sciences Journal 19 (5): 710-713, 2012. ISSN 1818-4952; © IDOSI Publications, 2012.

[5] Stanislav Böhm, Marek Běhálek, "Usage of petri nets for high performance computing". FHPC '12 Proceedings of the 1st ACM SIGPLAN workshop on Functional high-performance computing. Pages 37-48. ACM New York, NY, USA ©2012. ISBN: 978-1-4503-1577-7.

[6] Bin Cheng, Weiqin Tong, and Xingang Wang, "Hybrid Performance modeling and analyzing of parallel systems". International Journal of numerical analysis and modeling, Volume 9, Number 2, Pages 232-246, © 2012 Institute for Scientific Computing and Information.

[7] Gianfranco Ciardo, Ludmila Cherkasova, Vadim Kotov, and Tomas Rokicki, "Modeling A Scalable High-Speed Interconnect with Stochastic Petri Nets". Department of Computer Science College of William and Mary USA, Hewlett-Parkard Labs. 1994.

18

Complex modeling of matrix parallel algorithms

Peter Hanuliak

Dubnica Technical Institute, Sladkovicova 533/20, Dubnica nad Vahom, 018 41, Slovakia

Email address:

phanuliak@gmail.com

Abstract: Parallel principles are the most effective way how to increase performance in parallel computing (parallel computers and algorithms too). In this sense the paper is devoted to a complex performance evaluation of matrix parallel algorithms (MPA). At first the paper describes the typical matrix parallel algorithms and then it summarizes common properties of them to complex performance modeling of MPA. To complex performance analysis we are able to take into account all overheads influence performance of parallel algorithms (parallel computer architecture, parallel computation, communication etc.). To be le to analyze MPA in their abstract form we have defined needed decomposition models of MPA. For these decomposition strategies we derived analytical relation for defined complex performance criterions including isoefficiency functions, which allow us to predict performance although for hypothetical parallel computer. In its experimental part the paper considers the achieved results using defined complex performance criterions including issoefficiency function for performance prediction also for hypothetical future parallel computers. Such idea of common abstract analysis could be very useful in deriving complex performance criterions for groups of other similar parallel algorithms (PA) as for example numerical integration PA, optimization PA etc.

Keywords: Parallel Computer, NOW, Grid, Parallel Algorithm (PA), Matrix PA, Decomposition Model, Performance Modeling, Optimization, Overhead Function H (S, P), Inter Process Communication IPC, Performance Prediction, Issoeficiency Function

1. Trends in Parallel Computing

Basic common properties in parallel computing (parallel computers, parallel algorithms) computing, which are reaching continuous demands to performance acceleration, are as follows

- embedded parallel principles on various levels of technical (hardware) and program support means (software) [8]
- using of homogenous shared resources so in computing nodes of parallel computers (processors, cores, computers) as in parallel algorithms too [24]
- using of high speed communication networks reducing communication latency [39]
- increased client/server computing on symmetrical multiple processors or cores (SMP)
- trends to unified modeling of parallel computers (shared memory, distributed memory) and in parallel algorithms (shared memory, distributed memory, hybrid)
- continuous demands to increase mobility and data migration [23]
- the development of hardware neutral parallel programming language, such as Java, provides a virtual computational environment in which computing nodes of parallel computer appear to be homogenous
- continuous improvements in network technology and communication middleware in order to use shared parallel resources in unified manner (cloud computing, Internet computing).

Current trends also in high performance computing (HPC) are to use networks of workstations (NOW, SMP) as a cheaper alternative to traditionally used massively parallel multiprocessors or supercomputers and to profit from unifying of both mentioned disciplines [19]. The individual powerful computing nodes (workstations) could be so single personal computer (PC) as parallel computers based on modern SMP parallel computers implemented within computing node of parallel computer [13, 15]. Based on such modular NOW modules there were are realized high integrated massive parallel computers named as Grid systems [38]. A member of NOW module or Grid could be any classic supercomputers [35].

2. Performance Evaluation in Parallel Computing

To performance evaluation of parallel computers and parallel algorithms we can use   evaluation methods as follows
- analytical
  - application of queuing theory results [11, 21]
  - order (asymptotic) analyze [12, 20]
  - Petri nets [7]
- simulation methods [25]
- experimental
  - benchmarks [28]
  - modeling tools [32]
  - direct measuring [9, 30].

Analytical method is a very well developed set of techniques which can provide exact solutions very quickly, but only for a very restricted class of models. For more general models it is often possible to obtain approximate results significantly more quickly than when using simulation, although the accuracy of these results may be difficult to determine.

Simulation is the most general and versatile means of modeling systems for performance estimation. It has many uses, but its results are usually only approximations to the exact answer and the price of increased accuracy is much longer execution times. They are still only applicable to a restricted class of models (though not as restricted as analytic approaches.) Many approaches increase rapidly their memory and time requirements as the size of the model increases.

Evaluating system performance via experimental measurements is a very useful alternative for computer systems. Measurements can be gathered on existing systems by means of benchmark applications that aim at stressing specific aspects of computers systems. Even though benchmarks can be used in all types of performance studies, their main field of application is competitive procurement and performance assessment of existing systems and algorithms.

3. Parallel Algorithms

In principal we can divide parallel algorithms (PA) to the following groups
- parallel algorithm using shared memory (PA_{sm}). These algorithms are developed for parallel computers with shared memory as actual modern symmetrical multiprocessors (SMP) or multicore systems on motherboard
- parallel algorithm using distributed memory (PA_{dm}). These algorithms are developed for parallel computers with distributed memory as actual NOW system and their higher integration forms named as Grid systems
- hybrid PA which combine using of both previous PA (PA_{hyb}). This trend support applied using of NOW consisted from computing nodes based on SMP parallel computers.

The main difference between PA_{sm} and PA_{dm} is in form of inter process communication (IPC) among created parallel processes [18, 33]. Generally we can say that IPC communication in parallel system with shared memory can use more communication possibilities (all the possibilities of communication in shared memory) than in distributed systems (only network communication).

2.1. Developing Steps of PA

The role of programmer is for the given parallel computer and for given application problem to develop the effective parallel algorithm. This task is more complicated in those cases, in which we have to create the conditions for any parallel activities in form of dividing the sequential algorithm to their mutual independent parts named parallel processes. Principally development of any parallel algorithms (shared memory, distributed memory, hybrid) includes performing of the following activities [29, 34].
- decomposition of a complex problem to a set of parallel processes including their data (decomposition model)
- mapping – distribution of decomposed parallel processes to computing nodes of used parallel computer
- inter process communication (IPC) to cooperation (data communications, synchronization, control) of performed parallel processes
- performance optimization (tuning) of developed parallel algorithm (effective PA).

The most important step is to choose for given complex problem optimal decomposition model. To do this there is necessary to understand given complex problem, shared data, applied sequential algorithms (SA) and the flow of SA control [4, 26].

3.1.1. Decomposition Models

Decomposition model defines distribution of given complex problem to its independent parts (parallel processes) in such a way, that they could be performed in a parallel way via computing nodes of used parallel computer. Optimal selection of decomposition model and degree of parallelism are critical conditions to develop effective parallel algorithm. Potential decomposition of given complex problem is crucial for effectiveness of parallel algorithm [16]. The chosen decomposition model then drives the rest of effective parallel program development. This is true is in case of developing new applied PA as in porting serial code. The decomposition model defines structure of PA codes and their data and estimate the optimal topology of needed communication network [27, 31]. The existed decomposition models we have been analyzed in [16].

3.1.2. Mapping

This step allocates created parallel processes to computing nodes of parallel computer for their parallel executions. There is necessary to achieve that every computing node should perform allocated parallel processes (one or more) with at least approximate input loads (load balancing) on real assumption of equal powerful computing nodes. Fulfillment of this condition contributes to optimal parallel execution time.

3.1.3. Inter process Communication

Inter process communication (IPC) represents a needed tool to cooperation of decomposed parallel processes. In general we can say that dominated parts of parallel algorithms are decomposed parallel processes (independent sequential parts) and inter process communication (IPC) among created parallel processes in performing of PA. We have been analyzed IPC communication in detail in [18].

3.1.4. Performance Optimization

After verifying developed parallel algorithm on used parallel computer the further step is performance modeling and its optimization in order to develop effective PA. This step contents analysis of previous steps in such a way to minimize whole execution time latency of parallel computing T(s, p). Performed optimization of T(s, p) for given parallel algorithm depends mainly from following factors

- allocation of balanced input load to used computing nodes of parallel computer (load balancing) [1, 36]
- minimization of accompanying overheads amounts (parallelization, IPC, synchronization control of PA) [14, 22].

To do load balancing we need in case of obvious using of equally powerful computing nodes of PC results of load allocation for given developed PA. In dominated parallel computers (NOW, Grid) there are necessary to reduce (optimize) mainly number of inter process communications IPC (communication complexity) for example by considering of alternative existing decomposition model.

3.2. Complex Performance Evaluation Metrics

To evaluating parallel algorithms we have been defined in [14] complex performance criterions of PA. Tradeoffs among these performance factors are often encountered in real applied PA. We summarize these criterions as follows

- complex parallel execution time T(s, p) including overhead function h(s, p)
- complex speed up S(s, p)
- complex efficiency E(s, p)
- issoeficiency w(s).

4. Typical Matrix Parallel Algorithms

Some of the typical matrix parallel algorithms we have been yet analyzed as follows

- parallel matrix multiplication [16]
- parallel fast discrete Fourier Transform (DFFT) in [14].

We will short describe further typical MPA.

4.1. System of Linear Equations

System of n linear equations (SLE) with n variables x_1, $x_2, x_3, ..., x_n$, in matrix form is defined as follows [3, 10]

$$A \cdot X = B$$

where the matrix A is a square matrix of coefficients, B is the vector of the right side and X is a vector of searching unknown as follows

$$A = \begin{pmatrix} a_{11}, a_{12}, a_{13}, \dots a_{1n} \\ a_{21}, a_{22}, a_{23}, \dots a_{2n} \\ \dots \\ \dots \\ \dots \\ a_{n1}, a_{n2}, a_{n3}, \dots a_{nn} \end{pmatrix} \quad B = \begin{pmatrix} a_{1,n+1} \\ a_{2,n+1} \\ \cdot \cdot \\ \cdot \cdot \\ a_{n,n+1} \end{pmatrix} \quad X = \begin{pmatrix} X_{1n} \\ X_2 \\ \cdot \cdot \\ \cdot \cdot \\ X_n \end{pmatrix}$$

4.1.1. Methods of SLE Solving

There is no known universal optimal method of solving systems of linear equations. There are several different ways of solving SLR whereby each of them at fulfillment of defined assumption implies the option of the solution method. In principle, we divide the available methods for exact (finite) and iterative. There exist many various ways how to solve system of linear equations. But there does not exist any optimal way of solving it. The existed methods can be divided into

- exact
  - Cramer rule
  - Gaussian elimination methods (GEM)
  - GEM alternatives
- iterative

4.1.2. Typical Decomposition Models

To parallel solution of SLE by preferred Gauss eliminated method (GEM) the decomposition models are as follows [16]

- allocation of block strips
  - gradually allocation of strips.

In the first allocation method strips are divided to set of strips and to every computing node is assigned one block. Illustration of these allocation methods is at Fig. 1.

Number of columns

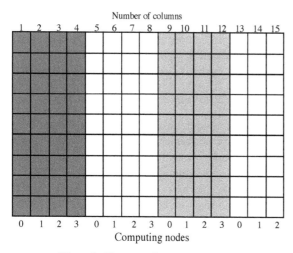

Computing nodes

Figure 1. Allocation of matrix data blocks.

Another alternative decomposition model with gradual allocation of columns they are allocated columns to individual computing nodes like the card are gradually passing out at games to game participants. Illustration of gradual assignment of columns is at Fig. 2.

Number of columns

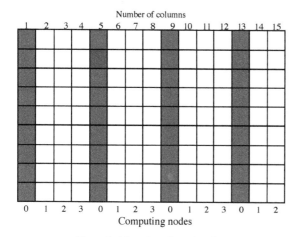

Computing nodes

Figure 2. Allocation of matrix columns.

4.2. Partial Differential Equations

Partial differential equations (PDE) are the equation involving partial derivates of an unknown function with respect to more than one independent variable. PDEs are of fundamental importance in modeling all types of continuous phenomena in nature. Typical examples are weather forecasting, optimization of aerodynamic shapes, fluid flow, and the like. Simple PDE can be solved directly, but in general it is necessary to approximate the solution on the extensive network of final points by iterative numerical methods [18]. We will confine our attention to PDE with two space independent variables x, y. The needed function we denote as u (x, y). The considered partial derivations we denote as u_{xx}, u_{xy}, u_{yy} etc. For practical use the most important PDE are two ordered equations as follows

- heat equation, $u_t = u_{xx}$
- wave equation, $u_{tt} = u_{xx}$

- Laplace equation $u_{xx} + u_{yy} = 0$.

These three types are the basic types of general linear second order PDR as in follows $u_{xx} + b\ u_{xy} + c\ u_{yy} + d\ u_x + e\ u_y + f\ u + g = 0$. This equation could be transformed by changing the variables to one of three basic equations, including the members of the lower rows, provided that the coefficients a, b, c are not all equal to zero. Variable $b^2 - 4\ ac$ is refer to as discriminant whereby its value determines the following basic groups PDR of second order

- $b^2 - 4\ ac > 0$, hyperbolic (typical equation for waves).
- $b^2 - 4\ ac = 0$, parabolic (typical of heat transfer)
- $b^2 - 4\ ac < 0$, elliptical (typical is the Laplace equation) [6, 37].

Classification of more general types of PDE is not so clear. When the coefficients are variable, then the type of equation can be modified by changes in the analyzed area and if it is intended at the same time with several equations, each equation can generally be of a different type. Simultaneously analyzed problem may be nonlinear or equation requires more than second order [2, 23]. Nevertheless, the basic used classification of PDE is also used when determining if it is not accurate. Specifically the following types of PDEs are as follows

- hyperbolic. This group is characterized by time dependent processes that are not stabilized at some steady state
- parabolic. Group characterized by a time dependent processes, which tend to the stabilization
- elliptical. Group describes the processes that have reached steady state and are therefore time independent. A typical example is the Laplace equation.

Here we show how to solve in parallel way specific PDE – Laplace equation in two dimensions – by means of a grid computation method that employs finite difference method. Although we focus on this specific problem, the same techniques are used for solving other PDE (Laplace - three dimensional, Poisson equation etc.), extensive approximations calculations on various parallel computers (supercomputers, massive, SMP, NOW, Grid) eventually solving another similar complex problems.

4.2.1. Parallel Application of Iterative Algorithms

Here we show how to solve in parallel way specific PDE – Laplace equation in two dimensions – by means of a grid computation method that employs finite difference method. Although we focus on this specific problem, the same techniques are used for solving other PDE (Laplace - three dimensional, Poisson equation etc.), extensive approximations calculations on various parallel computers (supercomputers, massive, SMP, NOW, Grid) eventually solving another similar complex problems. Laplace equation is a practical example of using iterative methods to its solution. The equation for two dimensions is following

$$\frac{\delta^2\Phi}{\delta x^2}+\frac{\delta^2\Phi}{\delta y^2}=0$$

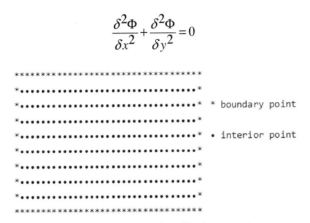

Figure 3. Grid approximation of Laplace equation.

Function $\Phi(x, y)$ could represent some unknown potential, such as heat, stress etc. Given a two-dimensional region and values for points of the region boundaries, the goal is to approximate the steady-state solution $\Phi(x, y)$ for points in the interior by the function $u(x, y)$. We can do this by covering the region with a grid of points (Fig. 1) and to obtain the values of $u(x_i, y_j) = u_{i,j}$.

Let us consider square region (a, b) x (a, b). For coordinates of grid points is valid $x_i = i*h$, $y_j = j*h$, $h = (b-a) / N$ for $i,j = 0, 1, ..., N$. We replace partial derivations of Φ ~ $u(x, y)$ by the differences of $u_{i,j}$. After substituting we obtain final iteration formulae as

$$X_{i,j}^{(t+1)} = (X_{i-1,j}^{(t)} + X_{i+1,j}^{(t)} + X_{i,j-1}^{(t)} + X_{i,j+1}^{(t)}) / 4$$

orits alternative version

$$X_{i,j}^{(t+1)} = (4 X_{i,j}^{(t)} + X_{i-1,j}^{(t)} + X_{i+1,j}^{(t)} + X_{i,j-1}^{(t)} + X_{i,j+1}^{(t)}) / 8$$

Each interior point is initialized to some value. The steady-state values of the interior points are then computed by repeated iterations. In each iteration the new point value is set to a combination of the previous values of neighboring points. The computation terminates either after a given number of iterations or when every new value is within some acceptable difference Epsilon > 0 of the previous value.

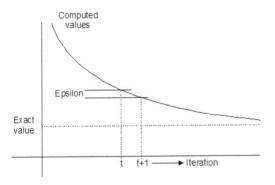

Figure 4. Convergence rate.

Illustration of convergence rate of iterative parallel algorithms is at Fig. 4. For the convergence of Gauss-Seidel iterative method is valid the same conditions as for Jacobi iterative method whereby the Gauss-Seidel method converges faster. Given condition is not only necessary but only a sufficient one for the convergence of both methods. In practice, we use both iterative methods also in case of not satisfying of this condition based on them that convergence is influenced also by selection of initial vector.

4.2.1.1. Communication Model

For Jacobi finite difference method a two-dimensional grid is repeatedly updated by replacing the value at each point with some function of the values at a small fixed number of neighboring points. The common approximation structure uses a four-point stencil to update each element $X_{i,j}$ (Fig. 5.).

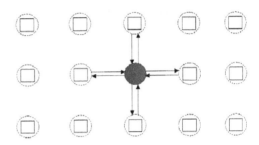

Figure 5. Communication model for 4 - points approximation

Similar for more accurate value of any point we can also used more precise multipoint approximation relation, and that for example through approximation of nine points according the pencil at Fig. 6 with following relation

$$X_{i,j}^{(t+1)} = (16 X_{i-1,j}^{(t)} + 16 X_{i+1,j}^{(t)} + 16 X_{i,j-1}^{(t)} + 16 X_{i,j+1}^{(t)}$$
$$- X_{i-2,j}^{(t)} - X_{i+2,j}^{(t)} - X_{i,j-2}^{(t)} - X_{i,j+2}^{(t)}) / 60$$

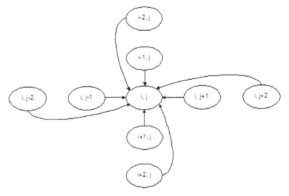

Figure 6. Stencil with nine points.

5. Complex analytical performance modeling of MPA

For a complex performance analysis (including overhead function) of matrix parallel algorithms (MPA) we will be consider general shape of a square n x n matrix at Fig. 7.

$$A = \begin{pmatrix} a_{11}, a_{12}, & \cdots & , a_{1n} \\ a_{21}, a_{22}, & \cdots & , a_{2n} \\ \cdot\cdot & & \cdot \\ \cdot\cdot & & \cdot \\ \cdot\cdot & & \cdot \\ a_{n1}, a_{n2}, & \cdots & , a_{nn} \end{pmatrix}$$

Figure 7. Square matrix n x n.

The reason for preferred square matrix reducing of parameter number (n = m) in derivation process of complex analytical performance relations (execution time, speed up, efficiency, isoefficiency etc.). Such more transparent approach is supported with following additional reasons

*   any rectangular matrix n x m could be transformed into a square matrix n x n either by extending the number of rows (where m < n) or column (if m > n)
*   derivation process of performance relations will be the same except for the fact that when considering the complexity of the matrix instead of n^2 (square matrix) we have to consider product n. m .

5.1. Basic Common Characteristics of MPA

Typical characteristics of matrix parallel algorithms are regularities both in the program (matrix computational activities) and also in data structure matrix (matrix elements). Such regularity we refer to as a domain. Matrix computational activities we will represent as $T(s, p)_{comp}$ latency. Considering square matrix n x n sequential computational complexity is given as n^2. Common characteristics of matrix parallel algorithms (MPA) are as follows

*   parallelization - matrix itself is well parallelized theoretically up to level of its single data element. But applying such a maximal degree of parallelization could not be effective because of low computation complexity for one matrix element. Therefore we will consider basic matrix decomposition models in group of matrix data elements
*   using of domain decomposition models in which domain is represented by data matrix elements (data domain)
*   applied matrix date domain decomposition models define that for parallel computation on decomposed parts of matrix data elements there is necessary to perform in a parallel way the whole computation as in sequential matrix algorithms
*   to do any computational operation on matrix there is necessary to do this operation on every matrix element or group of them. From performed analysis comes out that at solving typical MPA parallel matrix computations are performed as follow
    *   allocated matrix data part of given computing node are repeatedly evaluated according used

PA (iteration PA). After every iterations step there is necessary to perform IPC communication to neighboring computing nodes of shared matrix data elements
*   allocated matrix data part of given computing node in one computing step are reduced according used PA to simpler matrix data part (for example GEM PA). After every reduction step there is necessary to perform IPC communication to all other used computing nodes.

Based on these conclusions to modeling of MPA there is necessary to derive needed computational and communication complexity.

5.1.1. Computational Matrix Complexity

5.1.1.1. Sequential

Sequential computational matrix complexity for considered square matrix n x n is given as n^2 (computation on each matrix element). Then the used asymptotic complexity is given as $O(n^2)$.

5.1.1.2. Parallel

Computational parallel matrix complexity $Z(s, k)$ is given as computational complexity of one decomposed parallel process in k computing steps where parameter s we have been defined [14] as working load of given problem. For square matrix $s = n^2$ (sequential matrix complexity). Through matrix decomposition models we are creating p parallel processes (matrix decomposition to p matrix parts) the $Z(s, k)$ will be given as a quotient of computational sequential matrix complexity n^2 and number of decomposed parallel processes p as follows

$$Z(s,k) = Z(s,1)\frac{n^2}{p}$$

where $Z(s, 1)$ represent computational complexity in one computation step. From derived relation the parallel computation time complexity $T(s, p)_{comp}$ is given through quotient of parallel computing time running time of one parallel process (product of its complexity $Z(s, 1)_{comp}$ and a constant t_{c1} as an average value of performed computation operations) through number of decomposed parallel processes p as follows

$$T(s,p)_{comp} = \frac{Z(s,1)_{comp} \cdot n^2 \cdot t_{c1}}{p}$$

In MPA we are oft using mapping under the condition n = p. Then we get for $T(s, p)_{comp}$ following simpler relation as

$$T(s,p)_{comp} = Z(s,1)_{comp} \cdot n \cdot t_{c1}$$

For this simpler relation asymptotic complexity is given as O (n). At the same time in relation to ideally parallelized MPA and under assumptions of theoretical unlimited number of computation nodes p mathematical limit of T(s, p)$_{comp}$ is given as

$$T(s, p)_{comp} = \lim_{p \to \infty} \frac{Z(s,1)_{comp} \cdot n^2 \cdot t_{c1}}{p} = 0$$

From this result we can see that a MPA the dominant influence will have mainly communication complexity. Therefore we will exanimate basic matrix decomposition models and their consequences to defined complex performance criterion.

5.2. Basic Matrix Decomposition Models

Supposed efficiency of parallel matrix algorithms (shared memory, distributed memory, hybrid) required to allocate a parallel process to more than one internal element of the square matrix (data elements). Then for decomposition of matrix elements to some groups of matrix data elements we have in principal two basic decomposition models as follows

- decomposition model of n x n matrix to square blocks of matrix elements (parallel process). Illustration example of matrix decomposition to p blocks (B$_1$, B$_2$, ..., B$_p$) is at Fig. 8 a. In this case the decomposed blocks consist from at least four matrix data elements.

- decomposition model of n x n to continual matrix strips of matrix elements. Continual strips consist of at least one matrix row or one matrix column. Illustration example of matrix decomposition to p strips (S$_1$, S$_2$, ..., S$_p$) is at Fig. 8 b. In this case the decomposed strips consist from at least one matrix row.

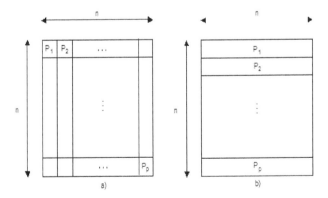

Figure 8. *Matrix decomposition models a) blocks b) strips.*

5.2.1. Decomposition Model to Blocks

For mapping matrix elements in blocks a inter process communication is performed on the four neighboring edges of blocks, which it is necessary in computation flow to exchange. Every parallel process therefore sends four messages and in the same way they receive four messages at the end of every calculation step (Fig. 9. a) supposing that all needed data at every edge are sent as a part of any message).

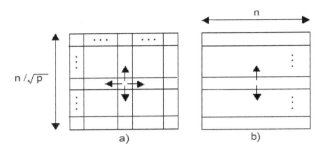

Figure 9. *Communication decomposition models a) blocks b) strips.*

Then the requested communication time for this decomposition model is given as

$$T(s, p)_{commb} = 8 \, (t_s + \frac{n}{\sqrt{p}} t_w)$$

using defined technical communication parameters [18] as follows
- t$_s$ is defined parameter for communication initialization
- t$_w$ is defined parameter for data unit latency.

This equation is correct for p ≥ 9, because only under this assumption it is possible to build at least one square because only then is possible to build one square block with for communication edges. Using these variables for the communication overheads in decomposition method to blocks is correct

$$T(s, p)_{comm} = T(s, p)_{commb} = h(s,p) = 8 \, (t_s + \frac{n}{\sqrt{p}} t_w)$$

Then the requested communication time for this decomposition method is given as

$$T_{comb} = 8 \, (t_s + \frac{n}{\sqrt{p}} t_w)$$

5.2.2. Matrix Decomposition to Strips

Decomposition method to rows or columns (strips) are algorithmic the same and for their practical using is critical the way how are the matrix elements putting down to matrix. For example in C language are array elements put down from right to left and from bottom to top (step by step building of matrix rows).

In this way it is possible send very simple through specification of the beginning address for a given row and through a number of elements in row (addressing with indexes). Let for every parallel process (strips) two messages are send to neighboring processors and in the same way two messages are received from neighboring processors (Fig. 8 b) supposing that it is possible to transmit for example one row to one message. Communication time for a calculation step T(s, p)$_{comms}$ is then given as

$$T(s, p)_{comms} = 4 \, (t_s + n \, t_w)$$

Using these variables for the communication overheads in decomposition method to strips is correct

$$T(s,p)_{comm} = T(s,p)_{comms} = h(s,p) = 4(t_s + n\,t_w)$$

The whole time to execute parallel algorithm T(s, p) for decomposition to strips is then given in general as

$$T(s,p) = \frac{n^2 . t_c}{p} + 4(t_s + n\,t_w)$$

In this case a communication time for one calculation step does not depend on the number of used calculation processors.

5.3. Complex Analytical Performance Modeling

To complex MPA performance modeling we have been defined as deriving of evaluation criterions of IPA including considering overhead function h(s, p). We summarized derived analytical results as following

Shared results for both decomposition models (blocks, strips)

- execution time of sequential square matrix algorithm T(s, 1)

$$T(s,1)_{comp} = n^2\,t_{c1}$$

- execution time for own parallel computation time of IPA parallel algorithms T(s, p)$_{comp}$

$$T(s,p)_{calc} = \frac{n^2 . t_{c1}}{p}$$

- optimal conditions to selection of matrix decomposition model for t$_s$,and for t$_w$ respectively

$$t_s > n\,(1 - \frac{2}{\sqrt{p}})\,t_w \quad t_w > n\,(1 - \frac{2}{\sqrt{p}})\,t_s .$$

Different results for basic matrix decomposition models (blocks, strips)

- overhead function for blocks h(s, p)$_b$ and for strips h(s, p)$_s$ as follows

$$h(s,p)_b = 8\,(t_s + \frac{n}{\sqrt{p}}\,t_w)$$

$$h(s,p)_s = 4\,(t_s + n\,t_w)$$

- complex parallel execution time for blocks T(s, p)$_{compb}$, and for strips T(s, p)$_{comps}$

$$T(s,p)_{calcb} = T(s,p)_{calc} + T(s,p)_{ipcb} = \frac{n^2 . t_{c1}}{p} + 8(t_s + \frac{n}{\sqrt{p}}\,t_w)$$

$$T(s,p)_{calcs} = T(s,p)_{calc} + T(s,p)_{ipcs} = \frac{n^2 . t_{c1}}{p} + 4(t_s + n\,t_w)$$

- parallel speed up for blocks S(s, p)$_b$, and for strips S(s, p)$_s$

$$S(s,p)_b = \frac{T(s,1)}{T(s,p)_{calcb}} = \frac{n^2\,p\,t_{c1}}{n^2 t_{c1} + 8\,(p\,t_s + \sqrt{p}\,n\,t_w)}$$

- efficiency for blocks E(s, p)$_b$, and for strips E(s, p)$_s$

$$E(s,p)_b = \frac{S(s,p)_b}{p} = \frac{n^2\,t_{c1}}{n^2 t_{c1} + 8\,(p\,t_s + \sqrt{p}\,n\,t_w)}$$

$$E(s,p)_s = \frac{S(s,p)_s}{p} = \frac{n^2\,t_{c1}}{n^2 t_{c1} + 4\,p\,(t_s + n\,t_w)}$$

- constant C (needed constant in deriving an isoefficiency function w (s)) and that for blocks as C$_b$, and for strips as C$_s$ in issoeficiency function

$$C_b = \frac{E(s,p)}{1 - E(s,p)} = \frac{n^2 t_{c1}}{8\,(p\,t_s + \sqrt{p}\,n\,t_w)}$$

$$C_s = \frac{E(s,p)}{1 - E(s,p)} = \frac{n^2 t_{c1}}{4\,p\,(t_s + n\,t_w)}$$

Fig. 10 illustrates growth dependencies of parallel computing time T(s, p)$_{comp}$, communication time T(s, p)$_{comm}$ and complex parallel execution time T(s, p)$_{complex}$ from input load growth n (Square matrix dimension) at constant number of computing node p = 256

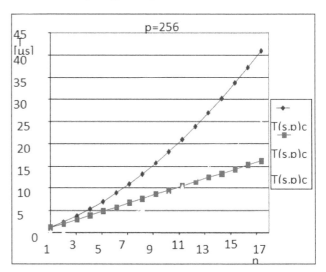

Figure 10. Dependencies of T(s, p)$_{complex}$, T(s, p)$_{comm}$ and T(s, p)$_{comp}$ from n (p=256).

Fig. 11 illustrates growth dependencies of parallel computing time T(s, p)$_{comp}$, communication time T(s, p)$_{comm.}$ and complex parallel execution time T(s, p)$_{complex}$ from increasing number of computing node p at constant input load n (Matrix dimension n = 512)

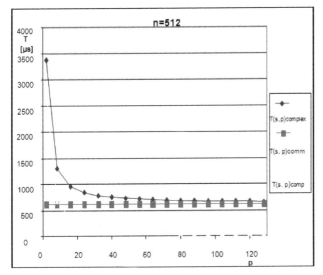

Figure 11. *Dependencies of $T(s, p)_{complex}$, $T(s, p)_{comm}$ and $T(s, p)_{comp}$ from p (n=512).*

5.3.1. Issoeficiency Functions

Issoeficiency function w(s) is very important for performance prediction of parallel algorithms PA. For modeling of performance prediction in PA we are going to derive for defined basic matrix decomposition models (blocks, strips) corresponding analytical issoeficiency functions $w(s)_b$ (Decomposition to blocks) and $w(s)_s$ (decomposition to strips). For asymptotic complexity of w(s) is valid following derived relation as

$$w(s) = \max\left[T(s, p)_{calc}, h(s, p)\right]$$

where defined workload s is a function of input load n. For IPA it is given as $s = n^2$. We have defined that for given value efficiency E(s, p) following quotient of efficiencies E(s, p) is constant

$$\frac{E}{1-E} = C$$

5.3.1.1. Canonical Matrix Decomposition Models

As basic matrix decomposition models we have defined such matrix decomposition models to which there is possible to reduce all other known matrix decomposition models. Then the canonical matrix decomposition models are as follows
- matrix decomposition model to blocks
- matrix decomposition model to strips.

For defined constants C_b (blocks) and C_s (strips) which are integral parts of issoeficiency functions $w(s)_b$ (blocks) and w(s) (strips) we have derived following relations

$$C_b = \frac{E(s,p)}{1-E(s,p)} = \frac{n^2 t_{c1}}{8\,(p\,t_s + \sqrt{p}\,n\,t_w)}$$

$$C_s = \frac{E(s,p)}{1-E(s,p)} = \frac{n^2 t_{c1}}{4\,p\,(t_s + n\,t_w)}$$

To win a closed form of issoeficiency function $w(s)_b$, $w(s)_s$ we have used an approach in which we performed at first the analysis of increasing input load influenced the analyzed expression contained t_s in relation to p so to keep this growth constant (we supposed that $t_w = 0$). Then for constants C_b, C_s we get following expressions

$$C_b = \frac{n^2 t_{c1}}{8\,p\,t_s} \quad C_s = \frac{n^2 t_{c1}}{4\,p\,t_s}$$

From these expressions we can derive for searched functions $w(s)_b = w(s)_s = n^2$ from relations for C_b, C_s following relations

$$w(s)_b = n^2 = 8\,C_b\,p\,\frac{t_s}{t_{c1}} \quad w(s)_s = n^2 = \frac{4\,C_s\,p\,t_s}{t_{c1}}$$

With a similar approach we can analyze the influence growth of input load caused another part of expression from t_s in relation to p so to keep this growth constant (we supposed that $t_s = 0$). Then after setting and performed needed adjustments we get for searched functions $w(s)_b$ and $w(s)_s$ following relations

$$w(s)_b = n^2 = 8\,C_b\,\sqrt{p}\,n\,\frac{t_w}{t_{c1}} \quad w(s)_s = 4\,C_s\,n\,p\,\frac{t_w}{t_{c1}}$$

Final derived analytical functions $w(s)_b$ and $w(s)_s$ are as follows

$$w(s)_b = \max\left[\frac{n^2 t_{c1}}{p}, 8\,C_b\,p\,\frac{t_s}{t_{c1}}, 8\,C_b\,n\,\sqrt{p}\,\frac{t_w}{t_{c1}}\right]$$

$$w(s)_s = \max\left[\frac{n^2 t_{c1}}{p}, 4\,C_s\,p\,\frac{t_s}{t_{c1}}, 4\,C_s\,n\,p\,\frac{t_w}{t_{c1}}\right]$$

5.3.1.2. Optimization of Issoefficiency Functions

Optimization of derived issoefficiency functions require to search for dominant expressions in derived final relations for $w(s)_b$ a $w(s)_s$. For this purpose we have been done comparison of individual expressions of $w(s)_b$ and $w(s)_s$ with following conclusions
- the first expressions of $w(s)_b$ and $w(s)_s$ are the same and therefore this expression will be the component of final optimized $w(s)_{opt}$. At the same time for this expression at performed asymptotical analysis in relation to parameter p the following limit is valid

$$\lim_{p \to \infty} \frac{n^2 . t_{c1}}{p} = 0$$

and therefore the similar first expressions of issoeficiency functions $w(s)_b$ and $w(s)_s$ we can omit from searched $w(s)_{opt}$
- in relation to the similarity of actually first expressions of $w(s)_b$ and $w(s)_s$ (After omitting expressions according previous conclusion) as

follows

$$8\,C_b\;p\,\frac{t_s}{t_{c1}} \geq 4\,C_s\;p\,\frac{t_s}{t_{c1}}$$

This condition after reducing of shared expression parts lead to inequality $2\,C_b \geq C_s$. After setting and following adjustments we get final condition as $p \geq 1$, which is valid on the whole range of spotted values of parameter p. It means the with this performed expression comparison we have got more dominated expression which we let to the next three comparisons (Two from $w(s)_b$ and one from $w(s)_s$)

- in an analogous way we do comparison of third expressions from original $w(s)_b$ and $w(s)_s$ issoeficiency functions and that as follows

$$4\,C_s\;n\,p\,\frac{t_w}{t_{c1}} \geq 8\,C_b n \sqrt{p}\,\frac{t_w}{t_{c1}}$$

These conditions after reducing of shared expression parts lead to following inequality $C_s\sqrt{p} \geq 2\,C_b$. After setting and performed adjustments we get final condition $p \geq 1$, which is fulfilled on the whole range of parameter p. With performed comparison we have ignored further less expression and final relation $w(s)_{opt}$ is actually as follows

$$w(s)_{opt} = \max\left[4\,C_s\;n\,p\,\frac{t_w}{t_{c1}}, 8\,C_b\;p\,\frac{t_s}{t_{c1}}\right]$$

- final comparison of remaining expression comes to following expression comparison

$$4\,C_s\;n\,p\,\frac{t_w}{t_{c1}} \geq 8\,C_b\;p\,\frac{t_s}{t_{c1}}$$

This condition after reducing of shared expression parts leads to following inequality $C_s\,n\,t_w \geq 2\,C_b t_s$. After setting and performed adjustments we come to following inequality $n^2.\,t_w^2 \geq \sqrt{p}\,t_s^2$. This inequality we are able to solve only for concrete values of parameters n, p, t_s, t_w. For example using following values of parameters $t_s = 35\ \mu s$, $t_w = 0{,}23\ \mu s$ and under assumption of in praxis frequent case of choosing n = p we get simpler expression to condition validity as n. $\sqrt{n} \geq 152{,}17^2$or p. $\sqrt{p} \geq 152{,}17^2$. The smallest integer number which satisfies given condition is p = n = 813. Satisfying this condition for n or p, the final issoeficiency function $w(s)_{opt}$ given with first expression and in opposite is given with second expression of following final optimized issoeficiency function $w(s)_{opt}$

$$w(s)_{opt} = \max\left[4\,C_s\;n\,p\,\frac{t_w}{t_{c1}}, 8\,C_b\;p\,\frac{t_s}{t_{c1}}\right].$$

5.3.1.3. Conclusions of Issoeficiency Functions

Then for the given concrete value of E(s, p) and for given values of parameters p, n we can in analytical way the thresholds, for which growth of isoefficiency function means decreasing of efficiency of given parallel algorithm with assumed typical decomposition strategies. This means the minor scalability of the assumed algorithm. In case of decomposition strategy the approach is similar to analyzed practical used decomposition matrix strategies.

Based on analysis of computer technical parameters t_s, t_w, t_c for some parallel computers in the world they are valid following inequalities $t_s >> t_w > t_c$. Alike is valid that $p \leq n$. Using these inequalities it is necessary to analyze dominancy influence of the all derived expressions.

Then the asymptotic issoeficiency function is limited through dominancy conditions of second and third expressions. From their comparison comes out

- based on real condition $t_w \geq t_s$ a third expression is bigger or equal than a second expression and an issoefficiency function is limited through the first expression of $w(s)_{opt}$. If we used following technical parameters $t_s = 35\ \mu s$, $t_w = 0{,}23\ \mu s$ this is true for n ≥ 813
- for n < 813 and for the same technical constants $t_s = 35\ \mu s$, $t_w = 0{,}23\ \mu s$ issoefficiency function is limited through a second expression of $w(s)_{opt}$.

6. Results

We illustrate some of chosen performed tested results. To practical illustrations we have used MPA algorithms for iterative solving of Laplace PDE equation defined as follows

- four point iteration relation in which in one iteration are performed five arithmertic operations ($t_{c1} = 5\ t_c$)
- communication model according Fig. 5.

For experimental testing we have used workstations of NOW parallel computer (workstations WS 1 – WS 5) and supercomputer as follows

- WS 1 – Pentium IV (f = 2,26 G Hz)
- WS 2 - Pentium IV Xeon (2 proc., f = 2,2 G Hz)
- WS 3 - Intel Core 2 Duo T 7400 (2 cores, f=2,16 GHz)
- WS 4 - Intel Core 2 Quad (4 cores, 2.5 GHz)
- WS 5 - Intel SandyBridge i5 2500S (4 cores, f=2.7 GHz)
- supercomputer Cray T3E in remote computing node.

Comparison of decomposition model influence (D1 - blocks, D2 - strips) is at Fig. 11. For comparison were measured values recomputed to one iteration step. Performed measurements have proved higher efficiency of decomposition model to blocks for tested parallel computer Cray T3E. Technical parameters of parallel computer Cray T3E ($t_s = 3\ \mu s$, $t_w = 0{,}063\ \mu s$, $t_c = 0{,}011\ \mu s$).

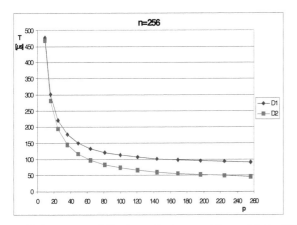

Figure 12. *Comparison of T(s, p)$_{complex}$ for decomposition models (n=256).*

Fig. 13 illustrate dependencies to optimal selection of decomposition strategy for technical parameter t_{si} (t_{s1}, t_{s2}) using verified technical parameters of supercomputer Cray T3E for t_w = 0,063 µs and n = 128, 256.

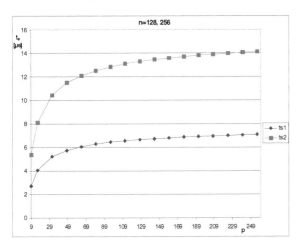

Figure 13. *Influences t_s for n = 128, 256*

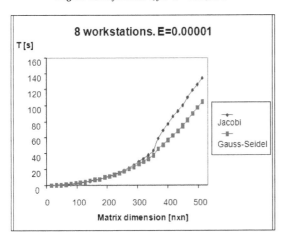

Figure 14. *Comparison of T(s, p)$_{complex}$ for Jacobi and Gauss-Seidel IPA for E=10^{-5}.*

At Fig. 14 we have presented measurement results of the whole solving time for both developed parallel algorithms (Jacobi, Gauss-Seidel) with the number of processors p = 8 and for various values of input workload n (Matrix

dimensions) for E=10^{-5}. From comparisons of these measurements come out that for great number of workstations (p=8) are the whole solving times approximately the same. The reasons are that lower computation complexity at Gauss-Seidel method (Computation) is eliminated through greater communication complexity in its parallel algorithm practically double higher than Jacobi IPA.

This figure illustrates continually percent spreading of the individual overheads (Initialization, computation, communication, gathering) for Jacobi parallel algorithm with the given number of workstations p = 4 for various values of workload n (matrix dimensions) and for accuracy E=0,001. From comparisons we can see raising trend of computation in dependence of accuracy E.

Generally for the problems with increasing communication complexity through using great number of processors p based on Ethernet NOW we come to the point (Threshold that parallel computing is no more effective, that means we are without any speed-up. It is evident that for the given problem, given parallel algorithms and given parallel computer to find such a threshold (no speed-up) is very important.

The individual parts of the whole execution parallel time are illustrated at Fig. 15 for Jacobi iterative parallel algorithm for 4 workstations and for E = 0,001.

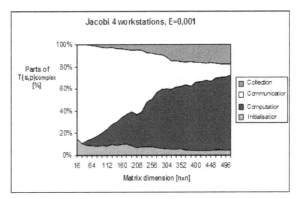

Figure 15. *Percentage comparison of T (s, p)$_{complex}$ for its components (E=0,001).*

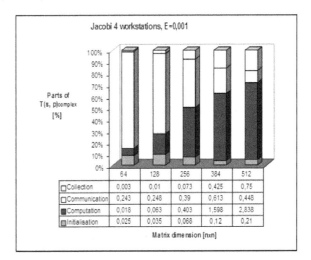

Figure 16. *Comparison of T (s, p)$_{complex}$ parts for p=4 (E=0,001).*

The influence of number of workstations at given accuracy E=0,001 to the individual parts of the whole solving time for pre Jacobi iterative parallel algorithm for various sizes of input workload (matrix dimensions from 64 x 64 to the size 512 x 512) illustrates Fig. 16 for the number of workstations p = 4. From the comparisons come out percent sinking of computations at the bigger number of workstations (parallel speed-up though the higher number of workstations) at the moderate percent raising of network communication overheads.

Fig. 17 illustrates the times of individual parts of the whole solving time as a function of input workload n (Square matrix dimensions) for number of workstations p=4 and for accuracy E=0,001. From comparison of these both graphs comes out higher contribution through the number of working stations than the raising overheads of the network communication.

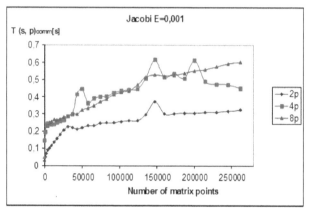

Figure 17. *Influence of computing nodes to T(s, p)$_{comm}$, (E=0,001).*

Fig. 18 illustrate influence of number of workstations NOW to quicker solutions of both distributed parallel algorithms (Gauss-Seidel parallel algorithm) for matrix dimensions 512x512 and various analyzed accuracies of Epsilon.

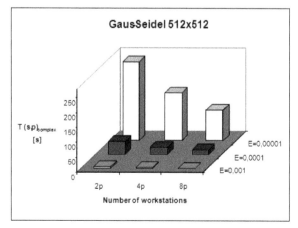

Figure 18. *Influence of workstation number for Gauss-Seidel IPA.*

Derived analytical issoeficiency functions allow us to predict parallel computer performance also for theoretical not existed ones. We have illustrated at Fig. 19

isoefficiency functions for individual constant values of efficiency (E = 0,1 to 0,9) for n < 152 using the published technical parameters t_c, t_s, t_w communication constants of used NOW (t_c= 0,021 μs, t_s = 35 μs, t_w= 0,23 μs,).

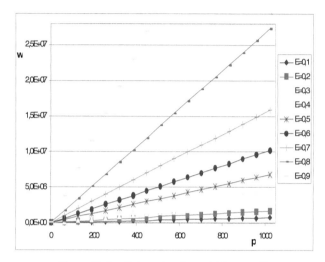

Figure 19. *Isoefficiency functions w(s) for n < 152*

Fig. 20 illustrates isoefficiency functions for individual constant values of efficiency (E = 0,1 to 0,9) for n = 1024 and for communication parameters of parallel computer Cray T3E (t_c= 0,011μs, t_s = 3 μs, t_w= 0,063 μs,).

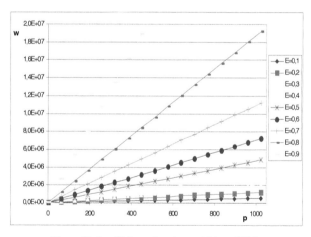

Figure 20. *Isoefficiency functions w(s) (n = 1024).*

From both pictures (Fig. 19 and Fig. 20) we can see that to keep a given value of efficiency we need step by step increasing number of computing processors and higher value of workload (useful computation) to balance higher communication overheads.

7. Conclusions

Performance evaluation as a discipline has repeatedly proved to be critical for design and successful use in parallel computing. At the early stage of design, performance models can be used to project the system scalability and evaluate alternative solutions. At the production stage, performance evaluation methodologies can be used to detect bottlenecks and subsequently suggests ways to alleviate them. Queuing networks has been established in modeling of parallel computers [5, 17]. Extensions of complexity theory to parallel computing have been successfully used for the evaluation of parallel algorithms and communication complexity too. Via extended form of isoefficiency concept for parallel algorithms we have demonstrated its applied using to performance prediction in typical matrix parallel algorithms (MPA).

To derive isoefficiency function in analytical way it is necessary to derive al typical used criterion for performance evaluation of parallel algorithms including their overhead function (parallel execution time, speed up, efficiency). Based on this knowledge we are able to derive issoefficiency function as real criterion to evaluate and predict performance of parallel algorithms also for future hypothetical parallel computers. So in this way we can say that this process includes complex performance evaluation including performance prediction.

Due to the dominant using of parallel computers based on NOW modules and their high integration named as Grid there has been great interest in performance prediction of parallel algorithms in order to achieve effective parallel algorithms (optimized). Therefore this paper summarizes the used methods for complexity analysis which can be applicable to all types of parallel computers (supercomputer, NOW, Grid).

This paper finalizes applying of complex analytical modeling to the whole group of matrix parallel algorithms which are characterized by domain decomposition models. To present this group of MPA we have modeled abstract matrix using basic decomposition models with supposed intensive communication complexity. In such a way performed complex modeling could be inspiring also to other PA or even a group of PA. The complex analyzed examples we have been evaluated so on classic massive supercomputers (hypercube, mesh) as on dominant parallel computers represented by NOW module.

Acknowledgements

This work was done within the project "Complex performance modeling, optimization and prediction of parallel computers and algorithms" at University of Zilina, Slovakia. The author gratefully acknowledges help of project supervisor Prof. Ing. Ivan Hanuliak, PhD.

References

[1] Arora S., Barak B., Computational complexity - A modern Approach, Cambridge University Press, pp. 573, 2009

[2] Bahi J. H., Contasst-Vivier S., Couturier R., Parallel Iterative algorithms: From Sequential to Grid Computing, CRC Press, USA, 2007

[3] Bronson R., Costa G. B., Saccoman J. T., Linear Algebra - Algorithms, Applications, and Techniques, 3rd Edition, Elsevier Science & Technology, Netherland, pp. 536, 2014

[4] Casanova H., Legrand A., Robert Y., Parallel algorithms, CRC Press, USA, 2008

[5] Dattatreya G. R., Performance analysis of queuing and computer network, University of Texas, Dallas, USA, pp.472, 2008

[6] Davis T. A., Direct methods for sparse Linear Systems, Cambridge University Press, United Kingdom, pp. 184, 2006

[7] Desel J., Esperza J., Free Choise Petri Nets, Cambridge University Press, United Kingdom, pp. 256, 2005

[8] Dubois M., Annavaram M., Stenstrom P., Parallel Computer Organization and Design, Cambridge university press, United Kingdom, pp. 560, 2012

[9] Dubhash D.P., Panconesi A., Concentration of measure for the analysis of randomized algorithms, Cambridge University Press, United Kingdom, 2009

[10] Edmonds J., How to think about algorithms, Cambridge University Press, United Kingdom, pp. 472, 2010

[11] Gelenbe E., Analysis and synthesis of computer systems, Imperial College Press, pp. 324, 2010

[12] Goldreich O., P, NP, and NP - Completeness, Cambridge University Press, United Kingdom, pp. 214, 2010

[13] Hager G., Wellein G., Introduction to High Performance Computing for Scientists and Engineers, CRC Press, USA, pp. 356, 2010

[14] Hanuliak P., Hanuliak J., Complex performance modeling of parallel algorithms , American J. of Networks and Communication, Science PG, Vol. 3, USA, 2014

[15] Hanuliak M., Modeling of parallel computers based on network of computing nodes, American J. of Networks and Communication, Science PG, Vol. 3, USA, 2014

[16] Hanuliak M., Hanuliak J., Decomposition models of parallel algorithms, American J. of Networks and Communication, Science PG, Vol. 3, USA, 2014

[17] Hanuliak M., Hanuliak I., To the correction of analytical models for computer based communication systems, Kybernetes, Vol. 35, No. 9, UK, pp. 1492-1504, 2006

[18] Hanuliak J., Modeling of communication complexity in parallel computing, American J. of Networks and Communication, Science PG, Vol. 3, USA, 2014

[19] Hanuliak M., Unified analytical models in parallel and distributed computing, AJNC (Am. J. of Networks and Comm.), SciencePG, Vol. 3, No. 1, USA, pp. 1-12, 2014

[20] Hanuliak J., Hanuliak I., To performance evaluation of distributed parallel algorithms, Kybernetes, Volume 34, No. 9/10, United Kingdom, pp. 1633-1650, 2005

[21] Hillston J., A Compositional Approach to Performance Modeling, University of Edinburg, Cambridge University Press, United Kingdom, pp. 172 pages, 2005

[22] Hwang K. and coll., Distributed and Parallel Computing, Morgan Kaufmann, USA, 472 pages, 2011

[23] Kshemkalyani A. D., Singhal M., Distributed Computing, University of Illinois, Cambridge University Press, United Kingdom, pp. 756 pages, 2011

[24] Kirk D. B., Hwu W. W., Programming massively parallel processors, Morgan Kaufmann, USA, pp. 280, 2010

[25] Kostin A., Ilushechkina L., Modeling and simulation of distributed systems, Imperial College Press, United Kingdom, pp. 440, 2010,

[26] Kshemkalyani A. D., Singhal M., Distributed Computing, University of Illinois, Cambridge University Press, UK, pp. 756, 2011

[27] Kushilevitz E., Nissan N., Communication Complexity, Cambridge University Press, United Kingdom, pp. 208, 2006,

[28] Le Boudec Jean-Yves, Performance evaluation of computer and communication systems, CRC Press, USA, pp. 300, 2011

[29] Levesque John, High Performance Computing: Programming and applications, CRC Press, USA, pp. 244, 2010

[30] Lilja D. J., Measuring Computer Performance, University of Minnesota, Cambridge University Press, United Kingdom, pp. 280, 2005

[31] McCabe J., D., Network analysis, architecture, and design (3rd edition), Elsevier/ Morgan Kaufmann, USA, pp. 496, 2010

[32] Meerschaert M., Mathematical modeling (4-th edition), Elsevier, pp. 384, 2013

[33] Misra Ch. S.,Woungang I., Selected topics in communication network and distributed systems, Imperial college press, United Kingdom, pp. 808, United Kingdom

[34] Peterson L. L., Davie B. C., Computer networks – a system approach, Morgan Kaufmann, USA, pp. 920, 2011

[35] Resch M. M., Supercomputers in Grids, Int. J. of Grid and HPC, No.1, pp. 1 - 9, 2009

[36] Riano l., McGinity T.M., Quantifying the role of complexity in a system´s performance, Evolving Systems, Springer Verlag, Germany, pp. 189 – 198, 2011

[37] Shapira Y., Solving PDEs in C++ - Numerical Methods in a Unified Object-Oriented Approach (2nd edition), Cambridge University Press, United Kingdom, pp. 800, 2012

[38] Wang L., Jie Wei., Chen J., Grid Computing: Infrastructure, Service, and Application, CRC Press, USA, 2009 www pages

[39] www.top500.org.

19

Complex performance modeling of parallel algorithms

Peter Hanuliak, Juraj Hanuliak

Dubnica Technical Institute, Sladkovicova 533/20, Dubnica nad Vahom, 018 41, Slovakia

Email address:

phanuliak@gmail.com (P. Hanuliak)

Abstract: Parallel principles are the most effective way how to increase parallel computer performance and parallel algorithms (PA) too. In this sense the paper is devoted to a complex performance evaluation of chosen PA. At first the paper describes very shortly PA and then it summarized basic concepts for performance evaluation of PA. To illustrate the analyzed evaluation concepts the paper considers in its experimental part the results for real analyzed examples of discrete fast Fourier transform (DFFT). These illustration examples we have chosen first due to its wide application in scientific and engineering fields and second from its representation of similar group of PA. The basic form of parallel DFFT is the one-dimensional (1-D), unordered, radix–2 algorithm which uses divide and conquer strategy for its parallel computation. Effective PA of DFFT tends to computing one – dimensional FFT with radix greater than two and computing multidimensional FFT by using the polynomial transfer methods. In general radix - q DFFT is computed by splitting the input sequence of size s into q sequences each of them in size n/q, computing faster their q smaller DFFT's, and then combining the results. So we do it for actually dominant asynchronous parallel computers based on Network of workstations (NOW) and Grid systems.

Keywords: Parallel Computer, NOW, Grid, Parallel Algorithm (PA), Matrix PA, Decomposition, Performance Modeling, Optimization, Issoeficiency Function, Numerical Integration, Discrete Fast Fourier Transform (DFFT), Overhead Function

1. Introduction

Parallel and distributed computing has been evolved as two separate research disciplines. Parallel computing has addressed problems of communication and intensive computation on highly-coupled computing nodes while distributed computing has been concerned with coordination, availability, timeliness, etc., of more likely coupled computing nodes. Current trends, such as parallel computing on networks of high performance computing nodes (workstations) and Internet computing, suggest the advantages of unifying these two research disciplines. In relation to these trends we have developed a flexible model of computation that supports both parallel and distributed computing [11].

Parallel and distributed computing share the same basic computational model consisting on physically distributed parallel processes that operate concurrently and interact with each other in order to accomplish a task as a whole. In parallel computing, processes are assumed to be placed closer to each other and they could communicate frequently and hence the ratio of computation/communication of parallel applications is usually much smaller than that in distributed applications. On the other hand, distributed computing focuses on parallel processes that could be allocated in a wide area i. e., communication between some parallel processes is assumed to be more costly than in parallel computing.

A number of recent trends point to a convergence of research in parallel and distributed computing. First, increased communication bandwidth and reduced latency make geographical distribution of computing nodes less of a barrier to parallel computing. Second, the development of architecture neutral programming language, such as Java, provides a virtual computational environment in which computing nodes appear to be homogenous. Finally, increased client/server computing is adopting symmetrical multiprocessor architecture (SMP), often multiple processors or cores with a shared memory in a single workstations. While such architectures are less scalable than networks of computers, some parallel programs with high communication traffic may execute on them more efficiently. Another important trend is a convergence of parallel and distributed computing is the potential of Internet computing. With improvements in network technology and communication middleware, one can view the Internet as a huge of parallel and distributed computers. Because connectivity on the Internet can be intermittent variable of the bandwidth, the ability of processes as well as data to migrate becomes critical. In turn, this requires a satisfactory treatment of mobility.

2. Architectures of Parallel Computers

It is very difficult to classify all existed parallel computers. We have tried to classify them from the point of program developer [1, 9] to the two following basic groups according Figure1.

- synchronous parallel computers. They are often used under central control, that means under the global clock synchronization (vector, array system etc.) or a distributed local control mechanism (systolic systems etc.). The typical architectures of this group of parallel computers illustrate Figure 1 on its left side
- asynchronous parallel computers. They are composed of a number of fully independent computing nodes (processors, cores, and computers). To this group belong mainly various forms of computer networks (cluster), network of workstation (NOW) or more integrated Grid modules based on NOW modules. The typical architectures of asynchronous parallel computers illustrate Figure 1 on its right side.

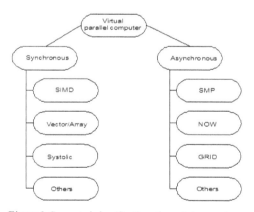

Figure 1. Suggested classification of parallel computers.

2.1. Dominant Parallel Computers

2.1.1. Network of Workstations

There has been an increasing interest in the use of networks (Cluster) of workstations connected together by high speed networks for solving large computation intensive problems [5]. This trend is mainly driven by the cost effectiveness of such systems as compared to massive multiprocessor systems with tightly coupled processors and memories (Supercomputers). Parallel computing on a cluster of workstations connected by high speed networks has given rise to a range of hardware and network related issues on any

given platforms [15, 26]. Load balancing, inter processor communication (IPC), and transport protocol for such machines are being widely studied [20, 21].

This trend is mainly driven by the cost effectiveness of such systems as compared to massive multiprocessor systems with tightly coupled processors and memories (Supercomputers). With the availability of high performance personal computers (workstation) and high speed communication networks (Infiniband, Quadrics, Myrinet), recent trends are to connect a number of such workstations to solve complex problems in parallel on such NOW modules [24, 28].

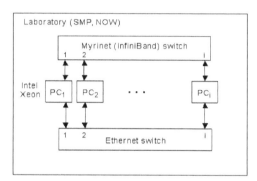

Figure 2. Illustration of NOW.

Principal example of NOW module is at Figure 2. The individual workstations PC_i are mainly powerful workstations based on symmetrical multiprocessor or multicore platform (SMP).

2.1.3. Grid Systems

Grid represents a new way of managing and organizing of individual resources (processors, memory modules, I/O devices etc.) [33]. Grids go out conceptually from a structure of virtual parallel computer based on NOW modules. We have illustrated at Figure 3 typical integrated Grid module based on NOW modules. Any classic parallel computer (massive multiprocessors as supercomputers etc.) could be a member of any NOW module [29].

3. Parallel Algorithms

In principal we can divide parallel algorithms (PA) to the following groups

- parallel algorithm using shared memory (PA_{sm}). These algorithms are developed for parallel computers with shared memory as actual modern symmetrical multiprocessors (SMP) or multicore systems on motherboard
- parallel algorithm using distributed memory (PA_{dm}). These algorithms are developed for parallel computers with distributed memory as actual NOW system and their higher integration forms named as Grid systems
- hybrid PA which combine using of both previous PA (PA_{hyb}). This trend support applied using of NOW consisted from computing nodes based on SMP parallel computers.

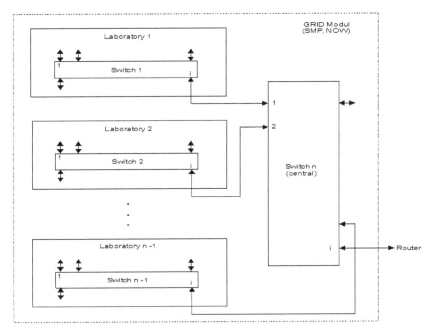

Figure 3. *Grid as integration of NOW network.*

The main difference between PA_{sm} and PA_{dm} is in form of inter process communication (IPC) among created parallel processes [12, 17]. Generally we can say that IPC communication in parallel system with shared memory can use more communication possibilities (all the possibilities of communication in shared memory) than in distributed systems (only network communication).

3.1. Developing Process of PA

The role of programmer is for the given parallel computer and for given application problem to develop the effective parallel algorithm. This task is more complicated in those cases, in which we have to create the conditions for any parallel activities in form of dividing the sequential algorithm to their mutual independent parts named parallel processes. Principally development of any parallel algorithms (shared memory, distributed memory) includes following activities [10, 18].

- decomposition - the division of the application problem into a set of parallel processes and their data
- mapping - the way how created parallel processes and data are distributed among the nodes of parallel computer
- inter-process communication as a way of parallel processes cooperation and synchronization
- tuning – performance optimization of developed parallel algorithm

The most important step is to choose the best decomposition method for given complex problem. To do this it is necessary to understand the concrete complex problem, shared data domain, the used algorithms and the flow of control in given complex problem [13, 23].

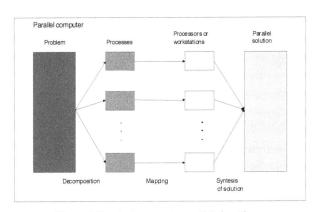

Figure 4. *Developing steps in parallel algorithms.*

3.1.1. Decomposition models

Decomposition strategy defines potential dividing of given complex problem to their independent parts (parallel processes) in such a way, that they could be performed in a parallel way through computing nodes of given parallel computer. Existence of some decomposition method is critical assumption to possible parallel algorithm. Potential decomposition degree of given complex problem is crucial for effectiveness of parallel algorithm [3, 7]. The chosen decomposition method drives the rest of program development. This is true is in case of developing new application as in porting serial code. The decomposition method tells us how to structure the code and data and defines the communication topology [10, 25]. The used decomposition models are as follows

- naturally parallel decomposition
- domain decomposition
- control decomposition
  - manager/workers

·   functional
- divide-and-conquer strategy
- decomposition of big problems
- object oriented programming (OOP).

To the illustration of developing effective parallel algorithm and the way of its complexity evaluating we used applied problem of discrete fast Fourier Transform (DFFT). In relation to it we illustrate the principle of divide and conquer decomposition model (DM), which is used to decompose DFFT.

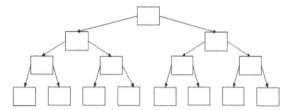

Figure 5. *Illustration of divide and conquer DM (n=8).*

4. The Role of Performance

Modeling techniques allow to model contention both at hardware and software levels by combining approximate solutions and analytical methods [30]. We would like to prefer analytical methods although the complexity of parallel computers and parallel algorithms could limit the applicability of these techniques. To the verification of analytical results we would like to use simulation method or in some cases experimental measuring.

4.1. Performance Evaluation Methods

To performance evaluation of parallel algorithms we can use analytical approach to get under given constraints analytical laws or some other derived analytical relations. We can use following solution methods to get a function of complex performance
- analytical
  ·   order (asymptotic) analysis [2, 12]
  ·   Petri nets [4]
  ·   queuing theory [8, 14]
- simulation [19]
- experimental
  ·   benchmarks [16]
  ·   modeling tools [22]
  ·   direct measuring [6, 27].

Analytical method is a very well developed set of techniques which can provide exact solutions very quickly, but only for a very restricted class of models. For more general models it is often possible to obtain approximate results significantly more quickly than when using simulation, although the accuracy of these results may be difficult to determine [31].

Simulation is the most general and versatile means of modeling systems for performance estimation. It has many uses, but its results are usually only approximations to the exact answer and the price of increased accuracy is much

longer execution times. They are still only applicable to a restricted class of models in spite of its computation and time requirements.

Evaluating system performance via experimental measurements is a very useful alternative for computer systems. Measurements can be gathered on existing systems by means of benchmark applications that aim at stressing specific aspects of computers systems. Even though benchmarks can be used in all types of performance studies, their main field of application is competitive procurement and performance assessment of existing parallel computers and parallel algorithms.

5. Performance Evaluation Criterions of PA

To evaluations of parallel algorithms there have been developed several fundamental concepts. Tradeoffs among these performance factors are often encountered in real-life applications.

5.1. Basic Performance Concepts

5.1.1. Parallel Execution Time

We have defined parallel execution time T(s, p) as the execution time performed by p computing nodes (processors, cores, workstations) of given parallel computer and s defines input size (load) of given problem. Then T(s, 1) defines execution time for classic sequential computer.

5.1.2. Speed Up

The speed up factor S(s, p) we can define as

$$S(s,p) = \frac{T(s,1)}{T(s,p)}$$

Speed up factor (dimensionless) is a measure obtained at given complex algorithm using p computing nodes solving given problem with its problem size s. Since S(s, p) ≤ p, we would like to design algorithms that achieve S(s, p) ≈ p.

5.1.3. Efficiency

The efficiency for processor system with p computing nodes is defined by

$$E(s,p) = \frac{S(s,p)}{p} = \frac{T(s,1)}{p\,T(s,p)}$$

The efficiency is always less than 1. A value of E(s, p) approximately equal to 1 for given p, indicates that such parallel computer, using p computing nodes runs approximately p times faster than it does on sequential computer.

5.1.4. The Isoefficiency Concept

The workload w of an algorithm often grows in the order O(s), where s is the problem size. Thus, we denote the workload w = w(s) as a function of problem size s. In parallel computing is very useful to define an isoefficiency function relating workload to parallel computer size p which is needed to achieve given fixed efficiency E (s, p). Let h(s, p) be the total overhead function consisted from existed overhead latencies involved in PA implementations. This overhead function is a function of both parallel computer size p and input problem size s. Then we can define efficiency E(s, p) of a parallel algorithm as

$$E(s,p) = \frac{w(s)}{w(s) + h(s,p)}$$

The workload w(s) corresponds to useful performed parallel computations while the overhead function h(s, p) represents latency times attributed to communication of parallel processes, synchronization, waiting to shared resources etc. In general, the overheads increase with respect to increasing both values of parameters p and s. The question is hinged on relative growth rates between w(s) and h(s, p). With a fixed problem size the efficiency E(s, p) decreases as p increase. The reason is that the overhead function h(s, p) increases with p. With a fixed parallel computer size, the overhead function h(s, p) grows slower than the workload w(s). Thus the efficiency E(s, p) increases with increasing problem size s for a fixed parallel computer size. Therefore, one can expect to maintain a constant efficiency E(s, p) if the workload w(s) is allowed to grow properly with increasing parallel computer size p.

5.2. Complex Performance Modeling of PA

Complex performance modeling of PA we qualify as modeling with considering overhead function h(s, p) in all defined fundamental performance concepts T(s, p), S(s, p), E(s, p) and w(s).

5.2.1. Complex Parallel Execution Time
Complex parallel execution time T(s, p)$_{complex}$ will be defined as the whole parallel execution time included overhead function h(s, p) as follows

$$T(s,p)_{complex} = T(s,p) + \text{h(s, p)}$$

5.2.2. Complex Speed Up Factor
The complex speed S(s, p)$_{complex}$ we can rewrite as

$$S(s,p)_{complex} = \frac{T(s,1)}{T(s,p)_{complex}}$$

5.2.3. Complex Efficiency
Similarly the complex efficiency E(s, p)$_{complex}$ we can rewrite as

$$E(s,p)_{complex} = \frac{S(s,p)_{complex}}{p} = \frac{T(s,1)}{p\,T(s,p)_{complex}}$$

5.2.4. Isoefficiency Function
We rewrite equation for efficiency concept E(s, p) as E(s, p) = 1/(1 - h(s, p) / w(s). In order to maintain a constant E(s, p), the workload w(s) should grow in proportion to the overhead h(s, p). This leads to the following relation

$$w(s) = \frac{E(s,p)}{1 - E(s,p)} h(s,p)$$

The factor C = E(s, p)/1 - E(s, p) is for a given efficiency E(s, p) constant. We can then define the isoefficiency function as follows

$$w(s) = C\ h(s,p)$$

5.2.5. Overhead Function
Overhead function h (s, p) defines coincident overhead latencies of given PA (shared memory, distributed memory, hybrid).The typical existed overheads of PA are as follows
- communication latency T(s, p)$_{comm}$
- parallelization latency T(s, p)$_{par}$
- synchronization latency T(s, p)$_{syn}$
- waiting latency to use shared resource T(s, p)$_{wait}$
- influence of parallel computer architecture T(s, p)$_{arch}$
- specific latency of given PA T(s, p)$_{spec}$.

Taking into account all named overhead latencies the overhead function h(s, p) is as follows

$$h(s,p) = \sum \left(T(s,p)_{comm}, T(s,p)_{par}, T(s,p)_{syn}, T(s,p)_{arch}, T(s,p)_{spec} \right)$$

and the complex parallel execution time will be defined as follows

$$T(s,p)_{complex} = T(s,p)_{comp} + h(s,p)$$

In general influence of at least most important overhead latencies is necessary to take into account in performance modeling of parallel algorithms because their influence to complex parallel execution time T(s, p)$_{complex}$ could be dominant. The illustration of such dominant influence of communication overhead latency T(s, p)$_{comm}$ to own parallel computation time T(s, p)$_{comp}$ is shown at Figure 6.

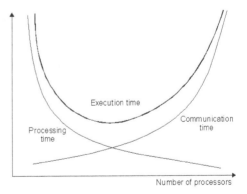

Figure 6. *Illustration of dominant influence of T(s, p)$_{comm}$ to T(s, p)$_{comp}$.*

5.2.6. Modeling of PA Latencies

5.2.6.1. Own Parallel Computation Time
Own computation time T (s, p)$_{comp}$ of PA is given through quotient of maximal time of running parallel processes i. e. as product of complexity C_{pp} (complexity of maximal parallel process) and a parameter t_c as an average value of defined performed computation units (instruction,

block of instructions etc.) divided by number of used computing nodes p of used parallel computer. Parallel computation execution time $T(s, p)_{comp}$ is then given as follows

$$T(s, \mathrm{p})_{comp} = \frac{C_{pp} \cdot t_c}{p}$$

In general execution time of sequential and parallel algorithms are given through multiplicity product of algorithm complexity C_{alg} (dimensionless number of considered computation units) and a parameter t_c as an average value of performed considered operations (instructions, computing steps etc.).

For asymptotic complexity of parallel execution time $T(s, p)_{comp}$ supposing well paralleled problems is then given as

$$T(s, \mathrm{p})_{comp} = \lim_{p \to \infty} \frac{C_{pp} \cdot t_c}{p} = 0$$

5.2.6.2. Modeling of Overhead Latencies

Overhead latencies are defined with its overhead function h(s, p). For every concrete parallel algorithm (shared memory, distributed memory, hybrid), or a similar group of PA, we are able to define their overhead function h(s, p) considering at least the most important overhead latencies of given PA. The individual latencies of given PA comes out from used decomposition model. Complex modeling based on considering overhead function h(s, p) of given PA opens unified complex modeling of any developed parallel algorithm.

5.2.6.2.1. Modeling of Communication Latency

Inter process communication of parallel processes (IPC) $T(s, p)_{comm}$ (communication latency) influences in a decisive degree used decomposition model of PA. Obviously it is higher in distributed computing than in parallel one. For example world known parallel computing model with shared memory PRAM (Parallel Random Access Machine) does not consider communication latency. To model communication latency we have applied theory of complexity to inter process communication of parallel processes in a similar way as in modeling computation latency focusing to a number of performed communication steps (communication complexity). Then communication latency $T(s, p)_{comm}$ is given through number of performed communication steps (communication complexity) for used decomposition model of given PA. Every communication step within parallel computer based on NOW module we can characterized through two basic communication parameters as follows

- communication parameter t_s defined as parameter for initialization of communication step (start up time)
- communication parameter t_w as parameter for transmission latency of considered data unit (typically word).

Illustration of defined communication parameters is at Figure 7. These communication parameters t_s, t_w are

constants for concrete parallel computer [10].

The whole communication overhead latency is given through two basic following functions

- function $f_1(t_s)$ which represents the whole number of communication initializations for given parallel process
- function $f_2(t_w)$ which correspondents to whole performed data unit transmission (usually time of word transmission for given parallel computer) in given parallel process.

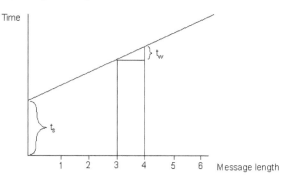

Figure 7. *Illustration of communication parameters.*

These two defined functions limit performance of used parallel computer on defined NOW module of parallel computer. Then using a superposition we can write for communication latency in NOW module $T(s, p)_{comm}$ as follows

$$T(s, p)_{commNOW} = f_1(t_s) + f_2(t_w)$$

To the practical illustration of communication overheads we used discrete fast Fourier transform (DFFT) representing typical matrix parallel algorithm with divide and conquer decomposition model.

To derive the whole communication latency $T(s, p)_{comm}$ in other dominant parallel computers based on integration of NOW modules (network of NOW modules named as Grid) we need to extend the considered two communication functions $f_1(t_s)$, $f_2(t_w)$ in NOW module by third function component $f_3(t_h)$, which will determine potential multiple crossing used NOW modules of integrated parallel computer. This third function is characterized through multiplying hops parameter l_h among NOW modules (generally NOW networks) and average communication latency time of jumped NOW modules with the same communication speed or a sum of individual communication latencies for jumped NOW modules with their different communication speed. Then in general to the whole communication latency in Grid is valid

$$T(s, p)_{commGRID} = f_1(t_s) + f_2(t_w) + \sum_{i=1}^{u} f_3(t_s, t_w, l_h)$$

In general communication latency time $f_3(t_s, t_w, l_h)$ is time to send data message with m words in one communication step among integrated NOW modules with

l_h hops. The communication time for one communication step is then given as $t_s + m\, t_w l_h t_h$, where the new parameters are.

- l_h is the number of network hops
- m is the number of transmitted data units (usually words)
- t_h is average communication time for one hop.

The new communication parameters t_h, l_h depend from a concrete architecture of Grid communication network and used routing algorithm. In [11] we have developed unified models which could help to establish these parameters for dominant parallel computers. For the complex analytical modeling there is necessary to derive for given parallel algorithm or a group of similar algorithms (matrix parallel algorithms) needed communication functions and that always individually for any decomposition strategy) isoefficiency function and defined technical parameters (computational, communication) for used parallel computer (NOW, Grid).

5.2.6.2.2. Modeling of Parallelization Latency

The parallelization latency $T(s, p)_{par}$ represents overhead latency of PA though parallelization of given complex problem. Its consequences are mostly projected as additional communication complexity to $T(s, p)_{comm}$.

5.2.6.2.3. Modeling of Synchronization Latency

The third part of h(s, p) function $T(s, p)_{syn}$ we can eliminate through optimization of load balancing among individual computing nodes of used parallel computer. For this purpose we would measure performance of used computing nodes for given developed parallel algorithm and then based on these achieved results we are able to redistribute better workload of computing nodes. This activity we can repeat until we have optimal redistributed input load (optimal load redistribution based on real performance results).

5.2.6.2.4. Modeling of Influence of Parallel Computer Architecture

The fourth part of h(s, p) overhead function $T(s, p)_{arch}$ (influence of parallel computer architecture) we will model with defined technical parameters t_c, t_s, t_w, which are constant for given parallel computer [10].

5.2.6.2.5. Modeling of Waiting Latency

The fifth part of h(s, p) overhead function $T(s, p)_{wait}$ represents whole waiting times to use shared resources of parallel computer (memory modules, communication channels, I/O devices etc.). This kind of overhead latency is typical in using shared technical resources in a massive way (shared memory modules, communication channels, I/O devices etc.). We can only limit it by optimal allocation of shared resources. To model this overhead latency in an analytical way is a very crucial problem. We have been modeling it in an analytical way applied queuing theory results in combination with experimental measurements [10, 11].

5.2.6.2.6. Modeling of Specified PA Latency

The sixth part of h(s, p) overhead function $T(s, p)_{spec}$ (influence of specified PA) represents any other possible specific overhead latency of given parallel algorithm.

5.2.6.3. Asymptotic Analytical Modeling of PA

Most of to this time known results in analytical modeling of PA in the world for extended classical parallel computers with shared memory (supercomputers, SMP and SIMD systems) or parallel computers with distributed memory based on some cluster of computing nodes mostly did not consider existed overhead influences of PA supposing that they are lower in comparison to the own computation latency $T(s, p)_{comp}$ of performed computations . In this sense analysis and modeling of complexity in parallel algorithms (PA) were rationalized to the analysis of only computation complexity of PA, that mean that the defined function of control and communication overheads h(s, p) were not a part of derived relations for the whole parallel execution time.

The complexity function in the relation for isoefficiency supposed, that dominate influence to the whole complexity of PA has computation complexity of performed massive computations. Such assumption has been proved to be true mainly in using classical parallel computers in the world (supercomputers, massive SMP – shared memory, SIMD architectures etc.). To map mentioned assumption to the relation for asymptotic isoefficiency w(s) means that

$$w(s) = \max\left[T(s,p)_{comp}, h(s,p) < T(s,p)_{comp}\right] = \max\left[T(s,p)_{comp}\right]$$

But in our unified complex modeling of PA possible influence of any part of defined overhead function h(s, p) could be dominant in a nonlinear way. Then for complex isoefficiency function it is necessary to consider it as follows

$$w(s) = \max\left[T(s, p)_{comp}, h(s, p)\right]$$

6. Discrete Fourier Transform

The discrete Fourier transform (DFT) has played an important role in the evolution of digital signal processing techniques. It has opened new signal processing techniques in the frequency domain, which are not easily realizable in the analogue domain. The DFT is a linear transformation that maps n regularly sampled points from a cycle of a periodic signal, like a sine wave, onto an equal number of points representing the frequency spectrum of the signal. The discrete Fourier transform (DFT) is defined as [3, 7]

$$Y_k = \frac{1}{N} \sum_{j=0}^{N-1} X_j e^{-2\pi i \left(\frac{jk}{N}\right)}$$

and the inverse discrete Fourier transform (IDFT) as

$$X_k = \sum_{j=0}^{N-1} Y_j e^{2\pi i\left(\frac{jk}{N}\right)}$$

for $0 \le k \le N-1$. For N real input values $X_0, X_1, X_2, \ldots, X_{N-1}$, transforms generates N complex values $Y_0, Y_1, Y_2, \ldots, Y_{N-1}$. If we use $w = e^{-2\pi i/N}$, that is w is N–th root of complex number i in complex plane we get

$$Y_k = \frac{1}{N}\sum_{j=0}^{N-1} X_j w^{jk}$$

and in inverse as

$$X_k = \sum_{j=0}^{N-1} Y_j w^{-jk}$$

Variable w is a basic part of DFFT computations and is named as twiddle factor. Defined transformation equations are on principle linear transformations.

A direct computation of the DFT or the IDFT requires N^2 complex arithmetic operations. For example the time required for just the complex multiplication in a 1024point DFT is $T_{mult}=1024 . 4 . T_{real}$, where we assumed that one complex multiplication corresponds to four real multiplications and the time required for one real multiplication (T_{real}) is known for given computer. But with this approach we could take into account only the computation times and not also the overheads delays connected with implantation on parallel way.

6.1. The Discrete Fast Fourier Transform

DFFT is a fast method of DFT computation with time complexity $O(N/2 \log_2(N))$ in comparison to sequential DFT algorithm complexity as $O(N^2)$. For a quick computation of the DFT is used adjustment of Cooley-Tukey [7]. To come to final adjustment we start with a modified form of the DFT.

$$Y_k = \frac{1}{N}\sum_{j=0}^{N-1} X_j w^{jk}$$

In general demanded sum is divided to two parts using decomposition strategy "divide-and-conquer". Its principle we describe with modification of origin sum to two following pairs as

$$Y_k = \frac{1}{N}\left(\sum_{j=0}^{\frac{N}{2}-1} X_{2j} w^{2jk} + \sum_{j=0}^{\frac{N}{2}-1} X_{2j+1} w^{(2j+1)k}\right),$$

where the first part of sum contents result part with even indexes and second part with odd indexes. In this sense we get

$$Y_k = \frac{1}{2}\left(\frac{1}{\left(\frac{N}{2}\right)}\sum_{j=0}^{\frac{N}{2}-1} X_{2j} w^{2jk} + w^k \frac{1}{\left(\frac{N}{2}\right)}\sum_{j=0}^{\frac{N}{2}-1} X_{2j+1} w^{2jk}\right) \text{ or}$$

$$Y_k = \frac{1}{2}\left(\frac{1}{\left(\frac{N}{2}\right)}\sum_{j=0}^{\frac{N}{2}-1} X_{2j} e^{-2\pi i\left(\frac{jk}{\left(\frac{N}{2}\right)}\right)} + w^k \frac{1}{\left(\frac{N}{2}\right)}\sum_{j=0}^{\frac{N}{2}-1} X_{2j+1} e^{-2\pi i\left(\frac{jk}{\left(\frac{N}{2}\right)}\right)}\right)$$

Every part of sum means DFT on N/2 values with even indexes and N/2 values with odd indexes. Then we can formally write $Y_k = \frac{1}{2}\left(Y_{even} + w^k Y_{odd}\right)$ for k=0, 1, 2, ..., N-1, whereby Y_{even} is N/2 – point DFT on the values with even indexes $X_0, X_2, X_4, \ldots$ and Y_{odd} is N/2 – point transformation on values $X_1, X_3, X_5, \ldots$.Supposed that k is limited to first 0, 1, ..., N/2–1, N/2 values from whole number of N values. The whole series we can divide to two following parts

$$Y_k = \frac{1}{2}\left(Y_{even} + w^k Y_{odd}\right)$$

and

$$Y_{k+\left(\frac{N}{2}\right)} = \frac{1}{2}\left(Y_{even} + w^{k+\left(\frac{N}{2}\right)} Y_{odd}\right) = \frac{1}{2}\left(Y_{even} - w^k Y_{odd}\right),$$

because $w^{k+N/2} = -w^k$, where $0 \le k < N/2$. In this way we can compute Y_k and $Y_{k+N/2}$ in a parallel way using two N/2 – point transformations according illustration at Figure 8.

Every from N/2 – point DFT we can again divide to next parts, that is to two N/4 – point DFT. Applied decomposition strategy could follows till to exhausting dividing possibility for given N. Dividing factor is named as radix - q, and that for dividing number higher than two.

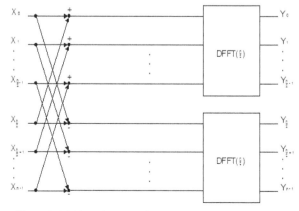

Figure 8. Divide and conquer decomposition strategy for DFFT.

The difference in parallel execution times between the direct implementation of the DFT and DFFT algorithm is significant for large N. Direct calculation of the DFT or IDFT, according to the following program requires N^2 complex arithmetic operations.

```
Program Direct_DFT;
var
  x, Y: array[0..Nminus1] of complex;
begin
  for k:=0 to N-1 do
  begin
    Y[k] :=x[0];
    for n:=1 to N-1 do
      Y[k] := Y[k] + Wnk * x[n];
  end;
end.
```

The difference in execution time between a direct computation of the DFT and the new DFFT algorithm is very high for large N. For example the time required for the complex multiplication in a 1024-point FFT is $T_{mult} = 0.5 \cdot N \cdot \log_2(N) \cdot 4 \cdot T_{real} = 0.5 \cdot 1024 \cdot \log_2(1024) \cdot 4 \cdot T_{real}$, where the complex multiplication corresponds approximately to four real multiplications.

6.2. Parallel Algorithms of Discrete Fast Fourier Transform

Cooley and Tukey have developed the fast DFFT algorithm which requires only $O(N.\log_2(N))$ operations. The difference in execution time between a direct computation of the DFT and the new DFFT algorithm is very large for large N. For example the time required for just the complex multiplication in a 1024-point FFT is $T_{mult} = 0.5.N \log_2(N).4.Treal = 0.5.1024. \log_2(1024).4. T_{real}$, where the complex multiplication corresponds approximately to four real multiplications. The basic form of parallel DFFT is the one-dimensional (1-D), unordered and radix–2 algorithms (using divide and conquer strategy according the principle at Figure 9). The effective parallel computing of DFFT tends to computing one dimensional DFFT with radix greater than two and computing multidimensional FFT by using the polynomial transfer methods. In general a radix-q DFFT is computed by splitting the input sequence of size s into a q sequences of size n/q each, computing faster the q smaller DFFT, and then combining the result. For example, in a radix-4 FFT, each step computes four outputs from four inputs, and the total number of iterations is $\log_4 s$ rather than $\log_2 s$. The input length should, of course, be a power of four. Parallel formulations of higher radix strategies (radix-3, radix-5) 1-D or multidimensional DFFT are similar to the basic form because the underlying ideas behind all sequential DFFT are the same [32]. An ordered DFFT is obtained by performing bit reversal (permutation) on the output sequence of an unordered DFFT. Bit reversal does not affect the overall complexity of a parallel implementation.

6.3. Two-Dimensional 2D DFFT

Processing of images and signals often requires the implementation of a multi-dimensional discrete fast Fourier transform (DFFT). Simplest method of computation of two dimensional DFFT (2D DFFT) is computation of one dimensional DFFT (1D DFFT) on each row and then follows computation of one dimensional DFFT for each column. This illustrates Figure 9.

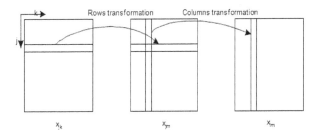

Figure 9. *Two dimensional DFFT.*

6.4. Analyzed Examples

6.4.1. One Element per Processor

This is the simplest example of complexity evaluation of the DFFT. In this case we consider a p=s parallel processor (d-dimensional hypercube architecture) to compute s-point DFFT. A hypercube is a multidimensional mesh of processors with exactly two processors in each dimension. A d-dimensional hypercube consists of p=2d processors. In a d-dimensional hypercube each processor is directly connected to d other processors. In this case we can simply derive that $T(s, 1) = s \log s$ and $T(s, p) = \log s$. Then speedup factor $S(s, p) = p$ and system efficiency $E(s, p) = 1$. Such a formulation of DFFT algorithm for a d-dimensional hypercube calculation is cost optimal but for the higher values of s and use of p=s processors could be only hypothetical.

6.4.2. Multiple Elements per Processor

This is very real case of practical DFFT parallel computations. In this example we examine implementing the binary exchange algorithm to compute an s-point DFFT on a hypercube with p processors, where p > s. Assume that both s and p are powers of two. According the Figure 10, we partition the sequences into blocks of s/p contiguous elements and assign one block to each processor. Assume that the hypercube is d-dimensional ($p=2^d$) and $s=2^r$. Figure 10 shows that elements with indices differing in their d (=2) most significant bits e mapped onto different processors. However, all elements with indices having the same r-d most significant bits are mapped onto the same processor. Hence, this parallel DFFT algorithm performs inter process communication only during the first d = log p of the log s iterations. There is no communication during the rested r – d iterations.

Each communication operation exchanges s/p words of data. Since all communications takes place between directly-connected processors, the total communication time does not depend on the type of routing. Thus the time spent in communication in the DFFT algorithm is $t_s \log p + t_w (s/p) \log p$, where t_s is the message start up time and t_w is the per-word transfer time. These times are known for the concrete parallel system. If a complex multiplication and addition pair takes time t_c, then the parallel run time for s-point DFFT on a p-processor hypercube is as follows

$$T(s,p)_{complex} = t_c \frac{s}{p} \log s + t_s \log p + t_w \frac{s}{p} \log p$$

The expressions for complex speedup $S(s,p)_{complex}$, efficiency $E(s,p)_{complex}$ and defined constant C (part of issoefficiency function) are given by the following equations

$$S(s,p)_{complex} = \frac{T(s,1)}{T(s,p)_{complex}} = \frac{t_c s \log s}{T(s,p)_{complex}} =$$

$$= \frac{p \, s \log s}{s \log s + (t_s/t_c) \, p \log p + (t_w/t_c) \, s \log p}$$

$$E(s,p)_{complex} = \frac{S(s,p)_{complex}}{p} =$$

$$= \frac{1}{1 + \{(\log p / \log s)[(t_s p / t_c s) + (t_w / t_c)]\}}$$

$$C = \frac{1 - E(s,p)_{complex}}{E(s,p)_{complex}} = \frac{t_s \, p \log p}{t_c \, s \log s} + \frac{t_w \log p}{t_c \log s}$$

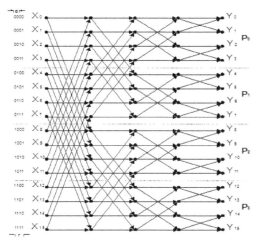

Figure 10. 16-point DFFT on four processors.

6.4.3. Multiple Elements per Processor with Routing

This is a most complicated, but very real, case in parallel computing. It is typical for the parallel architectures, in which the processors do not have enough direct connected processor to compute the given parallel algorithm. Then the communication of not direct connected processors or computers is realized through a number (hop) of other processors or communication switches according some routing algorithm.

The time for sending a message of size m between processors, that are l_h hops apart is given by $t_s + t_h l_h$ where t_s is the starting message time and t_h is the overhead time for one hop. The values t_s, t_h, l_h depend from the architecture of parallel system (mainly its interconnection

network) and routing strategy. If we can define these values for a concrete parallel system and routing strategy, cost performance tradeoffs are to be analyzed in a similar way than in previous case.

6.4.3. Multiple Elements per Processor in NOW

An example of our experimental parallel computer based on NOW we have illustrated at Figure 2. Based on this used NOW module communication overheads depends on topology of communication network and its communication parameters (transmission speed, bandwidth, transmission control etc.). To verify derived analytical results we have performed some simulation experiments in NOW module with DFFT's PA. From performed experiments comes out that for effective parallel computing of DFFT the use of centralized massively parallel system should by preferred to asynchronous parallel computers based on NOW modules. At experimental testing we have used workstations of NOW parallel computer as follows

- WS 1 – Pentium IV (f = 2.26 GHz)
- WS 2 - Pentium IV Xeon (2 proc., f = 2.2 GHz)
- WS 3 - Intel Core 2 Duo T 7400 (2 cores, f=2.16 GHz)
- WS 4 - Intel Core 2 Quad (4 cores, f = 2.5 GHz)
- WS 5 - Intel SandyBridge i5 2500S (4 cores, f=2.7 GHz).

7. Results

As scalable parallel computer we have defined any parallel computer in which the efficiency can be kept fixed as the number of computing nodes is increased, provided that the problem size is also increased. The scalability of a PA/parallel computer combination determines its capacity to use an increased number of computing nodes p effectively. We have considered the Cooley-Tukey algorithm for one dimensional s-point DFFT to maintain the same efficiency. Figure 11 illustrates the efficiency $E(s,p)_{complex}$ of the binary exchange DFFT parallel algorithm as a function of s on a 512 processors (computing nodes) hypercube parallel computer with its following technical parameters t_c=2μs, t_w= 4 μs and t_s =25 μs. The threshold point is given as $t_c / (t_c + t_w) = 0,33$. The efficiency initially increases rapidly with the problem size to the threshold, but then the efficiency curve platens out beyond the threshold. The binary exchange algorithm yields good performance on a hypercube provided that the communication bandwidth and the processing speed of the computing nodes are balanced. Efficiencies below a certain threshold can be maintained while increasing the problem size at a moderate rate with an increasing number of processors. We can say that the use of transpose algorithm will have much higher overhead than the binary exchange algorithm due to message start up time t_s, but has a lower overhead due to per-word transfer time t_w. As a result, either of the two algorithm formulations may be faster depending on the relative values of t_s and t_w. In principle supercomputers and other architectures with common memory have t_s very low in comparison to typical NOW or Grid.

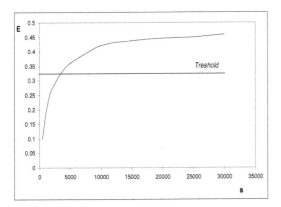

Figure 11. *The efficiency $E(s, p)_{complex}$ of the DFFT binary exchange parallel algorithm.*

However, this threshold is very low if the communication bandwidth of the hypercube is low, compared to the speed of its processors. Therefore it is necessary to describe a different parallel formulation of DFFT for interconnection network of computer network. Such a formulation involves matrix transposition for an array of s-input points and hence we called it the transpose parallel algorithm. In principal we can say that the use of transpose PA will have a much higher overhead than the binary exchange algorithm due to message start up time t_s, but has a lower overhead due to per-word transfer time t_w. As a result, either of the two algorithm formulations may be faster depending on the relative values of t_s and t_w. In principle parallel computers with shared memory have t_s very low in comparison to typical dominant parallel computers based on NOW and Grid.

The influence of matrix dimension for DFFT to the communication overheads in computer network of personal computers is illustrated at the Figure 12. From this picture we can see that to more effective DFFT computing in computer network there is necessary another transpose algorithm.

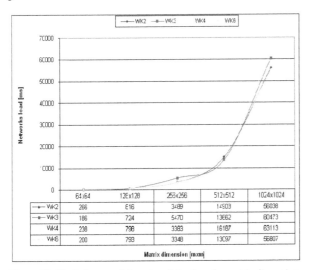

Figure 12. *The influence of the network load on the matrix dimension.*

The performance results in parallel computer NOW based on Ethernet we have graphically illustrated for 2D FFT at the Figure 13. These results of 2D FFT parallel algorithm document increasing of both computation and communication parts in geometrically way with the quotient value nearly four for analyzed matrix dimensions and that in increasing matrix dimension mean to do twice more computation on columns and twice more on rows.

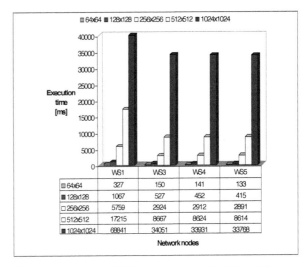

Figure 13. *Results in NOW for $T(s, p)_{complex}$ of 2DFFT for Ethernet.*

At Figure 14 we have illustrated parallel speed up $S(s, p)_{complex}$ of 1D DFFT parallel algorithm with binary data exchange for defined workload s = 65 536 ($s=n^2 / p$) as function of number of computing nodes p. Character of $S(s, p)_{complex}$ is sub linear i. e. always less as illustrated p curve (ideal speed up without overheads) as a consequence of overhead latencies (architecture, communication, synchronization etc.) of used parallel computer.

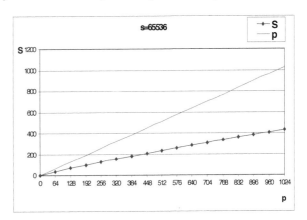

Figure 14. *Illustration of $S(s, p)_{complex}$ as factor of computing nodes number p.*

Figure 15 illustrates isoefficiency functions $w(s)_{complex}$ of 1D DFFT parallel algorithm. For lower values of $E(s, p)$ (0,1; 0,2; 0,3; 0,4; 0,45) to the threshold (E = 0,33) we have used based on performed analysis in theoretical part computed using following expression

$$w(s)_{complex} = s \log s = C \frac{ts}{tc} p \log p$$

and the values above the threshold of efficiency function were computed on following relation

$$w(s)_{complex} = C \frac{tw}{tc} p^{C \, tw/tc} \log p$$

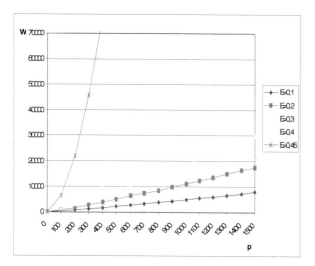

Figure 15. *Issoefficiency function w(s)ₐₒₘₚₗₑₓ of 1D DFFT (E =0,1; 0,2; 0,3; 0,4; 0,45).*

Figure 16 illustrates issoefficiency functions w(s)_complex 1D DFFT on hypercube parallel computer for the values of E(s, p)_complex = 0,5 and E(s, p)_complex =0,55. From illustrated curves at Figure 18 we can see in theoretical part of this section predicted stormy growth of issoeficiency function w(s)_complex i.e stormy tendency of algorithm scalability for analyzed parallel algorithm 1D DFFT with binary data exchange past the threshold.

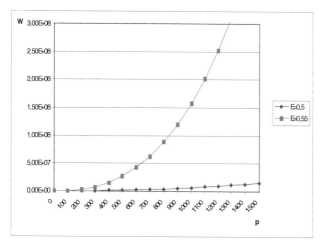

Figure 16. *Issoefficiency function w(s)ₐₒₘₚₗₑₓ of 1D DFFT (E =0,5; 0,55).*

Figure 17 illustrates influence of parallel computer architecture to its algorithm scalability based on given efficiency for E(s, p)_complex = 0,4 according the derived conclusions in theoretical part of this section. We

considered the Cooley-Tukey algorithm for one dimensional s-point DFFT parallel algorithm with binary data exchange. We can see that this parallel algorithm is essentially better scalable for the hypercube parallel architecture as for the mesh parallel computer. This implies better architecture of communication network in hypercube for parallel solution of 1D DFFT with binary data exchange (lover communication latency). This verified resulting idea documents important role of parallel computer architecture to the whole performance of given parallel algorithm. The scalability of an algorithm-architecture combination determines its capacity to use effectively an increased number of computing nodes p.

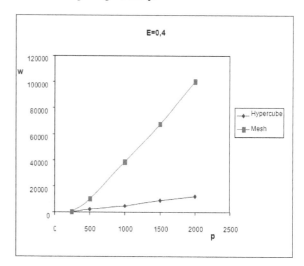

Figure 17. *Influence of parallel computer architecture to scalability.*

8. Conclusions

Performance evaluation as a discipline has repeatedly proved to be critical for design and successful use of parallel computers and parallel algorithms too. At the early stage of design, performance models can be used to project the system scalability and evaluate design alternatives. At the production stage, performance evaluation methodologies can be used to detect bottlenecks and subsequently suggests ways to alleviate them. Analytical methods (order analysis, queuing theory systems and Petri nets), simulation, experimental measurements, and hybrid modeling methods have been successfully used for the evaluation of system and its components too. Via the extended form of complex isoefficiency concept we have illustrated its concrete using to predicate the performance in applied matrix parallel algorithms.

To derive complex isoefficiency function in analytical way it is necessary to derive al typical used criterion for performance evaluation of parallel algorithms including their overhead function h(s, p). Based on these relations we are able to derive complex issoefficiency function as real criterion to evaluate and predict performance of parallel algorithms also for theoretical (not existed) parallel computers. So in this way we can say that this process includes complex performance evaluation including performance prediction.

This paper continues in applying complex analytical modeling to another group of matrix parallel algorithms (MPA) which is characterized by decreasing number of decomposed matrix elements (decreasing of complexity) in process of parallel execution. To present this group of MPA we have chosen discrete fast Fourier transform (DFFT) as Pa with its intensive communication complexity. On various real examples of the DFFT we have described complexity determination of PA not only for illustrated DFFT PA but also for other similar MPA. The considered complex analyzed examples we have been evaluated so on classic massive parallel computers (hypercube, mesh) as on actually dominant parallel computers NOW and Grid. It is obvious that in some cases using of network of workstations (NOW) or its higher integration parallel computers named as Grid (integrated network of NOW networks) could be less effective than on innovated classic massive parallel computers but NOW and Grid belong to more flexible and perspective parallel computers.

Acknowledgements

This work was done within the project "Modeling, optimization and prediction of parallel computers and algorithms" at University of Zilina, Slovakia. The author gratefully acknowledges help of project supervisor Prof. Ing. Ivan Hanuliak, PhD.

References

[1] Abderazek A. B., Multicore systems on chip - Practical Software/Hardware design, Imperial college press, pp. 200, 2010

[2] Arora S., Barak B., Computational complexity - A modern Approach, Cambridge University Press, pp. 573, 2009

[3] Casanova H., Legrand A., Robert Y., Parallel algorithms, CRC Press, 2008

[4] Desel J., Esperza J., Free Choise Petri Nets, Cambridge University Press, UK, pp. 256, 2005

[5] Dubois M., Annavaram M., Stenstrom P., Parallel Computer Organization and Design, pp. 560, 2012

[6] Dubhash D.P., Panconesi A., Concentration of measure for the analysis of randomized algorithms, Cambridge University Press, UK, 2009

[7] Edmonds J., How to think about algorithms, Cambridge University Press, UK, pp. 472, 2010

[8] Gelenbe E., Analysis and synthesis of computer systems, Imperial College Press, pp. 324, 2010

[9] Hager G., Wellein G., Introduction to High Performance Computing for Scientists and Engineers, pp. 356, July 2010

[10] Hanuliak P., Analytical method of performance prediction in parallel algorithms, The Open Cybernetics and Systemics Journal, Vol. 6, Bentham, UK, pp. 38-47, 2012

[11] Hanuliak M., Unified analytical models of parallel and distributed computing, American J. of Networks and Communication, Science PG, Vol. 3, No. 1, USA, pp. 1-12, 2014

[12] Hanuliak M., Hanuliak I., To the correction of analytical models for computer based communication systems, Kybernetes, Vol. 35, No. 9, UK, pp. 1492-1504, 2006

[13] Hanuliak M., Performance modeling of Nov and Grid parallel computers, AD ALTA – Vol. 3, issue 2, Hradec Kralove, Czech republic, pp. 91-96, 2013

[14] Harchol-BalterMor, Performance modeling and design of computer systems, Cambridge University Press, UK, pp. 576, 2013

[15] Hwang K. and coll., Distributed and Parallel Computing, Morgan Kaufmann, pp. 472, 2011

[16] John L. K., Eeckhout L., Performance evaluation and benchmarking, CRC Press, 2005

[17] Kshemkalyani A. D., Singhal M., Distributed Computing, University of Illinois, Cambridge University Press, UK, pp. 756, 2011

[18] Kirk D. B., Hwu W. W., Programming massively parallel processors, Morgan Kaufmann, pp. 280, 2010

[19] Kostin A., Ilushechkina L., Modeling and simulation of distributed systems, Imperial College Press, pp. 440, 2010

[20] Kumar A., Manjunath D., Kuri J., Communication Networking , Morgan Kaufmann, pp. 750, 2004

[21] Kushilevitz E., Nissan N., Communication Complexity, Cambridge University Press, UK, pp. 208, 2006

[22] Kwiatkowska M., Norman G., and Parker D., PRISM 4.0: Verification of Probabilistic Real-time Systems, In Proc. of 23rd CAV'11, Vol. 6806, Springer, pp. 585-591, 2011

[23] Le Boudec Jean-Yves, Performance evaluation of computer and communication systems, CRC Press, pp. 300, 2011

[24] McCabe J., D., Network analysis, architecture, and design (3rd edition), Elsevier/ Morgan Kaufmann, pp. 496, 2010

[25] Meerschaert M., Mathematical modeling (4-th edition), Elsevier, pp. 384, 2013

[26] Misra Ch. S.,Woungang I., Selected topics in communication network and distributed systems, Imperial college press, pp. 808, 2010

[27] Miller S., Probability and Random Processes, 2nd edition, Academic Press, Elsevier Science, pp. 552, 2012

[28] Peterson L. L., Davie B. C., Computer networks – a system approach, Morgan Kaufmann, pp. 920, 2011

[29] Resch M. M., Supercomputers in Grids, Int. J. of Grid and HPC, No.1, pp. 1-9, 2009

[30] Riano l., McGinity T.M., Quantifying the role of complexity in a system´s performance, Evolving Systems, Springer Verlag, pp. 189–198, 2011

[31] Ross S. M., Introduction to Probability Models, 10th edition, Academic Press, Elsevier Science, pp. 800, 2010

[32] Takahashi D., Kanada Y.: High-performance radix-2, 3 and 5 parallel 1-D complex FFT algorithms for distributed-memory parallel computers, J. of Supercomputing, 15, Kluwer Academic Publishers, The Netherlands, pp. 207-228, 2000

[33] Wang L., Jie Wei., Chen J., Grid Computing: Infrastructure, Service, and Application, CRC Press, 2009 www pages

[34] www.top500.org

[35] www.intel.com

[36] www.spec.org.

20

Modeling of communication complexity in parallel computing

Juraj Hanuliak

Dubnica Technical Institute, Sladkovicova 533/20, Dubnica nad Vahom, 018 41, Slovakia

Email address:

hanuliak@dti.sk

Abstract: Parallel principles are the most effective way how to increase parallel computer performance and parallel algorithms (PA) too. Parallel using of more computing nodes (processors, cores), which have to cooperate each other in solving complex problems in a parallel way, opened imperative problem of modeling communication complexity so in symmetrical multiprocessors (SMP) based on motherboard as in other asynchronous parallel computers (computer networks, cluster etc.). In actually dominant parallel computers based on NOW and Grid (network of NOW networks) [31] there is necessary to model communication latency because it could be dominant at using massive (number of processors more than 100) parallel computers [17]. In this sense the paper is devoted to modeling of communication complexity in parallel computing (parallel computers and algorithms). At first the paper describes very shortly various used communication topologies and networks and then it summarized basic concepts for modeling of communication complexity and latency too. To illustrate the analyzed modeling concepts the paper considers in its experimental part the results for real analyzed examples of abstract square matrix and its possible decomposition models. These illustration examples we have chosen first due to wide matrix application in scientific and engineering fields and second from its typical exemplary representation for any other PA.

Keywords: Parallel Computer, NOW, Grid, Shared Memory, Distributed Memory, Parallel Algorithm, MPI, Open MP, Model, Decomposition, Communication, Complexity, Modeling, Optimization, Overhead

1. Introduction

Communications in parallel and distributed computing has been considered as two separate research disciplines. Parallel computing has addressed problems of communication and intensive computation on highly coupled computing nodes while distributed computing has been concerned with coordination, availability, timeliness, etc., of more likely coupled computing nodes. Current trends, such as parallel computing on networks of high performance computing nodes (workstations) and Internet computing, suggest the advantages of unifying these two research disciplines.

Parallel and distributed computing share the same basic computational model consisting on physically distributed parallel processes that operate concurrently and interact with each other in order to accomplish a task as a whole. In parallel computing, processes are assumed to be placed closer to each other and they could communicate frequently and hence the ratio of computation/communication of parallel applications is usually much smaller than that in distributed applications. On the other hand, distributed computing focuses on parallel processes that could be allocated in a wide area i. e., communication between some parallel processes is assumed to be more costly than in parallel computing.

A number of recent trends point to a convergence of communication research in parallel and distributed computing [9, 15]. First, increased communication bandwidth and reduced latency make geographical distribution of computing nodes less of a barrier to parallel computing.

2. Communications in Parallel Computing

From the point of necessary communication modeling in parallel computing we can divide communications as

follows
- communications in parallel computers
- communications in parallel algorithms.

2.1. Communications in Parallel Computers

Communications in parallel computers we can divide as follows
- communications in parallel computers with shared memory
- communications in parallel computers with distributed memory.

2.1.1. Communication Networks with Shared Memory

To parallel computers with shared memory belong parallel computers as follows
- classic parallel computers
  - multiprocessors
  - massive parallel computers (supercomputers) [17, 28]
- modern symmetrical multiprocessor systems (SMP)
  - SMP multiprocessors
  - SMP multicores
  - mixed (processors, cores).

Typical actual example of SMP multiprocessor systems (Intel Xeon) illustrates Fig. 1.

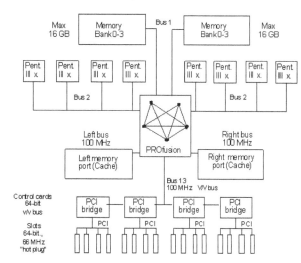

PROfusion - cross switch of 3 bus and 2 memory ports (parallel)

PCI cards - type Enhanced PCI (64 bit, 66 MHz, "Hot Plug" - on-line exchange)

Figure 1. *Architecture of SMP parallel computer (8-Intel processor).*

From illustrated Fig. 1 we can see that parallel using of computing nodes (processors) requires at least one communication network (at Fig. 1 PROfusion) to realize computing nodes cooperation solving any complex problem in a parallel way. Concretely it means two basic types of communications and that
- inter process communications (IPC) of processors via shared memory
- access of computing nodes to shared input/output (I/O) devices (I/O communications).

To this time various realized communication network

(switches) mainly in classic parallel computers with shared memory have used topologies or communication networks as follows [2, 25, 32]
- deterministic
  - bus (multibus)
  - multistage
  - array
  - crossbar
  - annulus (ring)
  - mesh, annuloid (torus)
  - boolean n-dimension cubes (hypercube)
  - butterfly
  - omega
  - shuffle (perfect, logarithmic, with exchange, k-routes)
  - De Bruin network
  - Banyan network
  - Batcher network
  - Benes network
  - ATM (asynchronous transfer mode)
  - FDDI (Fiber Distributed Data Interface)
  - trees (X-tree, H-tree, fat tree, hyper tree)
- stochastic
- hash networks.

2.1.2. Communication Networks for Parallel Computers with Distributed Memory

For parallel computers with distributed memory (computer networks, cluster, NOW, Grid) the typical used topologies or communication networks are as follows [27, 33]
- bus, multibus
- star
- tree
- ring
- Ethernet (Fast, Giga, 10 Giga)
- high speed communication networks
  - Myrinet
  - Infiniband
  - Quadrics.

2.2. Communications in Parallel Algorithms

In principal we can divide communication in parallel algorithms (PA) to the following groups
- inter process communications in parallel algorithm using shared memory (PA_{sm}). Shared memory (at least a part) allow to use it for communications via I/O instructions of given computing node (processor) or supported parallel developing standards
- inter process communications in parallel algorithm using distributed memory (PA_{dm}). All needed cooperation of parallel processes have to use only asynchronous data message communication via parallel supported developing standards
- inter process communications in hybrid PA which combine using of both previous PA (PA_{hyb}).

The main difference between PA_{sm} and PA_{dm} is in form of inter process communication (IPC) among created parallel processes [5, 17]. Generally we can say that IPC communication in parallel system with shared memory can use more existed communication possibilities (I/O instructions, communication services in shared memory) than in distributed systems (only network communication).

2.2.1. Inter Process Communication

In general we can say that dominated elements of parallel algorithms are their sequential parts (Parallel processes) and inter process communication (IPC) among performed parallel processes.

2.2.1.1. Inter Process Communication in Shared Memory

Inter process communication (IPC) for parallel algorithm with shared memory (PA_{sm}) is defined within supporting developing standards as following
- OpenMP
- OpenMP threads
  - Pthreads
  - Java threads
  - other.

The concrete communication mechanisms use existence of shared memory which allows every parallel process to story communicating data at some addressed memory place and then another parallel process to read stored data.

2.2.2.2. Inter Process Communication in Distributed Memory

Inter process communication (IPC) for parallel algorithm with distributed memory (PA_{dm}) is defined within supporting developing standards as following
- MPI (Message passing interface)
  - point to point (PTP) communication commands
    - send commands
    - receive commands
  - collective communication commands
    - data distribution commands
    - data gathering commands
- PVM (Parallel virtual machine)
- Java (Network communication services)
- other.

Typical MPI network communication is at Fig. 2. Based on existed communication links MPI contains mentioned collective communication commands.

2.3. Influence of Communications to Performance Tuning

Performance tuning means performance modeling and optimization of PA (effective PA). This step contents modeling and analysis in such a way to minimize the whole execution time of parallel computing. To achieve effective PA depends mainly from following factors
- optimal selection of communication networks in parallel computers
- minimization of needed inter process communication and other accompanying overheads

(parallelization, control of PA, waiting times) [16].

In actually dominated asynchronous parallel computers (NOW, Grid) there are necessary to reduce (optimize) mainly number of inter process communications IPC (Communication complexity) for example through possible using of alternative decomposition model.

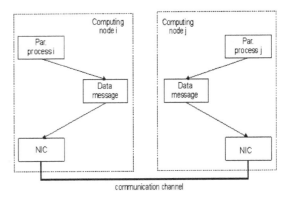

Figure 2. *Typical MPI network communication.*

3. Parallel Computing Models

Parallel computational model is an abstract model of parallel computing, which should include overhead and accompanying delays. Model is characterized by the possibility of parallel computers, which are for parallel computing deterministic. Abstraction degree should characterize also communication structure and contemporary permit at least approximation of its basic parameters (complexity, performance etc.) [14, 29]. On the other hand approximation accuracy is limited by requirement that abstract communication models have to represent similar parallel computer architectures and parallel algorithms [18]. It is clear that for every specific parallel computer and parallel algorithm too we are able to create their own communication model, which characterizes in detail their specific characteristics. Parallel communication models can be classified according various criterions. One of most used criteria is presentation way of model parameters. Typical used communication parameters can be divided into two groups as follows
- semantic
  - communications network architecture (architecture, channels, control)
  - communication methods (communication protocols)
  - communication delay (latency)
- performance (complexity, efficiency). Typical parameters are
  - working load s for given PA
  - size of the parallel system p (number of processors)
  - workload w - number of operations
  - sequential program execution time T(s, 1)
  - the computation execution time $T(s, p)_{comp}$
  - the whole execution time of a parallel algorithm T(s, p)

- parallel speed up $S(s, p)$
- efficiency $E(s, p)$
- isoefficiency $w(s)$
- average time of computation unit t_c (instruction, defined computing step etc.)
- communication technical parameters
  - average time to initialize communication (startup time) – t_s
  - average time to transmit data unit (data word) - t_w.

3.1. Communication Model with Shared Memory

3.1.1. PRAM Model

Model of parallel computer with shared memory PRAM (Parallel Random Access Machine) was previously used for its high degree of universality and abstractness. PRAM model still represents an idealized model, because it is not considering any delay. Although this approach has an important role in the theoretical design and development of parallel computers and parallel algorithms but for real modeling it is necessary to complete it by modeling at least of communication delays. Typical PRAM model illustrates Fig. 3.

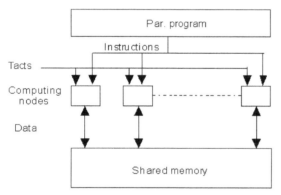

Figure 3. *PRAM model.*

In PRAM model computing nodes communicate via shared memory whereby every addressed place according PRAM model is available at the same time. Computing nodes are at their activities synchronized and communicate via shared memory. For practical design of a parallel algorithm programmer specifies sequences of parallel operations using shared memory. In performing parallel processes may come to long waiting delays which are increased proportionally to number of used parallel processes [4]. These time delays is necessary to model, analyze their behavior and make their real evaluation (removing of idealized PRAM model assumption).

3.1.1.1. Fixed Communication Model

Fixed communication model GRAM (Graph Random Access Machine) was one of the ways how to solve problem of waiting delays in PRAM model using distributed memory with precisely defined structure of its communication network in which symbol G determines topology graph of used communication network [25]. As examples we can name two dimensional communication networks and hypercube topology.

3.2. SPMD Model

Parallel computing model SPMD (Single Process Multiple Data) corresponded to classical parallel computers with shared memory (supercomputers, massive SMP) which were primary focused on massive data parallelism. Illustration of this model is shown at Fig.4. Such an orientation program assumes mostly following decomposition models [1, 17]

- domain decomposition
- manager / worker.

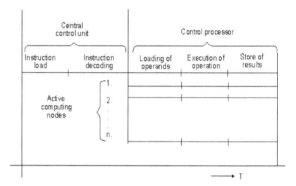

Figure 4. *SPMD model.*

3.3. Flexible Models

Previous models were not sufficiently precise, because increasing robustness of parallel computers cause also rises of communication overheads in parallel algorithms. The precise developed parallel computer represented robustness by number of computing nodes with parameter p, whereby every computing node was ready to work with n / p parallel processes. Parallel algorithm then consisted of sequence of defined parallel steps named as super steps, in which were done needed local calculations followed by communication exchange of data messages. It is obvious that such implemented parallel algorithm, in which number of super steps was small and independent of input load n, will be effective in any parallel computer providing efficient implementation just of communications procedures.

3.3.1. Flexible GRAM Model

Basic difference between fixed and flexible GRAM model is in number of computing nodes (Processors), which was considered with defined parameter p. At every stage of communication phase computing node could send data messages with their variable length to its neighbor computing nodes. Communication prize could be also subject of modeling and included following parts

- communication section to initialize communication (Startup time)
- own transmission part of communication defined as number of transmitted considered data units (Words).

3.3.2. BSP Model

Communication model BSP (Bulk Synchronous Parallel) is a realistic alternative of PRAM model (Fig. 5.). Number of parallel super steps (input load n) was divided to p computing processors. Updates of this communication model have used instead of synchronization after every performed instruction only synchronization at the end of performed partial computation referred as super step. Super step consisted from defined number of instructions (bulk). Each super step consisted of three following phases

- own partial computation
- global communications of processors
- barrier synchronization.

At super step used processors are performing their instructions asynchronously, whereby all read operations of collective memory of every processor were performed before performing the first write operation to shared memory. Existing delays of parallel algorithm were defined as follows

- parallel computation time were given by the maximum number of computation cycles w
- synchronization delay has its lower bound as the waiting time for transmission of minimal communication data messages through communications network
- • communication delay was given as the product g.h cycles, where parameter g characterizes throughput of communication network. Parameter h specified number of cycles for communication of maximal data message at super step. To avoid conflicts due to asynchronous communication network activities, send data message in stage by some processor is not dependent on received messages in the same phase of communication
- execution time for one super step is then given as the sum of the partly considered sub delays and that w + g.h + l.

BSP model does not exclude overlapping of individual super step activities. In the case of overlapping of defined actions execution time of super step were given as max (w, g.h, l).

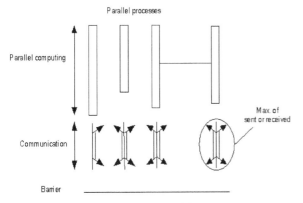

Figure 5. BSP model.

3.3.2.1. Adjusted BSP Model

Further innovation of BSP model includes adjustments PRAM and BSP models in such a way that modified model could be precisely characterize behavior of real parallel computers. These innovations were based on following

- parallel algorithm is performed in sequence of phases. There were following three types of phases and that namely
  - parallelization overheads $T(s, p)_{overh}$
  - own parallel computing $T(s, p)_{comp}$
  - interaction of computing nodes $T(s, p)_{interact}$ (communication, synchronization).
- for given computation phase were determinate input load with parameters that indicate average value of performed operations $t_c(p)$
- different interaction imposes different execution times. Execution time could be computed according following relation

$$T_{interact}(m, p) = t_s(p) + \frac{m}{r_\infty(p)} = t_s(p) + m\, t_c(p).$$

In this relationship m indicates data message length in bytes, $t_s(p)$ is communication start up time and $r \propto (p)$ is bandwidth limit of used communication channels.

3.3.2.2. CGM (Coarse Grained Multicomputer) Model

This model is based on BSP model and is represented by p processors whereby each of them with O(n / p) local memories for which every super step has h = O(n / p) communication cycles. The aim is to concentrate on a proposal with fewer super steps in order to achieve higher effectiveness of developed parallel algorithms. The ideal situation means to perform constant number of super steps as it was done in developed parallel algorithms as sorting, image processing, optimization problems etc.

3.3.2.3. Log P Model

Log P model is based on BSP model and focuses on a looser bound parallel computer architectures (Asynchronous parallel computers).The emphasis is on a parallel computer with distributed memory with parameters according Fig. 6 where

- L: time for communication initialization (Startup time)
- o: overflow due to communication activities. It is defined as the time interval during which computing node performs only control of performed communication
- g:gap between two consecutive transmitted data messages. It is defined as the inversion of bandwidth of the communication control processor
- p: number of computing nodes of parallel computer.

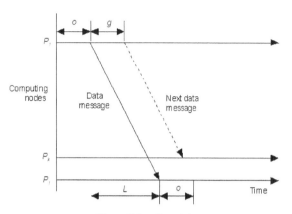

Figure 6. Log P model.

In this model they are considered the resources with their limited capacity. Consequently then only L / g data messages can be at a given time in communication network. Price for basic communication data block between two computing nodes is L + 2 o. If we require acknowledgement (ACK) then price is given as 2 L 2 + o.

3.4. Communication Models with Distributed Memory

3.4.1. Model MPMD

Computational model MPMD (Multiple Process Multiple Data) is associated with computer networks mainly in asynchronous parallel computers. As network topologies in computer networks (LAN, WAN) there is typically used following topological structure [6]

- bus
- star
- tree
- ring.

Suitable decomposition models are those which tend to functional parallelism, that mean to create of parallel processes, which can then perform allocated part of parallel algorithms on corresponding data. Typical decomposition models are as following [8]

- functional decomposition
- manager / server (Server / client, master / worker)
- object oriented programming OOP.

4. Complexity in Communication Networks

Typical communication network using single shared communication channel is illustrated at Fig. 7. The main disadvantage of such communication network is a serial communication among connected computer nodes. To analyze communication complexity we can apply analytical method of complexity theory. Then upper limit of communication complexity at Ethernet is given as O (p) for supposed network connection according Fig. 7. Communication network with this communication complexity limits development of effective parallel algorithms using serial communications as in case of

Ethernet network.

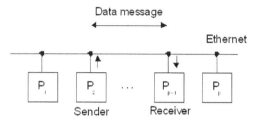

Figure 7. Communication in Ethernet network.

Typical communication network in NOW is based on Ethernet network. The communication principles in this network are illustrated at Fig. 7 where P_1, P_2, ... P_{p-1}, P_p could by common powerful single workstations or SMP parallel computers. Generally implementing computational model MPMD brings different overhead delays as follows

- parallelization of complex problem
- synchronization of decomposed parallel processes
- inter process communication (IPC) delay.

All these delays in parallel computers with distributed memory are reflected to communication complexity of used communication network.

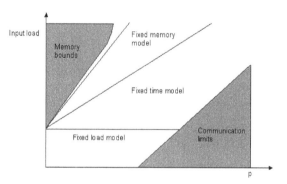

Figure 8. Real communication models.

Real application models should take into account potential lack of limited communication channels at implementation of parallel algorithms (technical communication limits) respectively other limited required technical resources [19]. Illustration of resource technical limits illustrates Fig. 8.

4.1. Modeling of Communication Complexity

To model communication complexity in actual parallel computing is of high importance from these causes [20]

- it plays important role in achieving high performance of all actual parallel computers (SMP, Now, Grid)
- to develop effective PA there is necessary to model and optimize inter process communications mainly for parallel algorithms with distributed memory [10].

Fig. 9 illustrates typical relation between parallel computation time $T(s, p)_{comp}$ (Processing time) in parallel computing and communication latency $T(s, p)_{comm}$ in parallel algorithms with intensive IPC communications.

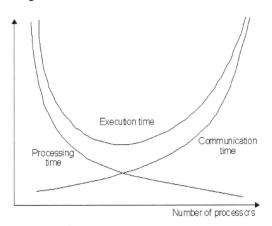

Figure 9. *Relations among parts of parallel execution time.*

We can easily show that limit of processing time $T(s, p)_{comp}$ with increasing number of computing nodes p goes to null using theory of complexity. Processing time complexity $T(s, p)_{comp}$ is given through quotient of running time of the greatest parallel process PP (product of its complexity Z_{pp} and a constant t_c as an average value of performed computation operations) through number of used computation nodes of the given parallel computer. Based on them we are able to derive for parallel computation time $T(s, p)_{comp}$ following relation

$$T(s, \mathrm{p})_{comp} = \frac{Z_{pp} \cdot t_c}{p}$$

Supposing ideally parallelized problem (for example matrix PA) and theoretical unlimited number of computation nodes p mathematical limit of $T(s, p)_{comp}$ is given as

$$T(s, \mathrm{p})_{comp} = \lim_{p \to \infty} \frac{Z_{pp} \cdot t_c}{p} = 0$$

For effective parallel algorithms we are seeking for the bottom part of whole execution time according Fig. 9.

4.2. Communication Latencies

Inter process communication of parallel processes (IPC) $T(s, p)_{comm}$ (communication latency) influences in a decisive degree used decomposition model of PA. Obviously it is higher in parallel algorithms with distributed memory PA_{dm} than in other ones. To model communication latency we have applied theory of complexity to inter process communication $T(s, p)_{comm}$ of parallel processes in a similar way as in modeling computation latency $T(s, p)_{comp}$ focusing to a number of performed communication steps (communication complexity). Then communication complexity $Z(s, p)_{comm}$ is given through number of performed communication steps (communication complexity) for used decomposition model of given PA. Every communication step within parallel computer based on NOW module we

can characterized through two basic communication parameters as follows

- communication parameter t_s defined as parameter for initialization of communication step (startup time)
- communication parameter t_w as parameter for transmission latency of considered data unit (typically word).

Illustration of defined communication parameters is at Fig. 10. These communication parameters t_s, t_w are constants for defined parallel computer [11]. Then the communication latency $T(s, p)_{comm}$ using communication complexity $Z(s, p)_{comm}$ and the defined communication parameters is given as follows

$$T(s, p)_{comm} = Z(s, p)_{comm}(t_s + t_w)$$

The whole communication latency is given through two basic following functions

- function $f_1(t_s)$ which represents the whole number of communication initializations for given parallel process
- function $f_2(t_w)$ which correspondents to whole performed data unit transmission (usually time of word transmission for given parallel computer) in given parallel process.

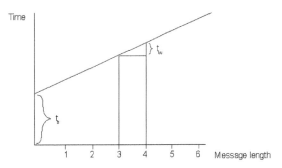

Figure 10. *Illustration of communication parameters.*

These two defined functions limit performance of used parallel computer on defined NOW module of parallel computer. Then using a superposition we can write for communication latency in NOW module $T(s, p)_{comm}$. as follows

$$T(s, p)_{commNOW} = f_1(t_s) + f_2(t_w)$$

The most difficult but in practice a common example of communication complexity for massive and Grid (network of NOW networks) is network communication included crossing through several communications networks (hops), which are interconnected by routers respectively other connecting communication elements (repeaters, switches, bridges, gates etc.).In such case, communication is done through number of control communications processors, or communication switches whereby in such transmission chain could occur communication networks with remote data transmission. Number of network crossings through various communication networks is defined as number of hops [21, 24].

To model communication latency we need to extend the considered two communication functions $f_1(t_s)$, $f_2(t_w)$ in NOW module by third function component $f_3(t_h)$, which will determine potential multiple crossing used NOW modules of integrated parallel computer. This third function is characterized through multiplying hops parameter l_h among NOW modules (generally u NOW networks) and average communication latency time of jumped NOW modules with the same communication speed or a sum of individual communication latencies for jumped NOW modules with their different communication speed. Then in general to the whole communication latency in Grid is valid

$$T(s,p)_{commGRID} = f_1(t_s) + f_2(t_w) + \sum_{i=1}^{u} f_3(t_s, t_w, l_h)$$

In general communication latency time $f_3(t_s, t_w, l_h)$ is time to send data message with m words in one communication step among integrated NOW modules with l_h hops. The communication time for one communication step is then given as $t_s + m\, t_w\, l_h\, t_h$, where the new parameters are

- l_h is the number of network hops
- m is the number of transmitted data units (usually words)
- t_h is average communication time for one hop.

The new communication parameters t_h, l_h depend from a concrete architecture of Grid communication network and used routing algorithm. In [9] we have developed unified models which could help to establish these parameters for dominant parallel computers. For the complex analytical modeling there is necessary to derive for given parallel algorithm or a group of similar algorithms (matrix parallel algorithms) needed communication functions and that always individually for any decomposition strategy) for known technical parameters (computational, communication) of used parallel computer (classic, NOW, Grid).

5. Communication Latencies of PA

To this time known results in complexity modeling on the in the world have used mainly classical parallel computers with shared memory (supercomputers) or massive multiprocessors with distributed memory which in most cases did not consider the influences of overheads in parallel computing (communication, synchronization, parallelization, waiting etc.) supposing that they would be lower in comparison to the latency of performed massive parallel computations [22].

In this sense analysis and modeling of complexity in parallel algorithms (PA) are to be reduced to only complexity analysis of own computations $T(s, p)_{comp}$, that mean that the function of all existed control and

communication overhead latencies $h(s, p)$ were not a part of derived relations for whole parallel execution time $T(s, p)$. In this sense the dominated function in the relation for used isoefficiency function $w(s)$ of the parallel algorithms is complexity of performed massive computations $T(s, p)_{comp}$. Such assumption has proved to be really true in using classical parallel computers (supercomputers, massive SMP, SIMD architectures etc.). Putting put on this assumption to the relation for asymptotic isoefficiency $w(s)$ we get $w(s)$ as follows

$$w(s) = \max\left[T(s,p)_{comp}, h(s,p)\right] < T(s,p)_{comp}$$

In opposite at parallel algorithms for actually dominant parallel computers based on NOW and Grid there is for complex modeling necessary to analyze   at least to evaluate most important overheads from all existed overheads which are [8, 10]

- architecture of parallel computer $T(s, p)_{arch}$
- own computations $T(s, p)_{comp}$
- communication latency $T(s, p)_{comm}$
  - start - up time (t_s)
  - data unit transmission (t_w)
  - routing
- parallelization latency $T(s, p)_{par}$
- synchronization latency $T(s, p)_{syn}$
- waiting caused by limiting shared technical resources $T(s, p)wait$ (memory modules, communication channels etc.).

All these named overhead latencies build in defined is efficiency function the whole overhead function $h(s, p)$. In general the influence of $h(s, p)$ is necessary to take into account in complex performance modeling of parallel algorithms or at least to evaluate their important individual components. The defined overhead function $h(s, p)$ is as follows

$$h(s,p) = \sum \left(T(s,p)_{arch}, T(s,p)_{par}, T(s,p)_{comm} T(s,p)_{syn}\right)$$

Overhead function $T(s, p)_{arch}$ is projected into used technical parameters t_c, t_s, t_w, which are constants for given parallel computer.

The second overhead function $T(s, p)_{par}$ depend from chosen decomposition strategy and their consequences are projected so to computation part $T(s, p)_{comp}$ as to communication part $T(s, p)_{comm}$.

The third overhead function $T(s, p)_{syn}$ we can eliminate through optimization of load balancing among individual computing nodes of used parallel computer. For this purpose we would measure performance of individual used computing nodes for given developed parallel algorithm and then based on done measured results we are able to redistribute better given input load. These activities we can repeat until we have optimal redistributed input load (Load balancing).

In general possible nonlinear influence of overhead function $h(s, p)$ should be taken into account in complex performance modeling of parallel algorithms. Then for asymptotic isoefficiency analysis of complex performance analysis we should consider w(s) as follows

$$w(s) = \max\left[T(s,p)_{comp}, h(s,p)\right]$$

, where the most important parts for dominant parallel computers (NOW, Grid) in overhead function h(s, p) are in relation to done analysis of individual overhead latencies the influence of communication latency of $T(s, p)_{comm}$.

Then the kernel of asymptotic analyze of h(s, p) is analysis of communication latency $T(s, p)_{comm}$ including projected consequences of used decomposition methods. In general derived isoefficiency function w(s) could have nonlinear character at gradually increasing number of computing nodes p. Analytical deriving of isoefficiency function w(s) including communication latency $T(s, p)_{comm}$ allow us to predict PA performance of given parallel algorithm so for real as for hypothetical parallel computers.

6. Applied Modeling of Communication Latency

We will illustrate modeling process of communication latency on approximation solution of steady state solution Φ (x, y) for points in the interior by the function u (x, y) according Fig. 11. Given a two dimensional region and values for points of the region boundaries, We can do this by covering the region with a grid of points and to obtain the values of u $(x_i, y_j) = u_{i,j}$ of the area. Each inner point is initialized to some initial value. Other stationary values of inner points will by computed applying iterative methods. In each iteration step, the new point value (next) will be defined as average of previous (old) or a combination of previous and new set of values of neighboring points. Iterative computation ends either after performed fixed number of iterations or after reaching defined precision acceptable difference E > 0 (epsilon value) for each new value. Epsilon accuracy is determined as desired difference between the previous and the new point value.

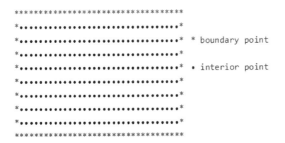

Figure 11. Grid approximation of two dimensional region.

These limits of points that indicate the boundary conditions as follows

- according to Dirichlet [10] giving the values of given function analyzed function at both ends of the field (at U = 0 for x = 0, 0 ≤ Y ≤ 1, U = 1 for X = 1,

0 ≤ Y ≤ 1)
- according to the Neumann [10] giving the values with solution derivations (for example u_y= 0 for Y = 0, 0 ≤ X ≤ 1, u_y = 0 for Y = 1 0 ≤ X ≤ 1).

To model communication complexity we will represent two dimensional grid of points with matrix in abstract form that means for empty matrix. Then we are considering the typical following n x m matrix

$$A = \begin{pmatrix} a_{11}, a_{12}, ..., a_{1n} \\ a_{21}, a_{22}, ..., a_{2n} \\ . \quad . \quad . \\ . \quad . \quad . \\ . \quad . \quad . \\ a_{m1}, a_{m2}, ..., a_{mn} \end{pmatrix}$$

To reduce number of variables in deriving modeling process we will consider matrix with m = n (square matrix). For this purpose there are also following causes

- we can transform any matrix n x m to n x n matrix through expanding rows (if m < n) or columns (if m > n)
- derivation process will be the same only when considering the workload instead of n^2 (square matrix) we should consider n x m (oblong matrix).

Communication model structure is involved in the communication complexity of the parallel algorithm. We would analyze the basic communication requirements potential using of iterative method. Analysis via iterative methods is based on iterative computation of the new iterative value of given internal point from fixed number of neighboring points according concrete iterative relationship. To compute each new value of one point of approximated network points they are setting to used iteration relation set of specified number of neighboring point values (iteration step). In our case it could be derived to compute the new value X I, j on a two dimensional network of points iterative relation where each new point value is computed as the arithmetic average of the four neighboring points as follows

$$X_{i,j}^{(t+1)} = (X_{i-1,j}^{(t)} + X_{i+1,j}^{(t)} + X_{i,j-1}^{(t)} + X_{i,j+1}^{(t)}) / 4$$

Iterative calculation according to the following iterative relationship and are repeated sequentially in each iteration step, the new gain values $X_{i,j}^{(1)}$, $X_{i,j}^{(2)}$...etc., while the name $X_{i,j}^{(t)}$ determines the value at a given point $X_{i,j}$ t - step. Suppose applying decomposition models create parallel processes for each point, respectively group of points as the two-dimensional network points.

For iterative finite difference method a two dimensional grid is repeatedly updated by replacing the value at each point with some function of the values at a small fixed number of neighboring points. The common approximation structure uses a four point stencil to update each element $X_{i,j}$ (Fig. 12.).

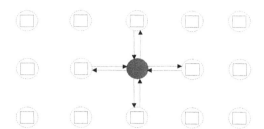

Figure 12. Communication for 4 - point approximation.

The needed communication process is as follows
for t = 0 to t–1 do
begin
send $X_{i,j}^{(t)}$ to each neighbor;
receive $X_{i-1,j}^{(t)}, X_{i+1,j}^{(t)}, X_{i,j-1}^{(t)}, X_{i,j+1}^{(t)}$ from the neighbors;
compute $X_{i,j}^{(t+1)}$ using the specified iteration relation for X_i;
end;

6.1. Matrix Decomposition to Strips

Decomposition method to rows or columns (Strips) are algorithmic the same and for their practical using is critical the way how are the matrix elements putting down to matrix.

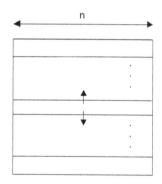

Figure 13. Communication consequences for decomposition to strips (rows).

In this way it is possible send very simple through specification of the beginning address for a given row and through a number of elements in row. Let for every parallel process (strips) two messages are send to neighboring processors and in the same way two messages are received from neighboring processors (Fig. 13) supposing that it is possible to transmit for example one row to one message. Communication time for a calculation step T(s, p)$_{coms}$ is then given as

$$T(s, p)_{comms} = 4\left(t_s + n\, t_w\right)$$

Using these variables for the communication overheads in decomposition method to strips is correct

$$T(s, p)_{comm} = T(s, p)_{comms} = h(s, p) = 4(t_s + n\, t_w)$$

In this case a communication time for one calculation step does not depend on the number of used calculation processors.

6.1.1. Matrix Decomposition to Blocks

For mapping matrix elements in blocks a inter process communication is performed on the four neighboring edges of blocks, which it is necessary in computation flow to exchange. Every parallel process therefore sends four messages and in the same way they receive four messages at the end of every calculation step (Fig. 14.) supposing that all needed data at every edge are sent as a part of any message).

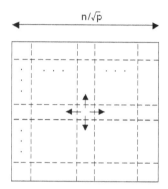

Figure 14. Communication model for decomposition to blocks.

Then the requested communication time for this decomposition method is given as

$$T(s, p)_{commb} = 8\left(t_s + \frac{n}{\sqrt{p}} t_w\right)$$

This equation is correct for p ≥ 9, because only under this assumption it is possible to build at least one square because only then is possible to build one square block with for communication edges. Using these variables for the communication overheads in decomposition method to blocks is correct

$$T(s, p)_{comm} = T(s, p)_{commb} = h(s, p) = 8(t_s + \frac{n}{\sqrt{p}} t_w)$$

Then the requested communication time for this decomposition method is given as

$$T_{comb} = 8\left(t_s + \frac{n}{\sqrt{p}} t_w\right)$$

Data exchange at all shared edges points for both decomposition strategies (blocks, strips) illustrates Fig. 15.

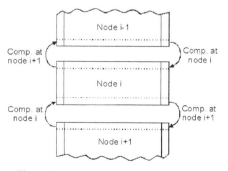

Figure 15. Data exchange among processors.

6.1.2. Optimization of Decomposition Model Selection

For comparison based on derived relations for communication complexity decomposition method to blocks demands higher communication time as decomposition to strips (more effective decomposition to strips) if

$$8\,(t_s + \frac{n}{\sqrt{p}}\,t_w) > 4\,(t_s + n\,t_w)$$

or after adjusting for technical parameter t_s.

$$t_s > n\,(1 - \frac{2}{\sqrt{p}})\,t_w.$$

or for second technical parameter t_w as follows

$$t_w > n\,(1 - \frac{2}{\sqrt{p}})\,t_s.$$

This relation is valid under assumption that $p \geq 9$, which is real condition in developed iterative parallel algorithms to build real square block. Fig. 16 illustrates choice optimization of suitable decomposition method based on derived dependences to establishing t_s for n = 256 and following values of t_w

- t_w = 230 ns = 0,23 μs (NOW IBM SP-2)
- t_w = 2,4 μs (NCUBE 2).

For higher values t_s as at given t_{si} (t_{s1} for t_w = 0, 23 μs and t_{s2} for t_w = 2, 4 μs) from the appropriate curve line for n = 256 is more effective decomposition method to strips. Limited values in choice optimal decomposition strategy are at given n for higher values t_w. Therefore in general decomposition to strips is more effective for higher values of t_s.

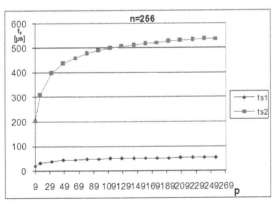

Figure 16. *Optimization of decomposition method.*

Threshold values t_s to select the optimal model of matrix decomposition strategy are for given n for larger values of t_w greater. Then in generally decomposed into strips (rows, columns) is more effective for higher values of t_s (NOW, Grid) and decomposition into blocks again for smaller values of t_s (classic massive SMP, supercomputers etc.). Generally the times of t_s are significant higher for parallel computers as NOW and Grid. For example NOW for FDDI

based optical cables with t_s = 1100 μs, t_w = 1.1 μs and for architecture with Ethernet t_s= 1500 μs, t_w= 5 μs) [8]. In such systems, the use of matrix decomposition method to strips then more effectively.

7. Measurement of Communication Latency

For experimental measurements of PA delays we can use available services of used parallel development environment (MPI services, Win API 32, Win API 64, etc.). For example to measurement of execution time of parallel processes we have been used following functions Win 32 API and Win API 64

- Query Performance Counter which returns actual value of counter
- Query Performance Frequency which defines counting frequency per second.

Values of both functions depend on used computer nodes. Using of above measuring time functions we can obtain execution times with high accuracy. For example for common Intel Pentium processors or higher it is 0.0008 ms which is sufficient for the time analysis of PA.

To measure communication latencies we can define function to measure time between two points of performed parallel algorithm respectively parallel process. An example of following pseudo code to measure time between two points T1 and T2 is as follows

```
T1:  time (&t1);    /*start of time*/
T2:  time (&t2);    /*stop time*/
measured_ time = difftime (t2,t1);   /*measured time = t2-t1*/
printf ("Measured time = % 5.2 f ms", measured_time);
```

Illustrated approach is for measurement of monitoring times universal that is we can use it so for measurements of parallel algorithms as for measuring monitoring overheads of parallel processes, or measuring delays which are typical to establish technical parameters of parallel computers.

7.1. Communication Technical Parameters

7.1.1. Classic Parallel Computers

We have been derived relations to communication comparisons of possible matrix decomposition models. We have been derived that decomposition model to strips (rows, columns) demands lower inter process communication delays than decomposition model to blocks with respecting derived condition that for blocks $p \geq 9$, for communication parameter t_s according following inequality

$$t_s > n\,t_w\,(1 - \frac{2}{\sqrt{p}})$$

At this time at general conclusion that decomposition model to strips (Rows, columns) is advantageous for higher values of technical parameter t_s, that using of this decomposition model is more effective. For applied matrix parallel algorithms MPA from this is resulting allocation of n = p. In case of massive equations in which n > p computing nodes will be repeatedly perform needed algorithm activities for remaining n > p strips (rows, columns). For example if we supposed for simplicity that the value of parameter n is divisible through parameter p without rest, used computing nodes of parallel computers will be repeatedly perform k times needed algorithm activities till exhausting quotient value of k = n/p. For both complexities (computation, communication) it mean k – multiplying factor of both mentioned complexities as derived complexities for pre n = p. From this fact it is clear that the base outgoing problem is to derive so computational as communication complexity for given n = p. Setting previous equality to the relation for technical parameter t_s after needed performed adjustments we get finally following relation

$$t_s > t_w (n - 2\sqrt{n})$$

In Tab. 1 we have computed the values t_s for known values of communication parameter t_w [7, 15].

Table 1. Optimization of decomposition method for t_s.

t_w [μs]	T(s, p)comm. – computed values		
	n=256	n=512	n=1024
	t_{s1} [μs]	t_{s2} [μs]	t_{s3} [μs]
0,063	14,11	29,40	60,48
0,070	15,68	32,67	67,20
0,080	17,92	37,34	76,80
0,230	50,60	107,35	220,8
0,440	97,68	205,37	422,4
0,540	119,88	252,04	518,4
1,100	244,20	513,4	1056
2,400	532,80	1 072,2	2 304
5,000	1 110,00	2 233,7	4 800

Graphical illustrations of optimal decomposition strategy for computed values in Tab. 1 are illustrated are at Fig. 17 (smaller values of t_s) and Fig. 18 (higher values of t_s).

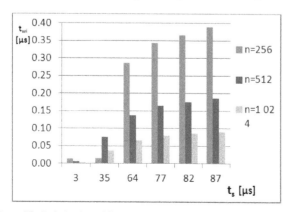

Figure 17. Optimization of decomposition methods for smaller values of t_s.

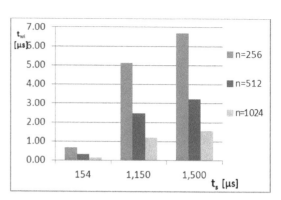

Figure 18. Optimization of decomposition methods for higher values t_s.

This inequality for t_s we are able modified for other input parameter (t_w, n) and compute remaining parameter. For example after adjustment for value of parameter t_w is valid

$$t_w < \frac{t_s}{n - 2\sqrt{n}}$$

At Tab. 2 we have been computed the values t_w for known values of communication parameter t_s [7, 15].

Table 2. Optimization of decomposition method for t_w.

t_s [μs]	T(s, p)comm. – computed values		
	n=256 (224)	n=512 (466,75)	n=1024 (960)
	t_{w1} [μs]	t_{w2} [μs]	t_{w3} [μs]
3	0,01339	0,0064	0,0031
35	0,01560	0,0750	0,0365
64	0,2857	0,1371	0,0666
77	0,3437	0,1650	0,0802
82	0,3661	0,1757	0,0854
87	0,3884	0,1864	0,0906
154	0,6875	0,3299	0,1604
1 150	5,1339	2,4638	1,1979
1 500	6,6964	3,2137	1,5625

Limiting values for optimal selection of decomposition strategy will be at given n for higher values of t_w also higher. Then decomposition model to strips will be more effective for higher values of t_s (NOW, Grid) and decomposition models to blocks for lower values of t_s.

7.2. NOW Parallel Computer

For analyzed decomposition models we are going to analyze generally their communication complexity. Communication function T(s, p)comm for given decomposition strategy is principally defined in NOW with defined two communications parameters and that namely
• communication latency due to parameter t_s
• communication latency due to parameter t_w.

Typical architecture of communication network used in NOW in our country (Slovakia) is still Ethernet. The bottleneck of this communication network is in its serial communication performing multiply point to point communications and this communication character remains unchanged also at collective communication mechanism but collective command Broadcast. We will therefore evaluate in an analytical way relations of all MPI collective communication commands in Ethernet. There is possible to evaluate it also for other concrete communication network architectures.

7.2.1. Evaluation of Collective Communication Mechanisms

For the following typical MPI communication mechanisms on the Ethernet network are valid the following relationships

- MPI command of data dispersion
  - MPI command Broadcast
  - MPI command Scatter
- MPI command of data collection
  - MPI command Gather
  - MPI command All gather
  - MPI command Reduce.

7.2.2. Collective Communication Mechanism Type Broadcast

Collective communication mechanism Broadcast is the only collective communication mechanism that can be effective also in Ethernet communications network. In this case of transmission of one of the same data unit (Byte, word) to all other p computing nodes (Processors) in terms of Ethernet network. Communication complexity $T(s, p)_{comm}$ of this command is $O(1)$ respectively using of established communication parameters t_s, t_w as

$$T(s,p)_{commbr} = t_s + t_w$$

To transmit data units various m always only one processor (Point to point) computational complexity is given as $O(m)$, respectively supporting communication parameters established by the following formula

$$T(s,p)_{commbr} = m(t_s + t_w)$$

For transmission of different data unit's m-1 remaining processor computational complexity is given by $O(p)$, respectively with the support of established communication parameters as

$$T(s,p)_{commbr} = \sum_{i=1}^{p-1} m(t_s + t_w)$$

The same communication delays are for any other collective communication mechanisms of Ethernet communication network. Communication complexity $T(s, p)_{commEth}$ common relationship will be determined as

$$T(s,p)_{commEth} = \sum_{i=1}^{p-1} m(t_s + t_w) = (p-1)\, m\,(t_s + t_w)$$

In general the value of the parameter t_s is highly significant for asynchronous parallel computers NOW and Grid. For example for parallel computer NOW based on fiber optic cable these communication parameters have

concrete values and that $t_s = 1100\ \mu s$ and $t_w = 1,1\ \mu s$ and for Ethernet communication network are these parameters even higher and that $t_s = 1500\ \mu s$ a $t_w = 5\ \mu s$. The realized measurements in our home conditions (DTI Dubnica and University of Zilina, Slovakia) of these communication parameters on an unloaded Ethernet network were higher than previous specified values. Causes for the substantial differences are mainly the following

- built sophisticated technical support to data block transmission based on direct memory access (DMA) for collective communication mechanisms
- multiple transmission channels based on multistage structure of communication network
- high speed communication network switches named as HPS (High Performance Switch)
- used high speed communications network as Infiniband, Quadrics and Myrinet [34].

8. Conclusions

Modeling of communication latency as a discipline has repeatedly proved to be critical for design and successful use of parallel computers and parallel algorithms too. At the early stage of design, communication models can be used to project the system performance, scalability and evaluate design alternatives [26, 30]. At the production stage, communication evaluation methodologies can be used to detect bottlenecks and subsequently suggests ways to alleviate them. Analytical methods (order analysis, queuing theory systems), simulation, experimental measurements, and hybrid modeling methods could be successfully used for the evaluation of system and its components too [14, 23]. Via the extended form of theory of complexity to modeling of communication latency we are able to predicate communication latency behavior also in other applied parallel algorithms than analyzed matrix decomposition models.

This paper continues in applying complex analytical modeling of PA including modeling communication complexity and latency too [12, 13]. To present used modeling concepts of communication complexity we have chosen abstract matrix with its possible decomposition models (rows or columns, blocks). Based on these decomposition models we have described communication complexity via deriving analytical relations including their comparisons and optimization. The both considered analyzed examples we have been evaluated so on classic supercomputers (hypercube, mesh, torus) as on actually dominant parallel computers NOW and Grid. It is obvious that in some cases using of network of workstations (NOW) or its higher integration parallel computers named as Grid (integrated network of NOW networks) could be less effective than on innovated classic massive parallel computers but NOW and Grid belong to more flexible and perspective parallel computers also for the future time.

Acknowledgements

This work was done within the project "Complex modeling, optimization and prediction of parallel computers and algorithms" at University of Zilina, Slovakia. The author gratefully acknowledges help of project supervisor Prof. Ing. Ivan Hanuliak, PhD.

References

[1]   Abderazek A. B., Multicore systems on chip - Practical Software/Hardware design, Imperial college press, UK, pp. 200, 2010

[2]   Arie M.C.A. Koster Arie M.C.A., Munoz Xavier, Graphs and Algorithms in Communication Networks, Springer Verlag, Germany, pp. 426, 2010

[3]   Arora S., Barak B., Computational complexity - A modern Approach, Cambridge University Press, UK, pp. 573, 2009

[4]   Barria A. J., Communication network and computer systems, Imperial College Press, UK, pp. 276, 2006

[5]   Casanova H., Legrand A., Robert Y., Parallel algorithms, CRC Press, USA, 2008

[6]   Dubois M., Annavaram M., Stenstrom P., Parallel computer organization and design, UK, pp. 560, 2012

[7]   Hager G., Wellein G., Introduction to High Performance Computing for Scientists and Engineers, CRC Press, USA, pp. 356, 2010

[8]   Hanuliak P., Analytical method of performance prediction in parallel algorithms, The Open Cybernetics and Systemics Journal, Vol. 6, Bentham, UK, pp. 38-47, 2012

[9]   Hanuliak M., Unified analytical models of parallel and distributed computing, American J. of Networks and Communication, Science PG, Vol. 3, No. 1, USA, pp. 1-12, 2014

[10] Hanuliak P., Hanuliak I., Performance evaluation of iterative parallel algorithms, Kybernetes, Vol. 39, No.1/ 2010, UK, pp. 107- 126

[11] Hanuliak M., Modeling of parallel computers based on network of computing nodes, American J. of Networks and Communication, Science PG, Vol. 3, USA, 2014

[12] Hanuliak P., Complex modeling of matrix parallel algorithms, American J. of Networks and Communication, Science PG, Vol. 3, USA, 2014

[13] Hanuliak M., Hanuliak I., To the correction of analytical models for computer based communication systems, Kybernetes, Vol. 35, No. 9, UK, pp. 1492-1504, 2006

[14] Harchol-Balter Mor, Performance modeling and design of computer systems, Cambridge University Press, UK, pp. 576, 2013

[15] Hwang K. and coll., Distributed and Parallel Computing, Morgan Kaufmann, USA, pp. 472, 2011

[16] Kshemkalyani A. D., Singhal M., Distributed Computing, University of Illinois, Cambridge University Press, UK, pp. 756, 2011

[17] Kirk D. B., Hwu W. W., Programming massively parallel processors, Morgan Kaufmann, USA, pp. 280, 2010

[18] Kostin A., Ilushechkina L., Modeling and simulation of distributed systems, Imperial College Press, USA, pp. 440, 2010

[19] Kumar A., Manjunath D., Kuri J., Communication Networking , Morgan Kaufmann, USA, pp. 750, 2004

[20] Kushilevitz E., Nissan N., Communication Complexity, Cambridge University Press, UK, pp. 208, 2006,

[21] Le Boudec Jean-Yves, Performance evaluation of computer and communication systems, CRC Press, USA, pp. 300, 2011

[22] McCabe J., D., Network analysis, architecture, and design (3rd edition), Elsevier/ Morgan Kaufmann, USA, pp. 496, 2010

[23] Meerschaert M., Mathematical modeling (4-th edition), Elsevier, pp. 384, 2013

[24] Misra Ch. S., Woungang I., Selected topics in communication network and distributed systems, Imperial college press, UK, pp. 808, 2010

[25] Mieghem P. V., Graph spectra for Complex Networks, Cambridge University Press, UK, pp. 362, 2010

[26] Park K., Willinger W., Self-Similar Network Traffic and Performance Evaluation, John Wiley & Sons, Inc., USA, pp. 558, 2000

[27] Peterson L. L., Davie B. C., Computer networks – a system approach, Morgan Kaufmann, USA, pp. 920, 2011

[28] Resch M. M., Supercomputers in Grids, Int. J. of Grid and HPC, No.1, Germany, pp. 1 - 9, 2009

[29] Riano l., McGinity T.M., Quantifying the role of complexity in a system´s performance, Evolving Systems, Springer Verlag, Germany, pp. 189 – 198, 2011

[30] Ross S. M., Introduction to Probability Models, 10th edition, Academic Press, Elsevier Science, Netherland, pp. 800, 2010

[31] Wang L., Jie Wei., Chen J., Grid Computing: Infrastructure, Service, and Application, CRC Press, USA, 2009

[32] Wolf Marilyn, High-Performance Embedded Computing (Second Edition), Morgan Kaufmann, USA, pp. 600, 2014

[33] Zhuge Hai., The Knowledge Grid, Imperial College Press, USA , pp. 360, December 2011
      www pages

[34] www.top500.org.

21

Modeling of parallel computers based on network of computing

Michal Hanuliak

Dubnica Technical Institute, Sladkovicova 533/20, Dubnica nad Vahom, 018 41, Slovakia

Email address:

michal.hanuliak@gmail.com

Abstract: The optimal resource allocation to satisfy such demands and the proper settlement of contention when demands exceed the capacity of the resources, constitute the problem of being able to understand and to predict system behavior. To this analysis we can use both analytical and simulation methods. Modeling and simulation are methods, which are commonly used by performance analysts to represent constraints and to optimize performance. Principally analytical methods represented first of all by queuing theory belongs to the preferred method in comparison to the simulation method, because of their potential ability of general analysis and also of their ability to potentially analyze also massive parallel computers. But these arguments supposed to develop and to verify suggested analytical models. This article goes further in applying the achieved analytical results in queuing theory for complex performance evaluation in parallel computing [9, 14]. The extensions are mainly in extending derived analytical models to whole range of parallel computers including massive parallel computers (Grid, meta computer). The article therefore describes standard analytical model based on M/M/m, M/D/m and M/M/1, M/D/1 queuing theory systems. Then the paper describes derivation of the correction factor for standard analytical model, based on M/M/m and M/M/1 queuing systems, to study more precise their basic performance parameters (overhead latencies, throughput etc.). All the derived analytical models were compared with performed simulation results in order to estimate the magnitude of improvement. Likewise they were tested under various ranges of parameters, which influence the architecture of the parallel computers and its communication networks too. These results are very important in practical use.

Keywords: Parallel Computer, Grid, Communication System, Correction Factor, Analytical Model, Jackson Theorem, NOW, Performance Modeling, Queuing System

1. Introduction

Performance of actually computers (sequential, parallel) depends from a degree of embedded parallel principles on various levels of technical (hardware) and program support means (software) [4]. At the level of intern architecture of basic module CPU (Central processor unit) of PC they are implementations of scalar pipeline execution or multiple pipeline (superscalar, super pipeline) execution and capacity extension of cashes and their redundant using at various levels and that in a form of shared and local cashes (L1, L2, L3). On the level of motherboard there is a multiple using of cores and processors in building multicore or multiprocessors system as SMP (symmetrical multiprocessor system) as powerful computation node, where such computation node is SMP parallel computer too [1]. On the level of individual computers the dominant trend is to use multiple number of high performed workstations based on single personal computers (PC) or SMP, which are connected in the network of workstations (NOW) or in a high integrated way named as Grid systems [34].

2. Architectures of Parallel Computers

We have tried to classify parallel computer from the point of system program developer to two following basic groups according Fig.1.

- synchronous parallel computers. They are often used under central control, that means under the global clock synchronization (vector, array system etc.) or a distributed local control mechanism (systolic systems etc.). The typical architectures of this group of parallel computers illustrate Fig. 1 on its left side
- asynchronous parallel computers. They are composed of a number of fully independent

computing nodes (processors, cores, and computers). To this group belong mainly various forms of computer networks (cluster), network of workstation (NOW) or more integrated Grid modules based on NOW modules [30]. The typical architectures of asynchronous parallel computers illustrate Fig. 1 on its right side.

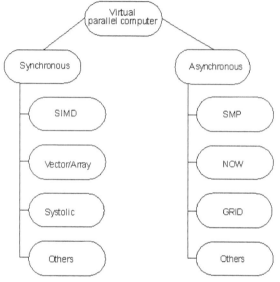

Figure 1. System classification of parallel computers.

3. Dominant Parallel Computers

3.1. Symmetrical Multiprocessor System

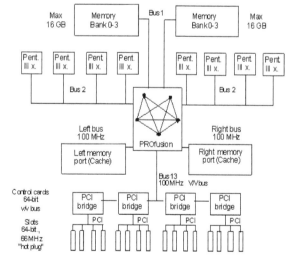

PROfusion - cross switch of 3 bus and 2 memory ports (parallel)

PCI cards - type Enthanced PCI (64 bit, 66 MHz, "Hot Plug"- on-line exchange)

Figure 2. Single computing node based on SMP (8-processors Intel Xeon).

Symmetrical multiprocessor system (SMP) is a multiple using of the same processors or cores which are implemented on motherboard in order to increase the whole performance of such system. Typical common characteristics are following

- each processor or core (computing node) of the multiprocessor system can access main memory (shared memory)
- I/O channels or I/O devices are allocated to individual computing nodes according their demands
- integrated operation system coordinates cooperation of whole multiprocessor resources (hardware, software etc.).

Real example of multiprocessor system illustrates Fig. 2.

3.2. Network of Workstations

Network of workstations belongs to actually dominant trends in parallel computing. This trend is mainly driven by the cost effectiveness of such systems as compared to massive multiprocessor systems with tightly coupled processors and memories (supercomputers). Parallel computing on a network of workstations connected by high speed networks has given rise to a range of hardware and network related issues on any given platform [20, 35]. With the availability of cheap personal computers, workstations and networking devises, the recent trend is to connect a number of such workstations to solve computation intensive tasks in parallel on such clusters. Network of workstations has become a widely accepted form of high performance computing (HPC). Each workstation in a NOW is treated similarly to a processing element in a multiprocessor system. However, workstations are far more powerful and flexible than processing elements in conventional massive multiprocessors (supercomputers).

Typical example of networks of workstations also for solving large computation intensive problems is at Fig. 3. The individual workstations are mainly extreme powerful personal workstations based on multiprocessor or multicore platform [8, 18].

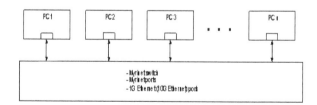

Figure 3. Typical architecture of NOW.

On such modular parallel computer we are able to study basic problems in parallel computing (parallel and distributed computing) as load balancing, inter processor communication IPC [22, 28], modeling and optimization of parallel algorithms etc. [10, 23, 26]. The coupled computing nodes PC_1, PC_2, ..., PC_i (workstations) could be single extreme powerful personal computers or SMP parallel computers. In this way parallel computing on networks of conventional PC workstations (single, multiprocessor, multicore) and Internet computing, suggest advantages of unifying parallel and distributed computing [9, 19].

3.3. Massive Parallel Computers

3.3.1. Grid Systems

Grid technologies have attracted a great deal of attention recently, and numerous infrastructure and software projects have been undertaken to realize various versions of Grids. In general Grids represent a new way of managing and organizing of computer networks and mainly of their deeper resource sharing. Conceptually they go out, similar like computer networks, from a structure of virtual computer based on computer networks.

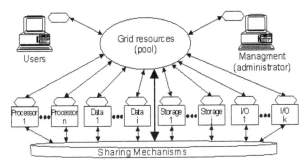

Figure 4. Architecture of Grid node.

Grid systems are expected to operate on a wider range of other resources as processors (CPU), like storages, data modules, network components, software (typical resources) and atypical resources like graphical and audio input/output devices, sensors and so one (Fig. 4.). All these resources typically exist within nodes that are geographically distributed, and span multiple administrative domains. The virtual machine is constituted of a set of resources taken from a resource pool. It is obvious that existed HPC parallel computers (supercomputers etc.) could be a member of such Grid systems too [31]. In general Grids represent a new way of managing and organizing of computer networks and mainly of their deeper resource sharing [34]. Grid systems are expected to operate on a wider range of other resources as processors (CPU), like storages, data modules, network components, software (typical resources) and atypical resources like graphical and audio input/output devices, sensors and so one (Fig. 4.). All these resources typically exist within nodes that are geographically distributed, and span multiple administrative domains. The virtual machine is constituted of a set of resources taken from a resource pool [34]. It is obvious that existed HPC parallel computers (supercomputers etc.) could be a member of such Grid systems too. In general Grids represent a new way of managing and organizing of computer networks and mainly of their deeper resource sharing.

3.3.2. Meta Computers

This term define massive parallel computer (supercomputer, Grid) with following basic characteristics [8, 34]
- wide area network of integrated free computing resources. It is a massive number of interconnected

networks, which are connected through high speed connected networks during which time whole massive system is controlled with network operation system, which makes an illusion of powerful computer system (virtual supercomputer)
- grants a function of meta computing that means computing environment, which enables to individual applications a functionality of all system resources
- system combines distributed parallel computation with remote computing from user workstations.

The best example of existed meta computer is Internet as massive international network of various computer networks according Fig. 5.

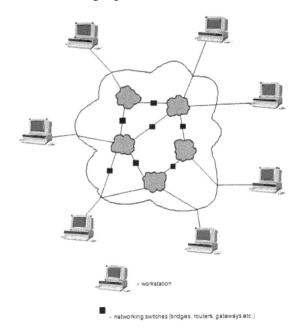

- workstation

- networking switches (bridges, routers, gateways etc.)

Figure 5. Internet as network of connected networks.

4. Analytical Performance Evaluation of Parallel Computers

To the behavior analysis of coupled computing nodes we can use various analytical models based on queuing theory results. Queuing theory is very good if you have to analyze a single independent computing node of sequential or parallel computers [3, 7]. But analysis of dominant parallel computers (NOW, Grid) lead to multiple connected computing nodes. The first problem, in comparison to a single node case, is existence of traffic dependency in any real network of computing nodes. If all the all node´s communication traffic has the property that it is Poisson, then even in a complicated network we can do under some conditions network analysis on a node-by-node basis [17, 29]. In fact, however, that is not yet true, because in communication networks of connected computing nodes the time a communication message spends in one node is related to the time it spends in another node, because the service one is looking for is network communication. That is one very nasty problem, but there have been developed some solutions.

The second serious problem is blocking as consequence of always real limited technical resources. If one node is blocked, the node feeding could not enter more data into that node. Consider a communication network in which you are given the location of computing nodes and the required communication traffic between pairs of computing nodes. Then according mentioned theorem says that if you have Poisson traffic into an exponential server you get Poisson traffic out; but a message maintains its length as it passes through the network, so the service times are dependent as it goes along its path. Thus, one thing we want to do is to get rid of that dependence. We can do this by making an independence assumption; we just assume that the dependence does not exist. We manage this by allowing the communication message to change its length as it passes through the communication network. Every time it hits a new computing node, we are going to randomly choose the message length so that we come up with an exponential distribution again. With that assumption, we can then solve the queuing problem of communication in parallel computers. Let us assume infinite storage at all points in the network of coupled computing nodes and refer to the problem M/M/1, where the question mark refers to the modified input process. We then run simulations, with and without the independence assumption for a variety of networks. The reason why it is good to do it is that a high degree of mixing takes place in a typical communication network; there are many ways into a node and many ways out of the node [6, 16].

The assumption of independence permits us to break also the massive parallel computer into independent computing nodes, and allowed all node analysis to take place. The reason we had to make that assumption was because the communication message maintains the same length as they pass through the network. If we accept the independence assumption, it turns out that the queuing theory contains a number of results for cases where the service at a node is an independent random variable in an arbitrary network of queues. A basic theorem is due to Jackson [17, 29]. Jackson's result essentially gives us the probability distribution for various numbers of messages at each of the nodes in such a network.

5. Application of Queuing Theory

The basic premise behind the use of queuing models for computer systems analysis is that the components of a computer system can be represented by a network of servers (resources) and awaiting lines (queues). A server is defined as an entity that can affect, or even stop, the flow of jobs through the system. In a computer system, a server may be the CPU, I/O channel, memory, or a communication port. Awaiting line is just that: a place where jobs queue for service. To make a queuing model work, jobs or customers or communication message (blocks

of data, packets) or anything else that requires the sort of processing provided by the server, are inserted into the network. A basic simple example could be the single server abstract model as single queuing theory system. In such model, jobs arrive at some rate, queue for service on a first-come first-served basis, receive service, and exit the system. This kind of model, with jobs entering and leaving the system, is called an open queuing system model [7, 27].

Queuing theory systems are classified according to various characteristics, which are often summarized using Kendall`s notation [3, 6]. The basic parameters of queuing theory systems are as following

- λ - arrival rate at entrance to a queue
- m- number of identical servers in the queuing system
- ρ - traffic intensity (dimensionless coefficient of utilization)
- q - random variable for the number of customers in a system at steady state
- w - random variable for the number of customers in a queue at steady state
- $E(t_s)$- the expected (mean) service time of a server
- $E(q)$- the expected (mean) number of customers in a system at steady state
- $E(w)$ - the expected (mean) number of customers in a queue at steady state
- $E(t_q)$ - the expected (mean) time spent in system (queue + servicing) at steady state
- $E(t_w)$ - the expected (mean) time spent in the queue at steady state.

Communication demands (parallel processes, IPC data) arrive at random at a source node and follow a specific route in the communication networks towards their destination node. Data lengths of communicated IPC data units (for example in words) are considered to be random variables following distributions according Jackson theorem. Those data units are then sent independently through the communication network nodes towards the destination node. At each node a queue of incoming data units is served according to a first-come first-served (FCFS) discipline.

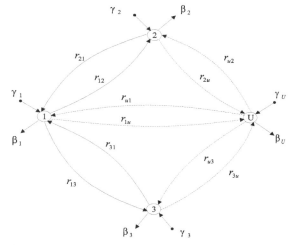

Figure 6. Communication network of connected computing nodes.

At Fig. 6 we illustrate generalization of any parallel computer including their communication network as following
- computing nodes u_i (i=1, 2, 3, ..., U) of any parallel computer are modeled as graph nodes
- network communication channels are modeled as graph edges r_{ij} (i≠j) representing communication intensities (relation probabilities).

The other used parameter of such abstract model are defined as following
- $\gamma_1, \gamma_2, \dots , \gamma_u$ represent the total intensity of input data stream to individual network computing nodes (the summary input stream from other connected computing nodes to the given i-th computing node. It is given as Poisson input stream with intensity λ_i demands in time unit

- r_{ij} are given as the relation probabilities from node i to the neighboring connected nodes j
- $\beta_1, \beta_2, ..., \beta_u$ correspond to the total extern output stream of data units from used nodes (the total output stream to the connected computing nodes of the given node).

The created abstract model according Fig. 6 belongs in queuing theory to the class of open queuing theory systems (open queuing networks). Formally we can adjust abstract model adding virtual two nodes (node 0 and node U+1 according Fig. 7 where
- virtual node 0 represent the sum of individual total extern input intensities $\gamma = \sum_{i=1}^{U} \gamma_i$ to computing nodes u_i
- virtual node U+1 represent the sum of individual total intern output intensities $\beta = \sum_{i=1}^{U} \beta_i$ from computing nodes u_i.

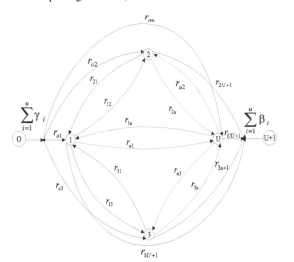

Figure 7. *Adjusted abstract model.*

Such a model corresponds in queuing theory to the model of open servicing network. Adjective "open" characterize the extern input and output data stream to the

servicing transport network [3, 29]. In common they are the open Markov servicing networks, in which the demand are mixed together at their output from one queuing theory system to another connected queuing theory system in a random way to that time as they are leaving the network. To the given i-th node the demand stream enter extern (from the network side), with the independent Poisson arrival distribution and the total intensity γ_i demands in seconds. After servicing at i-th node the demand goes to the next j-th node with the probability r_{ij} in such a way that the demand walks to the j-th node intern (from the sight network). At this time the demand departures from i-th node to the other nodes are defined with probability

$$1 - \sum_{j=1}^{U} r_{ij}$$

6. Modeling of the NOW and Grid

NOW is a basic module of any Grid parallel computer. Structure of essential parts in any workstation (i-th node) of NOW based on single processor (m=1) or multiprocessor system (m - processors or cores) is illustrated at Fig. 8. Inter process communication (IPC) represents all needed communication in NOW as
- communication among parallel processes
- control communication.

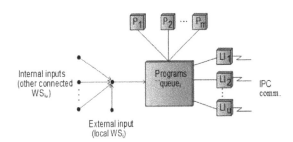

Figure 8. *Structure of i – th computing node (WSi).*

In principle we are assumed any constraints on structure of communication system architecture. Then we are modeling one workstation as a system with two dominant overheads
- computation execution time [2]
- communication latency [23].

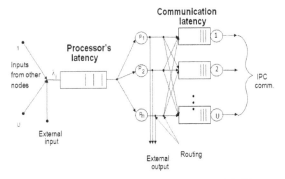

Figure 9. *Mathematical model of i – th node of NOW.*

To model these overheads through applying queuing theory we created mathematical model of one i-th computing node according Fig. 9, which models

- computation activities (processor's latency) as one queuing theory system
- every communication channel of i-th node LI_i i = 1, 2, ..., U as second queuing theory systems (communication latency).

6.1. Standard Analytical Model Based on M/M/M Queuing Systems

Let U be a node number of the whole transport system. For every node of NOW (i-th node according Fig. 10) we define the following parameters

- λ_i - the whole number of incoming demands to the i-th node, that is the sum both of external and internal inputs to the i-th node $\gamma = \sum_{i=1}^{U} \gamma_i$ represent the sum of individual total extern intensities in the NOW
- λ_{ij} - the whole input flow to the j-th communication channel at i-th node
- $E(t_q)_i$ - the average servicing time in the program queue (the waiting in a queue and servicing time) in the i-th node
- $E(t_q)_{ij}$ - the average servicing time of the j-th queue of the communication channel (the queue waiting time and servicing time) at i-th node.

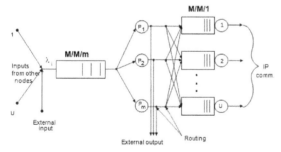

Figure 10. Standard analytical model of i-th computing node.

Then the whole extern input flow to the transport network is given as

$$\gamma = \sum_{i=1}^{U} \gamma_i \quad \text{and} \quad \lambda_i = \sum_{i=1}^{u} \lambda_{ij} + \beta_i$$

where β_i represents the intern output from i-th node (finished parallel programs in this node) which is not further transmitted and is therefore not entering to the $(LO)_i$. Then the whole delay we can modeled as

$$E(t_q)_{now} = \frac{1}{\gamma} \left[\sum_{i=1}^{U} \left(\lambda_i \cdot E(t_q)_i + \sum_{j=1}^{u_i} \lambda_{ij} \cdot E(t_q)_{ij} \right) \right]$$

, where $\frac{\lambda_i \cdot E(t_q)_i}{\gamma}$ and $\frac{\lambda_{ij} \cdot E(t_q)_{ij}}{\gamma}$ define individual contribution of computation queue delay (M/M/m) and communication channel delay (M/M/1) of

every node to the whole delay. For establishing $E(t_q)_i$ for computation queue delay it is necessary to know λ_i as the whole intensity of the input flow to the message queue where $\lambda_i = \gamma_i +$ all intern inputs flow to i-th node. The intern input flow to i-th node is defined as the input from other connected nodes. We can express it in two ways

- through solving a system of linear equations in matrix form as $\bar{\lambda} = \gamma + \bar{\lambda} \cdot R$
- using of two data structures in form of tables and that is the routing table (RT) and destination probability tables (DPT).

In related model the routing table creates deterministic logical way from source i to the destination j. Concretely RT(i,j) has index (1, ..., N) of the next node on the route from i to j. This assumption of the fixed routing is not rare. We have proved also experimental, that the fix routing produces good analytical results in comparison to the alternate adaptive routing in a concrete communication network. The destination probability table destiny for each i, j pair the probability, that the message which outstands in node i is destined for node j. This table with n x n dimension and elements DPT(i, j) terminates which fraction of the whole extern input γ_i has the destination j, that is $\gamma_i \cdot DTP(i, j)$. A path through the transport network we can define as the sequence $(x_1, x_2, ... , x_m)$ in which

- exist physical communication channel, which connects x_k a x_{k1}, k = 1,2, ... , m - 1
- x_j a $x_k, \forall j, k \; j \neq k$ (they do not exist loops).

We can define path with record "path (j→k), i)" as expression of the ordered sequence nodes, which are on the route from node j to the node k and they pass step by step through nodes i. That is x_i=j, x_m=k, x_p=i and $1 < p \leq m$.

We define then $\sum_{k \in path(j \to k, i)}^{U}$ as the summation over the set of all destination nodes k so that node i lies on the route from the source node j. Then we get the relation for the intern input flow to the i-th computing node λ_i as follows

$$\sum_{j=1}^{U} \sum_{k=1}^{U} \gamma_j \cdot DTP(j, k), \quad \text{for} \quad j \neq i, k \in \text{path} (j \to k, i)$$

and whole input flow to node i as

$$\lambda_i = \gamma_i + \sum_{j=1}^{U} \sum_{k=1}^{U} \gamma_i \cdot DTP(j, k) for \; j \neq i, k \in path(j \to k, i).$$

We supposed also that the incoming demands are exponential distributed and that queue servicing algorithm is FIFO (First In First Out). The program queue PQ_i is servicing through one or more the same computation processors, which performed incoming demands (parallel processes). In demand servicing in a given node could be two possibilities

- demand will be routed to another node of the transport networks by their placing to the one of the used communication channel (IPC communication)

• demand is in the addressed node and she will leave communication network.

To every communication channel is set the queue of the given communication lines (LQ), which stores the demands (their pointers) who are awaiting the communication through this communication channel. Also in this case we supposed its unlimited capacity, exponential inter arrival time distribution of input messages and the servicing algorithm FIFO. Every communication line queue has its communication capacity S_{ij} (in data units per second). Because we supposed the exponential demand length distribution the servicing time is exponential distributed too with average servicing time $1/\mu S_{ij}$, where μ is the average message length and S_{ij} is the communication capacity of node i and of communication channel j. For simplicity we will assume, as it is obvious, that S_{ij} is a part of μ. To find the average waiting time in the queue of the communication system we consider the model of one communication queue part node as M/M/1 queuing theory system according Fig. 11.

M/M/1

Figure 11. Model of one M/M/1 communication channel of the i-th node.

The total incoming flow to the communication channel j at node i which is given through the value λ_{ij} and we can determine it with using of routing table and destination probability table in the same way as for the value λ_i. Then ρ_{ij} as the utilization of the communication channel j at the node i is given as

$$\rho_{ij} = \frac{\lambda_{ij}}{\mu \, S_{ij}}$$

The total average delay time in the queue $E(t_q)_i$ is

$$E(t_q)_{ij} = \frac{1}{\mu_{ij} - \lambda_{ij}}$$

If we now substitute the values for T_i and T_{ij} to the relation for T we can get finally the relation for the total average delay time of whole transport system as

$$E(t_q)_{now} = \frac{1}{\gamma}\left[\sum_{i=1}^{U}\left(\lambda_i \cdot \frac{1}{\mu_i - \lambda_i} + \sum_{j=1}^{u_i}\lambda_{ij} \cdot \frac{1}{\mu_{ij} - \lambda_{ij}}\right)\right]$$

6.2. Model with M/D/m and M/D/1 Systems

The used model were built on assumptions of modeling incoming demands to program queue as Poisson input stream and of the exponential inter arrival time between communication inputs to the communication channels. The idea of the previous models were the presumption of decomposition to the individual independent channels

together with the independence presumption of the demand length, that is demand lengths are derived on the basis of the probability density function $p_i = \mu \, e^{-\mu t}$ for $t > 0$ and $f(t) = 0$ for $t \le 0$ always at its input to the node. On this basis it was possible to model every used communication channel as the queuing theory system M/M/1 and to derive the average value of delay individually for every channel too. The whole end-to-end delay was then simply the sum of the individual delays of the every used communication channel.

These conditions are not fulfilled for every input load, for all architectures of node and for the real character of processor service time distributions. These changes could cause imprecise results. To improve the mentioned problems we suggested the behavior analysis of the modeled NOW module improved analytical model (Fig. 12), which will be extend the used analytical model to more precise analytical model supposing that

• we consider to model computation activities in every node of NOW network as M/D/m system
• we consider an individual communication channels in i-th node as M/D/1 systems. In this way we can take into account also the influence of real non exponential nature of the inter arrival time of inputs to the communication channels.

These corrections may to contribute to precise behavior analysis of the NOW network for the typical communication activities and for the variable input loads. According defined assumption to modeling of the computation processors we use the M/D/m queuing theory systems according Fig. 12. To find the average program queue delay we have used the approximation formula for M/D/m queuing theory system as follows

$$E(t_w)(M/D/m_i) =$$
$$\left[1+(1-\rho_i)\cdot(m_i-1)\cdot\frac{\sqrt{45m_i}-2}{16\rho_i m_i}\cdot\frac{E(t_w)\,(M/D/1)}{E(t_w)\,(M/M/1)}\cdot E(t_w)\,(M/M/m_i)\right]$$

, in which

• ρ_i - is the processor utilization at i-th node for all used processors
• m_i - is the number of used processors at i-th node
• $E(t_w)(M/D/1)$, $E(t_w)\,(M/M/1)$ and $E(t_w)\,(M/M/m)$ are the average queue delay values for the queuing theory systems M/D/1, M/M/1 and M/M/m respectively.

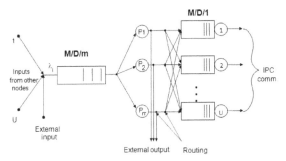

Figure 12. Precise mathematical model of i-th node.

The chosen approximation formulae we selected from two following points

- for his simply calculation
- if the number of used processors equals one the used relation gives the exact solution, that is W(M/D/1) system. Such number of processors is often used in praxis
- if the number of processors greater than one ($m_i > 1$) the used relation generate a relative error, which is not greater as 1%. This fact we verified and confirmed through simulation experiments.

Let $\overline{x_i}$ define the fixed processing time of the i-th node processors and $E(t_w)_i$ (PQ) the average program queue delay in the i-th node. Then ρi, as the utilization of the i-th node, is given as

$$\rho_i = \frac{\lambda_i . \overline{x_i}}{m_i}$$

Then the average waiting time in PQ queue $E(t_w)_i(M/D/m_i)$ is given trough the following relations

$$E(t_w)_i(M/D/1) = \frac{\rho_i \cdot \overline{x_i}}{2(1 - \rho_i)}$$

$$E(t_w)_i(M/M/1) = \frac{\rho_i \cdot \overline{x_i}}{1 - \rho_i}$$

$$E(t_w)_i(M/M/m_i) = \frac{\dfrac{(m_i \cdot \rho_i)^{m_i}}{m_i!(1-\rho_i)}}{\displaystyle\sum_{j=0}^{m_i-1}\left[\dfrac{(m_i \cdot \rho_i)^j}{j!} + \dfrac{(m_i \cdot \rho_i)^{m_i}}{m_i!(''1-\rho_i)}\right] \cdot \dfrac{\overline{x_i}}{m_i}}{(1-\rho_i)}$$

By substituting relations for ρ_i, $E(t_w)_i$ (M/D/1), $E(t_w)_i$ (M/M/1) and $E(t_w)_i$ (M/M/m_i) in the relation for $E(t_w)_I$ (M/D/m_i) we can determine $E(t_w)_I$ (PQ). Then the total average delay for the communication activities in i-th node is simply the sum of average message queue delay (MQ) plus the fixed processing time

$$E(t_w)_i = E(t_w)_i(PQ) + \overline{x_i}$$

To find the average waiting time in the queue of the communication system we consider the model of one communication queue part node as M/M/1 queuing theory system according

Fig. 11. Let $\overline{x_{ij}}$ determine the average servicing time for channel j at the node i. Then ρ_{ij} as the utilization of the communication channel j at the node i is given as

$$\rho_{ij} = \frac{\lambda_{ij} \overline{x_{ij}}}{S_{ij}}$$

where S_{ij} is the communication channel speed of j-th node. For simplicity we will assume that $S_{ij} = 1$. The total

incoming flow to the communication channel j at node i which is given through the value λ_{ij} and we can determine it with using of routing table and destination probability table in the same way as for a value λ_i. Let $E(t_w)_{ij}$ (LQ) be the average waiting queue time for communication channel j at the node i. Then

$$E(t_w)_{ij}(LQ) = \frac{\rho_{ij} \cdot \overline{x_{ij}}}{(1 - \rho_{ij})}$$

The total average delay value is the queue $E(t_w)_{ij}$ is given then as

There If we now substitute the values for $E(t_q)_i$ and $E(t_q)_{ij}$ to the relation for $E(t_q)_{now}$ we can get finally the relation for the total average delay time of whole NOW model is given as

$$E(t_q)_{now} = \frac{1}{\gamma}\left[\sum_{i=1}^{U}\left(E(t_w)_i(PQ) + \overline{x_i}\right) + \sum_{j=1}^{u_i}\left(E(t_w)_{ij}(LQ) + \overline{x_{ij}}\right)\right]$$

6.3. Mixed Analytical Models

6.3.1. Analytical Model with M/M/m and M/D/1 Queuing Systems

This model is mixture of analyzed model. The first part of final total average time $E(t_q)_i$ we get from chapter 6.1 and second part from 6.2.1 one. Then for $E(t_q)_{now}$ we can get finally

$$E(t_q)_{now} = \frac{1}{\gamma}\left[\sum_{i=1}^{U}\left(\lambda_i \cdot \frac{1}{\mu_i - \lambda_i} + \sum_{j=1}^{u_i}\left(E(t_w)_{ij}(LQ) + \overline{x_{ij}}\right)\right)\right]$$

6.3.2. Model with M/D/m and M/M/1 Queuing Systems

In this model the first part of final total average time $E(t_q)_i$ we can also get from chapter 6.2.1 and second part from 6.1 respectively. Then for $E(t_q)_{now}$ we get for this model finally

$$E(t_q)_{now} = \frac{1}{\gamma}\left[\sum_{i=1}^{U}\left(\left(E(t_w)_i(PQ) + \overline{x_i}\right) + \sum_{j=1}^{u_i}\lambda_{ij} \cdot \frac{1}{\mu_{ij} - \lambda_{ij}}\right)\right]$$

6.3.3. Analytical Model of Massive Grid Parallel Computers

We have defined Grid system as network of NOW network modules. Let N is the number of individual NOW networks or similar clusters. Then final total average time $E(t_q)_{grid}$

$$E(t_q)_{grid} = \frac{1}{\alpha}\left[\sum_{i=1}^{N}E(t_q)_{i\,now}\right]$$

where

- $\alpha = \sum_{i=1}^{N}\gamma_i$ represent the sum of individual total extern intensities to the i-th NOW module in the Grid

- $E(t_q)_{i\ now}$ correspondent to individual average times in i- th NOW module (i=1, 2, ... N).

The intern input flow to i-th node is defined as the input from all other connected computing nodes. We can express it in two following ways

- through solving a system of linear equations in matrix form as $\bar{\lambda} = \gamma + \bar{\lambda} \cdot R$
- using of two data structures in form of tables and that is the routing table (RT) and destination probability tables (DPT).

To improve the mentioned problems we suggested improved analytical model, which extends the used standard analytical model to more precise analytical model (improved analytical model) supposing that

- we consider to model computation activities in every node of NOW network as M/D/m system (assumption input of balanced parallel processes to every node)
- we consider an individual communication channels in i- th node as M/D/1 systems. In this way we can take into account also the influence of real non exponential nature of the inter arrival time of inputs to the communication channels.

Both analyzed analytical models are not fulfilled for every input load, for all parallel computer architectures and for the real character of computing node service time distributions. These changes may cause at some real cases imprecise results. Another survived problem of the used standard analytical model is assumption of the exponential inter arrival time between message inputs to the communication channels in case of unbalanced communication complexity of parallel processes. To remove mentioned changes we derived a correction factor to standard analytical model.

7. Corrected Standard Analytical Model

The derived standard analytical model supposes that the inter arrival time to the node's communication channels has the exponential distribution. This assumption is not true mainly in the important cases of high communication utilization. The node servicing time of parallel processes (computation complexity) could vary from nearly deterministic (in case of balanced parallel processes) to exponential (in case of unbalanced ones). From this in case of node's high processors utilization the outputs from individual processor of node's multiprocessor may vary from the deterministic interval time distribution to exponential one. These facts violate the assumption of the random exponential distribution and could lead to erroneous value of whole node's delay calculation. Worst of all this error could the greater the higher is the node utilization. From these causes we have derived the correction factor which accounts the measure of violation for the exponential distribution assumption.

The inter arrival input time distribution to each node's communication channel depends on ρ_i, where ρ_i is the overall processor utilization at the node i. But because only the part λ_{ij} from the total input rate λ_i for node i go to the node's communication channel j, it is necessary to weight the influence measure of the whole node's processors utilization trough the value $\lambda_{ij} / \lambda_{ij}$ for channel j as

$$\rho_i \cdot (\lambda_{ij} / \lambda_i)$$

To clarify the node's processor utilization influence to the average delay of communication channel we have tested the 7-noded experimental parallel computer. The processing time was varied to develop the various workloads of node's processors.

Extensive testing have proved, that if we increase utilization of communication channel and that develops saturation of communication channel queue then average queue waiting time is less sensitive to the nature of inter arrival time distributions. This is due to the fact that the messages (communicating IPC data) wait longer in the queue what significantly influenced the increase of the average waiting time and the error influence of the non-exponential inter arrival time distribution is decreased. To incorporate this knowledge for the correlation factor we investigated the influence of the weighting $p_i(\lambda_{ij} / \lambda_i)$ through the value $(1 - \rho_{ij})^x$ for various values x. The performed experiments showed the best results for the value x = 1. Derived approximation of the average queue waiting time of the communication channel j at the node i, which eliminates violence of the exponential inter arrival time distribution is then given as

$$\frac{\rho_i \cdot (1 - \rho_{ij}) \cdot \lambda_{ij}}{\lambda_i}$$

The finally correction factor of the communication channel j at the node i, which we have named as c_{ij} is as following

$$c_{ij} = 1 - \frac{\rho_i \cdot (1 - \rho_{ij}) \cdot \lambda_{ij}}{\lambda_i}$$

With the derived correction factor c_{ij} we can define now the corrected average queue waiting time as:

$$W_{ij}'(LQ) = c_{ij} \cdot W_{ij}(LQ)$$

The standard analytical model we can simply correct in such a way that instead of $W_{ij}(LQ)$ we will consider its corrected value $W_{ij}'(LQ)$. In this way derived improved standard analytical model we have defined as corrected standard analytical model. From the performed tests it is also remarkable that decreasing of the node's processors workload the assumption of the exponential inter arrival message time distribution to the communication channel is more effective. The achieved results are summarized at Tab. 1 for one of communication channels at the node 1. Graphical illustration of achieved results is at Fig. 9.

Table 1. *Achieved results for correction factor*

Processor utilization at node 1	Average channel delay at node 1 – simulation [msec]	Standard analytical model		Corrected analytical model	
		Average channel delay [msec]	Relative error [%]	Average channel delay [msec]	Relative error [%]
0,6	21,97	22,27	1,4	22,03	0,3
0,7	21,72	22,27	2,5	21,92	0,9
0,8	21,43	22,27	3,9	21,70	1,3
0,9	21,05	22,27	5,8	21,45	1,9
0,95	20,91	22,20	6,5	21,31	1,9

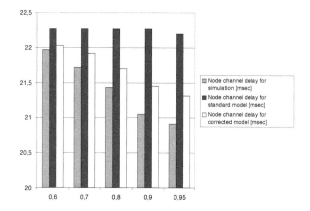

Figure 13. *The influence of the exponential time distribution and its correction.*

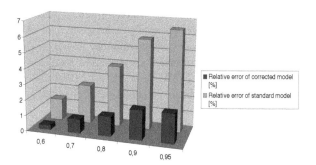

Figure 14. *Comparison of relative errors.*

The average delay values of the node's communication channel achieved through simulation are compared with the results of the standard analytical model (exponential inter arrival time distribution) and with the results of the corrected standard model. Comparison of the relative errors is illustrated in the Fig. 14.

At Table 2 there are results of the channel utilization influence to the average waiting time for the communication channel of 7 - noded communication network. For this case the channel utilization was influenced through communication speed changes.

Table 2. *The results of the channel influence*

Processor utilization at node 1	Average channel delay for node 1 using simulation [msec]	Standard analytical model		Corrected analytical model	
		Average channel delay [msec]	Relative error [%]	Average channel delay [msec]	Relative error [%]
0,6	8,89	9,25	4,1	8,68	2,4
0,7	15,92	16,38	2,9	15,91	0,06
0,8	31,04	31,94	2,9	31,39	1,1
0,9	79,76	81,08	1,7	80,38	0,8

The achieved results in Table 2 are illustrated at Fig. 15 including their relative errors related to simulation results.

The influence of communication channel utilization to the result accuracy of the analytical models is at the Fig. 16. From these achieved results follow that decreasing of the node's communication channel utilization the difference between simulated results and the standard analytical model increases.

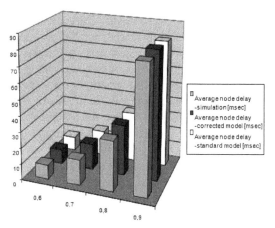

Figure 15. *The channel utilization influence to the total node delay.*

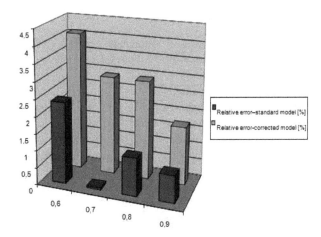

Figure 16. Influence of channel utilization to the accuracy of analytical models.

8. Other Achieved Results

Table 3 represents results and relative errors for the average value of the total message delay in the 5 nodes communication network so for classical analytical model (M/M/m + M/M/1) as for developed more precise analytical model (M/D/m + M/D/1) in which for multiprocessor's node activities we consider very real fixed latency. The same fixed delay was included to the average communication delay at each node and in simulation model too. These assumptions correspondence to the same communication speeds in each node's communication channel. If used communication channels do not have the same communication speeds then communication latencies are different constants. In both considered analytical models (M/M/m + M/M/1, M/D/m + M/D/1) performed experiments have proved that decreasing of processor utilization ρ cause decreasing of total average delay in NOW module $E(t_q)_{now}$. Therefore parallel processes are waiting in parallel processes queues shorter time. In contrary decreasing of node's communication channel speed increase communication channel utilization and then data of parallel processes have to wait longer in communication channel queues and increase the total node's latency. Tested results have also proved the influence of real non exponential nature of the input inter-arrival time to node's communication channels. In relation to it the analytical model M/D/m + M/D/1 provides best results and the analytical model M/M/m + M/M/1 the worst ones. The results for other possible mixed analytical models (M/M/m + M/D/1, M/D/m + M/M/1) provide results between the best and worst solutions. For simplicity deterministic time to perform parallel processes at node's multiprocessor activities (the servicing time of PQ queue) was settled to 8µs and the extern input flow for each node was the same constant too.

Table 3. Comparison of considered analytical models

Processor utilization	Whole delay for simulation [msec]	Standard analytical model		Corrected analytical model	
		End -to- end delay [msec]	Relative error [%]	End –to- end delay [msec]	Relative error [%]
0,2	21,45	20,06	6,48	20,83	2,89
0,3	23,53	21,58	8,29	22,85	2,89
0,4	26,24	23,49	10,48	25,51	2,78
0,5	30,16	26,51	12,10	29,44	2,39
0,6	34,69	29,79	14,12	33,92	2,22
0,7	41,67	35,19	15,55	41,38	0,70
0,8	54,25	44,08	18,75	54,43	0,33
0,9	80,01	60,38	24,53	84,47	6,82

To vary the processor utilization we modified the extern input flow in the same manner for each used node. Comparison of whole delay illustrates for both tested analytical models (standard, corrected) in relation to simulated results are presented at Fig. 17.

To vary node's processor utilization we modified the extern input flow in the same manner for each node of NOW module. For both analytical models (the best and the worst cases) are at Fig. 18 the relative errors in relation to simulation results. The best analytical model (M/D/m + M/D/1) provides very precision results in the whole range of input workload of multiprocessors and every communication channel's utilization with relative error, which does not exceed 6.2% and in most cases are in the range up to 5%. This is very important to project heavily loaded NOW network module (from about 80 to 90%), where the accurate results are to be in bad need of to avoid any bottleneck congestions or some other system instabilities.

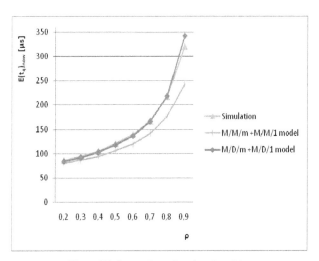

Figure 17. Comparison of analyzed models.

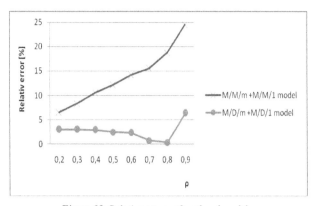

Figure 18. Relative errors of analyzed models.

The relative errors of worst analytical model are from 7 to 25%. This is due influences of processes queues delays, the nature of inter arrival input to the communication channel in the case of high processor utilization. In contrary the corrected analytical model in all cases has the relative number not greater than 7%. The achieved results in Table 3 indicate also other important critical fact. The derived corrected model produces more precise results in the whole range of node's processor utilization including the range of their higher utilization (in range 0,5 – 0,9) which are the most interesting to practical use. All developed analytical models could be applied also for large NOW networks practically without any increasing of the computation time in comparison to simulation method because of their explained module's structure based on NOW module. Simulation models require oft three orders of magnitude more computation time for testing massive meta computer. Therefore limiting factor of the developed analytical models will not be computation complexity, but space complexity of memories for needed RT and DPT tables. These needed RT and DPT tables require $O(n^2)$ memory cells, thus limiting the network analysis to the number of N nodes about 100 - 200 for the common SMP multiprocessor. In case of possible solving system of linear equations to find in analytical way node's λ_i and λ_{ij}, most parallel algorithms use to its solution Gauss elimination method (GEM). Used GEM parallel algorithms have computation complexity as $O(n^3)$ floating point multiplications and a similar number of additions [2, 15]. These values are however adequate to handle most existing communication network. In addition to it also for any future massive meta computers we would be always used hierarchically modular architecture, which consist on such simpler NOW modules.

We also point out, that accuracy contribution of corrected analytical model was achieved without the increasing the computation time in comparison to standard analytical model. It is also remarkable to emphasize increasing influence of the simulation complexity for the analysis of real massive parallel computers including their communication networks. The simulation models require three orders of magnitude more computation time for testing such complex parallel systems.

9. Conclusion

Performance evaluation of computers generally used to be a very hard problem from birthday of computers. This involves the investigation of the control and data flows within and between components of computers including their communication networks. The aim is to understand the behavior of the systems, which are sensitive from a performance point of view [32, 33]. It was, and still remains, not easy to apply any analytical method (queuing theory, theory of complexity, Petri nets) to performance evaluation of parallel computers because of their high number of not predictable parameters [21, 25]. Using of actual parallel computers (SMP -multiprocessor, multicore, NOV, Grid) open more possibilities to apply a queuing theory results to analyze more precise their performance. This imply existence of many inputs streams (control, data), which are inputs to modeled queuing theory systems and which are generated at various used resources by chance (assumption for good approximation of Poisson distribution). Therefore we could model computing nodes of parallel computers as M/D/m or M/M/m and their communication channels as M/D/1 or M/M/1 queuing theory systems in any existed parallel computer (SMP, NOW, Grid, meta computer).

Applied using of such flexible analytical modeling tool based on queuing theory results) shows real paths to a very effective and practical performance analysis tool including massive parallel computers (Grid, meta computers). In summary developed more precise analytical models could be applied to performance modeling of dominant parallel computers and that in following typical cases

- single computing nodes based on SMP parallel computer (multiprocessors, multicores, mix of them)
- NOW based on workstations (single, SMP)
- Grid (network of NOW modules)
- mixed parallel computers (SMP, NOW, Grid)
- meta computer (massive Grid).

From a point of user application of any analytical method is to be preferred in comparison with other possible methods, because of its universal and transparent character. Therefore the developed analytical models we can apply to performance modeling of any parallel computer or some parallel algorithms too (overheads). To practical applied using of developed analytical model we would like to advise following

- running of unbalanced parallel processes where λ is a parameter for incoming parallel processes with their exponential service time distribution as $E(t_s) = 1/\mu$ (corrected standard model)
  - in case of potential considering incoming units of parallel processes (data block, packet etc.) at using model based on M/M/m and M/M/1 queuing theory systems it would be necessary to recalculate at entrance incoming parallel processes to wanted data units. The way how to

recalculate them to such units at first node entrance we would like to refer in next paper

- running of parallel processes (λ parameter for incoming parallel processes with their deterministic service time $E(t_s) = 1/\mu =$ constant). The deterministic servicing times are a very good approximation of balanced parallel processes (M/D/m) with nearly equal amount of communication data blocks for every parallel process (M/D/1)
  - in case of using analytical model using M/D/m and M/D/1 we can consider λ parameter also for incoming units of parallel processes (data block, packet etc.) with their average service time for considered unit t_i, where $E(t_s) = 1/\mu = t_i =$ constant.

Using developed analytical models we are able to apply them so to both traditionally parallel computers (massive SMP) as distributed computers (NOW, Grid, meta computer). In such unified parallel computer models we are able better to study load balancing, mixed inter process communication IPC (shared and distributed memory), communication transport protocols, performance optimization and prediction in parallel algorithms etc. We would also like to analyze nasty problems in parallel computing as follows

- blocking problem (exhausted limited shared resources)
- waiting time $T(s, p)_{wait}$ as blocking consequence [11, 12]
- influence of routing algorithms
- to prove, or to indicate experimentally, the role of the independence assumption, if you are looking for higher moments of delay
- to verify the suggested model also for node limited buffer capacity and for other servicing algorithms than assumed FIFO (First In First Out)
- unified grouped decomposition models for parallel and distributed computing [13, 15]
- intensive testing, measurement and analysis to estimate technical parameters of used parallel computers [5, 24].

Acknowledgements

This work was done within the project "Complex modeling, optimization and prediction of parallel computers and algorithms" at University of Zilina, Slovakia. The author gratefully acknowledges help of project supervisor Prof. Ing. Ivan Hanuliak, PhD.

References

[1] Abderazek A. B., Multicore systems on chip - Practical Software/Hardware design, Imperial college press, United Kingdom, pp. 200, 2010

[2] Arora S., Barak B., Computational complexity - A modern Approach, Cambridge University Press, UK, pp. 573, 2009

[3] Dattatreya G. R., Performance analysis of queuing and computer network, University of Texas, Dallas, USA, pp. 472, 2008

[4] Dubois M., Annavaram M., Stenstrom P., Parallel Computer Organization and Design, Cambridge university press, United Kingdom, pp. 560, 2012

[5] Dubhash D.P., Panconesi A., Concentration of measure for the analysis of randomized algorithms, Cambridge University Press, UK, 2009

[6] Gelenbe E., Analysis and synthesis of computer systems, Imperial College Press, UK; pp. 324, 2010

[7] Giambene G., Queuing theory and telecommunications, Springer, pp. 585, 2005

[8] Hager G., Wellein G., Introduction to High Performance Computing for Scientists and Engineers, CRC Press, USA, pp. 356, 2010

[9] Hanuliak Peter, Hanuliak Michal, Modeling of single computing nodes of parallel computers, American J. of Networks and Communication, Science PG, Vol.3, USA, 2014

[10] Hanuliak J., Hanuliak I., To performance evaluation of distributed parallel algorithms, Kybernetes, Volume 34, No. 9/10, UK, pp. 1633-1650, 2005

[11] Hanuliak P., Analytical method of performance prediction in parallel algorithms, The Open Cybernetics and Systemics Journal, Vol. 6, Bentham, UK, pp. 38-47, 2012

[12] Hanuliak P., Hanuliak I., Performance evaluation of iterative parallel algorithms, Kybernetes, Volume 39, No.1/ 2010, United Kingdom, pp. 107- 126, 2010

[13] Hanuliak J., Modeling of communication complexity in parallel computing, American J. of Networks and Communication, Science PG, Vol. 3, USA, 2014

[14] Hanuliak P., Complex analytical performance modeling of parallel algorithms, American J. of Networks and Communication, Science PG, Vol. 3, USA, 2014

[15] Hanuliak P., Hanuliak J., Complex performance modeling of parallel algorithms, American J. of Networks and Communication, Science PG, Vol. 3, USA, 2014

[16] Harchol-BalterMor, Performance modeling and design of computer systems, Cambridge University Press, United Kingdom, pp. 576, 2013

[17] Hillston J., A Compositional Approach to Performance Modeling, University of Edinburg, Cambridge University Press, United Kingdom, pp. 172, 2005

[18] Hwang K. and coll., Distributed and Parallel Computing, Morgan Kaufmann, USA, pp. 472, 2011

[19] Kshemkalyani A. D., Singhal M., Distributed Computing, University of Illinois, Cambridge University Press, United Kingdom , pp. 756, 2011

[20] Kirk D. B., Hwu W. W., Programming massively parallel processors, Morgan Kaufmann, USA, pp. 280, 2010

[21] Kostin A., Ilushechkina L., Modeling and simulation of distributed systems, Imperial College Press, United Kingdom, pp. 440, 2010

[22] Kumar A., Manjunath D., Kuri J., Communication Networking , Morgan Kaufmann, USA, pp. 750, 2004

[23] Kushilevitz E., Nissan N., Communication Complexity, Cambridge University Press, United Kingdom, pp. 208, 2006

[24] Kwiatkowska M., Norman G., and Parker D., PRISM 4.0: Verification of Probabilistic Real-time Systems, In Proc. of 23rd CAV'11, Vol. 6806, Springer, Germany, pp. 585-591, 2011

[25] Le Boudec Jean-Yves, Performance evaluation of computer and communication systems, CRC Press, USA, pp. 300, 2011

[26] McCabe J., D., Network analysis, architecture, and design (3rd edition), Elsevier/ Morgan Kaufmann, USA, pp. 496, 2010

[27] Meerschaert M., Mathematical modeling (4-th edition), Elsevier, Netherland, pp. 384, 2013

[28] Misra Ch. S.,Woungang I., Selected topics in communication network and distributed systems, Imperial college press, UK, pp. 808, 2010

[29] Natarajan Gautam, Analysis of Queues: Methods and Applications, CRC Press , USA, pp. 802, 2012

[30] Peterson L. L., Davie B. C., Computer networks – a systemapproach, Morgan Kaufmann, USA, pp. 920, 2011

[31] Resch M. M., Supercomputers in Grids, Int. J. of Grid and HPC, No.1, pp. 1 - 9, 2009

[32] Riano l., McGinity T.M., Quantifying the role of complexity in a system´s performance, Evolving Systems, Springer Verlag, Germany, pp. 189 – 198, 2011

[33] Ross S. M., Introduction to Probability Models, 10th edition, Academic Press, Elsevier Science, Netherland, pp. 800, 2010

[34] Wang L., Jie Wei., Chen J., Grid Computing: Infrastructure, Service, and Application, CRC Press, USA, 2009
www pages

[35] www.top500.org.

22

Modeling of single computing nodes of parallel computers

Peter Hanuliak, Michal Hanuliak

Dubnica Technical Institute, Sladkovicova 533/20, Dubnica nad Vahom, 018 41, Slovakia

Email address:

phanuliak@gmail.com (P. Hanuliak), michal.hanuliak@gmail.com (M. Hanuliak)

Abstract: The paper describes analytical modeling for single computing nodes of parallel computers. At first the paper describes very shortly the developing steps of parallel computer architectures and then he summarized the basic concepts for performance evaluation. To illustrate theoretical evaluation concepts the paper considers in its experimental part the achieved results on concrete analyzed examples and their comparison. The suggested analytical models consider for single computing node based on processor or core and SMP modeling of own computer node´s activities and node´s communication channels of performed data communications within computing node queuing theory systems M/D/m or M/D/. In case of using SMP parallel system as node computer the suggested models consider for own node's activities M/M/m or M/D/m queuing theory systems. Although we are able to use other more complicated queuing theory systems we prefer modeling with mentioned models because achieved results for these models we can use in decomposed modeling of coupled computing nodes as network of workstations (NOW) or network of massive NOW modules (Grid). The achieved results of the developed analytical models we have compared with the results of tested computing nodes with other alternative evaluation method based on suitable benchmarks to verify developed analytical models. The developed analytical models could be used under various ranges of input analytical parameters, which influence the architecture of analyzed computing nodes which are interested for the praxis.

Keywords: Parallel Computer, Computing Node, Network of Workstation (NOW), Grid, Analytical Modeling, Queuing Theory, Performance Evaluation, Queuing Theory System, Benchmark

1. Developing Periods of Parallel Computers

In the first period of parallel computers between 1975 and 1995 dominated scientific supercomputers, which were specially designed for the high performance computing (HPC). These parallel computers have been mostly used computing models based on data parallelism. Those systems were way ahead of standard common computers in terms of their performance and price. Increased processor performance was caused through massive using of various parallel principles in all forms of produced processors. Parallel principles were used so in single PC's and workstations (scalar or super scalar pipeline, symmetrical multiprocessor systems SMP) [5] so as on extreme powerful PC as in various connected network of workstations (NOW, cluster). Gained experience with the implementation of parallel principles and intensive extensions of computer networks, leads to the use of connected computers for parallel solution. This period we can name as the second developing period. Their large growth since 1980 have been stimulated by the simultaneous influence of three basic factors [10, 29]

- high performance processors and computers
- high speed interconnecting networks
- standardized tools to development of parallel algorithms (OpenMP, MPI, Java).

Developing trends are actually going toward building of wide spread connected NOW networks with high computation and memory capacity (Grid). Conceptually Grid comes to the definition of meta computer [20], where meta computer could be understood as big computer network consisting on massive number of computing nodes, memories and other needed resources together creating an illusion of one single powerful supercomputer. These high integrated forms of NOW's create various Grid systems or meta computers we could define as the third period of parallel computers.

2. Basic Modules of Parallel Computers

Basic technical components of parallel computers illustrate Fig. 1 as follows
- modules of processors, cores of mix of them
- modules of computers (Sequential, parallel)
- memory modules
- input/output (I/O) modules.

These modules are connected through intern high speed communication networks (within concrete module) and extern among used computing modules via high speed communication networks [25, 35].

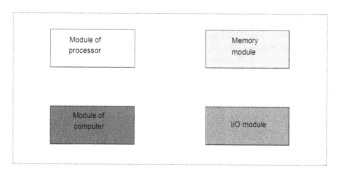

Figure 1. Basic building modules of parallel computers.

2.1. Classification of Parallel Computer

It is very difficult to classify all to this time realized parallel computers. The basic classification is from the point of realized memory as follows
- parallel computers with shared memory (multiprocessors, multicores)
- parallel computers with distributed memory (mainly based on computer networks)
- others.

2.1.1. Parallel Computers with Shared Memory
We can name realized parallel computers with shared memory as follows
- switched system
- multi bus system
- vector processor
- array processor
- associative processor
- transputer
- pipeline system
- systolic system
- wave front array system
- cellular system
- n-dimension cubes
- algorithm structured
- supercomputers
- connection machines
- super reliable
- neural networks.

2.1.2. Parallel Computers with Distributed Memory
To this group of parallel computers belong mainly parallel computers based on some form of network connection as follows
- localcomputer networks (LAN)
  - network of workstations (NOW)
  - PC farms, clusters
  - others
- wide area networks (WDN)
  - network of NOW networks (Grid)|
  - meta computers (Internet)
  - others.

2.2. Classification from the Point of Programmer

But from the point of programmer we divide them to the two following different groups
- synchronous parallel architectures. These are used for performing the same or very similar process (independent part) on different sets of data (data parallelism) in active computing nodes of parallel system. They are often used under central control that means under the global clock synchronization (vector, array system etc.) or a distributed local control mechanism (systolic systems etc.). This group consists mainly of parallel computers (centralized supercomputers) with any form of shared memory. Shared memory defines typical system features and in some cases can in considerable measure reduces developing of some parallel algorithms. To this group belong actually dominated parallel computers based on multiply cores, processors or mix of them and most of realized massive parallel computers (classic supercomputers) [5, 31]. Basic common characteristics are as following
  - shared memory (at least a part of memory)
  - using shared memory for communication
  - supported developing standard OpenMP, OpenMP Threads, Java
- asynchronous parallel computers. They are composed of a number of fully independent computing nodes (processors, cores or computers) which are connected through some communication network. To this group belong mainly various forms of computer networks (cluster), network of powerful workstation (NOW|) or more integrated network of NOW networks (Grid). Any cooperation and control are performed through inter process communication mechanisms (IPC) per realized remote or local communication channels. According the latest trends asynchronous parallel computers based on PC computers (single, SMP) are dominant parallel computers. Basic common characteristics are as following [10, 29]
  - no shared memory (distributed memory)
  - computing node could have some form of local memory where this memory in use only by connected computing node
  - cooperation and control of parallel processes

only using asynchronous message communication
- ■ supported developing standard
  - o MPI (Message passing interface)
  - o PVM (Parallel virtual machine)
  - o Java.

3. Typical Architectures of Modern Parallel Computers

3.1. Symmetrical Multiprocessor System

Symmetrical multiprocessor system (SMP) is a multiple using of the same processors or cores which are implemented on motherboard in order to increase the whole performance of such system. Typical common characteristics are following
- each processor or core (computing node) of the multiprocessor system can access main memory (shared memory)
- I/O channels or I/O devices are allocated to individual computing nodes according their demands
- integrated operation system coordinates cooperation of whole multiprocessor resources (hardware, software etc.).

Concept of multiprocessor system illustrates Fig. 2.

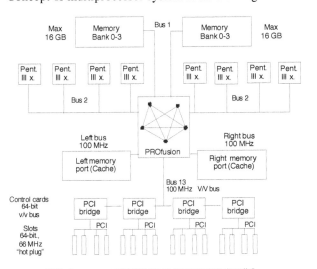

PROfusion - cross switch of 3 bus and 2 memory ports (parallel)

PCI cards - type Enhanced PCI (64 bit, 66 MHz, "Hot Plug" - on-line exchange)

Figure 2. Single computing node based on SMP (8-processors Intel Xeon).

3.2. Network of Workstations

There has been an increasing interest in the use of networks of workstations (NOW) connected together by high speed networks for solving large computation intensive problems. This trend is mainly driven by the cost effectiveness of such systems as compared to massive multiprocessor systems with tightly coupled processors and memories (supercomputers). Parallel computing on a network of workstations connected by high speed networks has given rise to a range of hardware and network related

issues on any given platform [3]. With the availability of cheap personal computers, workstations and networking devises, the recent trend is to connect a number of such workstations to solve computation intensive tasks in parallel on such clusters. Network of workstations [17, 19] has become a widely accepted form of high performance computing (HPC). Each workstation in a NOW is treated similarly to a processing element in a multiprocessor system. However, workstations are far more powerful and flexible than processing elements in conventional multiprocessors (supercomputers). To exploit the parallel processing capability of a NOW, an application algorithm must be paralleled. A way how to do it for an application problem builds its decomposition strategy. We will refer to it in [13].

Typical example of networks of workstations also for solving large computation intensive problems is at Fig. 3. The individual workstations are mainly extreme powerful personal workstations based on multiprocessor or multicore platform [1, 36]. Parallel computing on a cluster of workstations connected by high speed networks has given rise to a range of hardware and network related issues on any given platform.

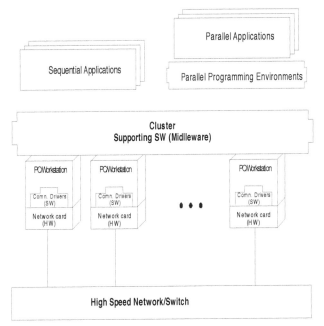

Figure 3. Typical architecture of NOW.

On such modular parallel computer we have been able to study basic problems in parallel computing (parallel and distributed computing) as load balancing, inter processor communication IPC [22, 32], modeling and optimization of parallel algorithms etc. [2, 14]. The coupled computing nodes PC_1, PC_2, ...,PC_i (workstations) could be single extreme powerful personal computers or SMP parallel computers. In this way parallel computing on networks of conventional PC workstations (single, multiprocessor, multicore) and Internet computing, suggest advantages of unifying parallel and distributed computing [30].

3.3. Grid Systems

Grid technologies have attracted a great deal of attention recently, and numerous infrastructure and software projects have been undertaken to realize various versions of Grids. In general Grids represent a new way of managing and organizing of computer networks and mainly of their deeper resource sharing [34]. Grid systems are expected to operate on a wider range of other resources as processors (CPU), like storages, data modules, network components, software (typical resources) and atypical resources like graphical and audio input/output devices, sensors and so one (Fig. 4.). All these resources typically exist within nodes that are geographically distributed, and span multiple administrative domains. The virtual machine is constituted of a set of resources taken from a resource pool [34]. It is obvious that existed HPC parallel computers (supercomputers etc.) could be a member of such Grid systems too. In general Grids represent a new way of managing and organizing of computer networks and mainly of their deeper resource sharing (Fig. 4).

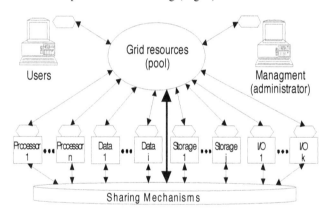

Figure 4. Architecture of Grid node.

Conceptually they go out from a structure of virtual parallel computer based on computer networks. In general Grids represent a new way of managing and organizing of resources like network of NOW networks. This term define massive computational Grid with following basic characteristics

- wide area network of integrated free computing resources. It is a massive number of inter connected networks, which are connected through high speed connected networks during which time whole massive system is controlled with network operation system, which makes an illusion of powerful computer system (virtual supercomputer)
- grants a function of meta computing that means computing environment, which enables to individual applications a functionality of all system resources
- Grid system combines distributed parallel computation with remote computing from user workstations.

3.4. Meta Computing

This term define massive parallel computer (supercomputer, Grid).

The best example of existingmeta computer is Internet as massive international network of various computer networks. Fig. 5 illustrates Internet as virtual parallel computer from sight of common Internet user.

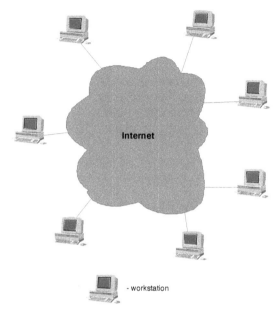

Figure 5. Internet as virtual parallel computer.

Another sight to Internet as network of connected individual computer networks is at Fig. 6. The typical networking switches are bridges, routers, gateways etc. which we denote with common term as network processors [27].

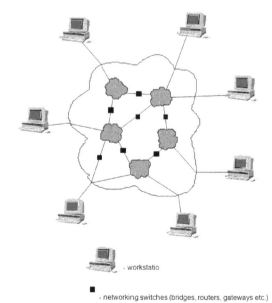

Figure 6. Internet as network of connected networks.

4. Modeling in Parallel Computing

Generally model is the abstraction of the system according Fig. 7 [15]. The functionality of the model represents the level of the abstraction applied. That means, if we know all there is about the system and are willing to pay for the complexity of building a true model, the role of abstraction is near nil. In practical cases we wish to abstract the view we take of a system to simplify the complexity of the real system. We wish to build a model that focuses on some basic elements of our interest and leave the rest of real system as only an interface with no details beyond proper inputs and outputs. A real system could be any parallel process or parallel computer that we are going to model [15]. In our cases they should be applied parallel algorithms (PA) or concrete parallel computers (SMP, NOW, Grid etc.).

The basic conclusion is that a model is a subjective view of modeler's subjective insight into modeled real system. This personal view defines what is important, what the purposes are, details, boundaries, and so one. Therefore the modeler must understand the system in order to guarantee useful features of the created model.

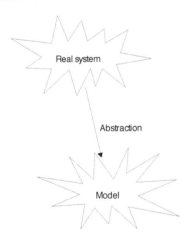

Figure 7. Modeling process.

4.1. Model Construction

Modeling is high creative process which incorporates following basic assumptions
- high ability of abstract thinking
- brain storming (creativity)
- alternating behavior and strategy
- logical hierarchical approaches to differ primary and secondary facts.

In general the development of model in any scientific area include the collection of following steps
- define the problem to be studied as well the criteria for analysis
- define and/or refine the model of the system. This include development of abstractions into mathematical, logical or procedural relationships
- collect data input to the model. Define the outside world and what must be fed to or taken from the

model to "simulate" that world
- select a modeling tool and prepare and augment the model for tool implementation
- verify that the tool implementation is an accurate reflection of the model
- validate that the tool implementation provides the desired accuracy or correspondence with the real world system being modeled
- experiment with the model to obtain performance measures
- analyze the tool results
- use findings to derive designs and improvements for the real world system.

Corresponding flow diagram of model development represents Fig. 8.

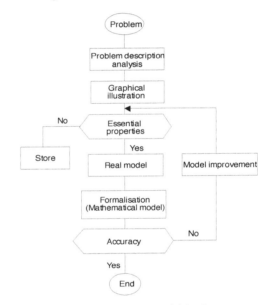

Figure 8. Flow diagram of model development.

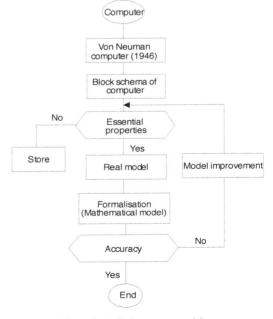

Figure 9. Applied computer modeling.

To practical illustration we have chosen applied modeling of classical sequential von Neumann computer according Fig. 9.

5. Abstract Models of Computing Nodes

5.1. Abstract model of SMP computing node with shared memory

Basic abstract model of parallel computer with shared memory is at Fig. 10.

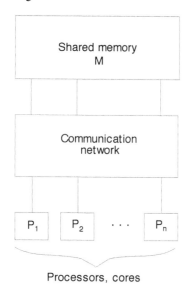

Figure 10. Abstract model of SMP.

5.2. Abstract Model of SMP with Distributed Memory

Basic abstract model of parallel computer with distributed memory is at Fig. 11.

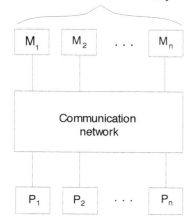

Figure 11. Abstract model of NOW.

6. The Role of Performance

Quantitative evaluation and modeling of hardware and software components of parallel systems are critical for the delivery of high performance. Performance studies apply to initial design phases as well as to procurement, tuning and capacity planning analysis. As performance cannot be expressed by quantities independent of the system workload, the quantitative characterization of resource demands of application and of their behavior is an important part of any performance evaluation study [24, 33]. Among the goals of parallel systems performance analysis are to assess the performance of a system or a system component or an application, to investigate the match between requirements and system architecture characteristics, to identify the features that have a significant impact on the application execution time, to predict the performance of a particular application on a given parallel system, to evaluate different structures of parallel applications [26].

6.1. Performance Evaluation Methods

The fundamental concepts have been developed for evaluating parallel computers. Trade-offs among these performance factors are often encountered in real-life applications. To the performance evaluation we can use following methods
- analytical methods
  - application of queuing theory [4, 8,16]
  - asymptotic (order) analysis [13, 14]
- simulation [21]
- experimental measurement
  - benchmarks [18]
  - modeling tools [23]
  - direct parameter measuring [6].

When we solve a model we can obtain an estimate for a set of values of interest within the system being modeled, for a given set of conditions which we set for that execution. These conditions may be fixed permanently in the model or left as free variables or parameters of the model, and set at runtime. Each set of m input parameters constitutes a single point in m-dimensional input space. Each solution of the model produces one set of observations. Such a set of n values constitutes a single point in the corresponding n-dimensional observation space. By varying the input conditions we hope to explore how the outputs vary with changes to the inputs.

6.1.1. Analytic Techniques

There is a very well developed set of techniques which can provide exact solutions very quickly, but only for a very restricted class of models. For more general models it is often possible to obtain approximate results significantly more quickly than when using simulation, although the accuracy of these results may be difficult to determine. The techniques in question belong to an area of applied mathematics known as queuing theory, which is a branch of stochastic modeling [7, 9]. Like simulation, queuing theory depends on the use of powerful computers in order to solve its models quickly. We would like to prefer techniques which yield analytic solutions.

6.1.2. The Simulation Method

Simulation is the most general and versatile means of modeling systems for performance estimation. It has many uses, but its results are usually only approximations to the exact answer and the price of increased accuracy is much longer execution times. To reduce the cost of a simulation we may resort to simplification of the model which avoids explicit modeling of many features, but this increases the level of error in the results. If we need to resort to simplification of our models, it would be desirable to achieve exact results even though the model might not fully represent the system. At least then one source of inaccuracy would be removed. At the same time it would be useful if the method could produce its results more quickly than even the simplified simulation. Thus it is important to consider the use of analytic and numerical techniques before resorting to simulation. This method is based on the simulation of the basic characteristics that are the input data stream and their servicing according the measured and analyzed probability values simulate the behavior model of the analyzed parallel system. Its part is therefore the time registration of the wanted interested discrete values. The result values of simulation model have always their discrete character, which do not have the universal form of mathematical formulas to which we can set when we need the variables of the used distributions as in the case of analytical models. The accuracy of simulation model depends therefore on the accuracy measure of the used simulation model for the given task.

6.1.3. Asymptotic (Order) Analysis

In the analysis of algorithms, it is often cumbersome or impossible to derive exact expressions for parameters such as run time, speedup, efficiency, issoefficiency etc. In many cases, an approximation of the exact expression is adequate. The approximation may indeed be more illustrative of the behavior of the function because it focuses on the critical factors influencing the parameter. We have used an extension of this method to evaluate parallel computers and algorithms in [13, 14].

6.1.4. Experimental Measurement

Evaluating system performance via experimental measurements is a very useful alternative for parallel systems and parallel algorithms. Measurements can be gathered on existing systems by means of benchmark applications that aim at stressing specific aspects of the parallel systems and algorithms. Even though benchmarks can be used in all types of performance studies, their main field of application is competitive procurement and performance assessment of existing systems and algorithms. Parallel benchmarks extend the traditional sequential ones by providing a wider a wider set of suites that exercise each system component targeted workload.

6.1.4.1. Benchmark

We divide used performance tests as following

- classical

  - Peak performance
  - Dhrystone
  - Whetstone
  - LINPAC
  - Khornestone
- problem oriented tests (Benchmarks)
  - SPEC tests [37]
  - PRISM [23].

6.1.4.2. SPEC Ratio

SPEC (Standard Performance Evaluation Corporation) defined one number to summarize all needed tests for integer number. Execution times are at first normalized through dividing execution time by value of reference processor (chosen by SPEC) with execution time on measured computer (user application program). The achieved ratio is labeled as SPEC ratio, which has such advantage that higher numerical numbers represent higher performance, that means that SPEC ratio is an inversion of execution time. INT 20xx (xx means year of latest version) or CFP 20xx result value is produced as geometric average value of all SPEC ratios. The relation for geometric average value is given as

$$n\sqrt{\prod_{i=1}^{n} normalised \ execution \ time_i}$$

, where normalized execution time is the execution time normalized by reference computer for $i - th$ tested program from whole tested group n (all tests) and

$$\prod_{i=1}^{n} a_i \ - product \ of \ individual \ a_i.$$

7. Application of Queuing Theory Systems

The basic premise behind the use of queuing models for computer systems analysis is that the components of a computer system can be represented by a network of servers (or resources) and waiting lines (queues). A server is defined as an entity that can affect, or even stop, the flow of jobs through the system. In a computer system, a server may be the CPU, I/O channel, memory, or a communication port. Awaiting line is just that: a place where jobs queue for service. To make a queuing model work, jobs (or customers or message packets or anything else that requires the sort of processing provided by the server) are inserted into the network. A simple example, the single server model, is shown in Fig. 12. In that system, jobs arrive at some rate, queue for service on a first-come first-served basis, receive service, and exit the system. This kind of model, with jobs entering and leaving the system, is called an open queuing system model.

We will now turn our attention to some suitable queuing systems, the notation used to represent them, the performance quantities of interest, and the methods for calculating them. We have already introduced many notations for the quantities of interest for random variables and stochastic processes.

7.1. Kendall Classification

Queuing theory systems are classified according to various characteristics, which are often summarized using Kendall`s notation [16, 28]. In addition to the notation described previously for the quantities associated with queuing systems, it is also useful to introduce a notation for the parameters of a queuing system. The notation we will use here is known as the Kendall notation in its extended form as

A/B/m/K/L/Z where
- A means arrival process definition
- B means service time distributions
- m is number of identical servers
- K means maximum number of customers allowed in the system (default = ∞)
- L is number of customers allowed to arrive (default = ∞)
- Z means discipline used to order customers in the queue (default = FIFO).

Three symbols used in a Kendall notation description also have some standard definitions. The more common designators for the A and B fields are as following
- M means Markovian (exponential) service time or arrival rate
- D defines deterministic (constant) service time or arrival rate
- G means general service time or arrival rate.

The service discipline used to order customers in the queue can be any of a variety of types, such as first-in first-out (FIFO), last in first out (LIFO), priority ordered, random ordered, and others. Next, we will apply several suitable queuing systems to model computer systems or workstations and give expressions for the more important performance quantities. We will suppose in Kendall notation default values that means we will use typical short Kendall notation.

7.2. Little's Laws

One of the most important results in queuing theory is Little's law. This was a long standing rule of thumb in analyzing queuing systems, but gets its name from the author of the first paper which proves the relationship formally. It is applicable to the behavior of almost any system of queues, as long as they exhibit steady state behavior. It relates a system oriented measure - the mean number of customers in the system - to a customer oriented measure - the mean time spent in the system by each customer (the mean end-to-end time), for a given arrival rate. Little's law says

$E(q) = \lambda \cdot E(t_q)$

or it's following alternatives
- $E(w) = \lambda \cdot E(t_w)$
- $E(w) = E(q) - \rho$ (single service where m=1)
- $E(w) = E(q) - m \cdot \rho$ (m – services).

We can use also following valid equation

$E(t_q) = E(t_w) + E(t_s)$.

where the named parameters are as
- λ - arrival rate at entrance to a queue
- m - number of identical servers in the queuing system
- ρ - traffic intensity (dimensionless coefficient of utilization)
- q - random variable for the number of customers in a system at steady state
- w - random variable for the number of customers in a queue at steady state
- $E(t_s)$ - the expected (mean) service time of a server
- $E(q)$ - the expected (mean) number of customers in a system at steady state
- $E(w)$ - the expected (mean) number of customers in a queue at steady state
- $E(t_q)$ - the expected (mean) time spent in system (queue + servicing) at steady state
- $E(t_w)$ - the expected (mean) time spent in the queue at steady state.

7.3. The M/M/1 Queue Model

To model a single workstation as single PC computer we give results needed results for M/M/1. There are many other kinds of queues, including those where FIFO strategy (First In First Out) is not assumed, but few yield easily usable analytic results. The M/M/1 queuing system is characterized by a Poisson arrival process and exponential service time distributions, with one server, and a FIFO queue ordering discipline. The system at Fig. 12 represents an input buffer holding incoming data bytes, with an I/O processor as the server. A few of the quantities that we will be interested in for this type of queuing system are the average queue length, the wait time for a customer in the queue, the total time a customer spends in the system, and the server utilization.

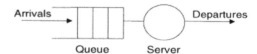

Figure 12. Queuing theory based model.

7.3.1. Poisson Distribution

The Poisson distribution models a set of totally independent events as a process, where each event is independent of all others. It is not the same as a uniform distribution. Where knowledge of past events does not allow us to predict anything about future ones, except that we know the overall average, the Poisson distribution represents the likelihood of one of a given range of numbers of events occurring within the next time interval. The definition of Poisson distribution is according following relation

$$p_i = \frac{\lambda^i}{i!} e^{-\lambda}$$

, where the parameter λ is defines as the average number of successes during the interval.

7.3.2. Exponential distribution

If the Poisson distribution represents the likely number of independent events to occur in the next time period, the exponential distribution is its converse. It represents the distribution of inter - arrival times for the same arrival process. Its mean is inter - event time, but it is often expressed in terms of the arrival rate, which is 1/inter - arrival time.

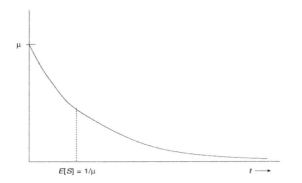

Figure 13. Graphic illustration of exponential distribution.

Exponential distribution function (Fig. 13) is defined as
$p_i = \mu \; e^{-\mu t}$ for t > 0 and f (t) = 0 for $t \le 0$.

and its mean value as
E (S) = E (t_s) = 1/μ.

The Poisson distribution models a set of totally independent events as a process, where each event is independent of all others. It is not the same as a uniform distribution. Where knowledge of past events does not allow us to predict anything about future ones, except that we know the overall average, the Poisson distribution represents the likelihood of one of a given range of numbers of events occurring within the next time interval.

7.3.3. The Derived Relation of M/M/1 Queue Model

We consider at first the M/M/1queue model. This represents a Poisson stream of independent arrivals into a queue whose single server has exponentially distributed service times. The queue is assumed to be unbounded and the population of potential customers to be infinite. Let λ be the (mean) rate of arrivals and μ be the (mean) rate of service (Fig. 14.).

Queue   Server rate = μ

Figure 14. M/M/1 queuing system model.

We derive a measure called traffic intensity, ρ as

$$\rho = \frac{\lambda}{\mu}$$

From this we can see that the mean and variance of the number of customers in the queue are

$$E(w) = \frac{\rho}{(1 - \rho)}$$

It also turns out that the end- to end delay (waiting time plus time being served) for each customer is exponentially distributed with parameter μ - λ. Thus the mean end to end delay is

$$E(q) = Var(q) = \frac{1}{(\mu - \lambda)}$$

Using Little's law we can get the end to end delay (waiting time plus time being served) for each

customer which is exponentially distributed with parameter μ - λ. Thus the mean end- to end delay is as

$$E(t_w) = \frac{\rho}{\mu - \lambda}$$

and waiting time in the queue as

$$E(t_q) = \frac{1}{\mu - \lambda}$$

7.3.4. M/M/m Queue Model

The illustration of M/M/m model for m=3 is at Fig. 15.

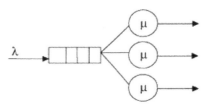

Figure 15. M/M/m (m=3) model of multiprocessor or multicore systems.

The basic needed derived relations for M/M/m queue model are following

$$\rho = \frac{\lambda}{\mu \cdot m} < 1$$

Average number of customer in the queue

$$E(w) = \frac{\rho \, (\rho \, m)^m}{m! \, (1 - \rho)^2} \, p_0$$

where the probability

$$p_0 = \left[\sum_{i=0}^{m-1} \frac{(m \, \rho)^i}{i!} + \frac{(m \, \rho)^m}{m!} \, \frac{1}{1 - \rho} \right]^{-1}$$

Average number of customer in the system is given as
$E(q) = E(w) + m \, \rho$

The further parameters E(t_q) and E(t_w) we can derive using the Little´s law.

7.4. Non Markovian models

7.4.1. M/D/1 Queue Model

In this queue model traffic intensity ρ is as

$$\rho = \frac{\lambda}{\mu} < 1$$

The service time is constant and is given as

$$E(t_s) = \frac{1}{\mu}$$

Then we can derive that the mean number of customers in the queue are

$$E(w) = \frac{\rho^2}{2(1-\rho)}$$

and the mean number of customers in the system is

$$E(q) = \frac{\rho(2-\rho)}{2(1-\rho)}$$

The waiting time in the queue for each customer is

$$E(t_w) = \frac{3\rho-2}{2\mu(1-\rho)}$$

and the end- to end delay (waiting time plus time being served) for each customer is

$$E(t_q) = \frac{\rho}{2\mu(1-\rho)}$$

7.4.2. M/D/m queue model

In this queue model traffic intensity ρ is as

$$\rho = \frac{\lambda}{\mu \cdot m} < 1$$

The service time is constant and is given as

$$E(t_s) = \frac{1}{\mu}$$

Then we can use for the mean number of customers in the queue following approximate relation

$$E(t_w)[M/D/m] \doteq$$

$$\left[1 + (1-\rho_i)\cdot(m-1)\frac{\sqrt{45m}-2}{16\rho_i\,m}\cdot\frac{E(t_w)[M/D/1]}{E(t_w)[M/M/1]}E(t_w)[M/M/m]\right]$$

Average number of customer in the system is given as
$$E(q) = E(w) + m\,\rho$$

The further parameters $E(t_q)$ and $E(t_w)$ we can derive using the Little's law.

8. Results

8.1. Application of THO Models

We have modeled single processor system as M/M/1 and multiprocessor system as M/M/m and M/D/m queuing models (two processor system as M/M/2 resp. M/D/2, four processor system as M/M/4 resp. M/D/4 etc.), where we were supposed parallel activity of used independent processors or cores. The differences between multiprocessor or multicore are in their performance (input parameters). Therefore we can model booth system with the same queuing theory system with appropriate input parameters. The individual result parameters are as follows

Input parameters
- λ - arrival rate at entrance to a queue
- ρ - traffic intensity
- m- number of identical servers in the queuing system.

Output parameters
- E(q) is the expected (mean) number of entities in a system
- E(w) is the expected (mean) number of entities in a queue
- $E(t_q)$ is the mean time spent in system (queue + servicing)
- $E(t_w)$ is the mean time spent in the queue
- $E(t_s)$ is the mean time of servicing.

Table 1 contains all the needed mean values of M/M/4 queuing system. The input parameter is ρ as input load. To compute the results we used concrete value of input intensity λ=3 and ρ = λ . $E(t_s)$ / 4 as input load intensity of four service equipment.

Table 1. Results for modeled 4 – multiprocessor system (λ=3)

P	E(w) [MIPS]	E(q) [MIPS]	E(t_w) [s]	E(t_q) [s]	E(t_s) [s]
0,1	0,000	0,400	0,000	0,133	0,133
0,2	0,002	0,802	0,001	0,267	0,267
0,3	0,016	1,216	0,005	0,405	0,400
0,4	0,060	1,660	0,020	0,553	0,533
0,5	0,174	2,174	0,058	0,725	0,667
0,6	0,431	2,831	0,144	0,944	0,800
0,7	1,000	3,800	0,333	1,267	0,933
0,8	2,386	5,586	0,795	1,862	1,067
0,9	7,090	10,690	2,363	3,563	1,200

Graphics illustration of results from Tab. 1 for modeled 4 – multiprocessor system (λ=3, ρ = λ . $E(t_s)$ / 4) are at Fig. 16, where x - axis contains values of parameter ρ (range of input load) and y - axis individual processing times per second.

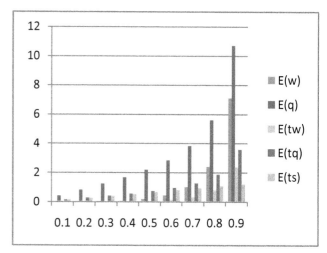

Figure 16. Illustration of modeling 4 – multiprocessor system (λ=3).

Graphic illustration at Fig. 17 illustrates average waiting times spent in system (queue + servicing) $E(t_q)$ of M/M/m queuing model (M/M/1, M/M/2, M/M/4) (λ=3, ρ = λ. $E(t_s)$ / m) for various number of services m, where x - axis contains values of parameter ρ (traffic intensity) and y - axis individual processing times per second.

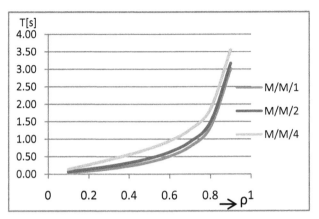

Figure 17. *Mean time in system T= E(t_q).*

Table 2. *Results for M/D/4 (λ=3, 4 – multiprocessor system)*

ρ	E(w) [MIPS]	E(q) [MIPS]	E(t_w) [s]	E(t_q) [s]	E(t_s) [s]
0,1	2,99	3,39	1,00	1,13	0,13
0,2	2,93	3,73	0,98	1,24	0,27
0,3	2,85	4,05	0,95	1,35	0,40
0,4	2,81	4,41	0,94	1,47	0,53
0,5	2,81	4,81	0,94	1,60	0,67
0,6	2,90	5,30	0,97	1,77	0,80
0,7	3,07	5,87	1,02	1,96	0,93
0,8	3,32	6,52	1,11	2,17	1,07
0,9	3,66	7,26	1,22	2,42	1,20

Graphics illustration of some results from Tab. 2 for modeled 4 – multiprocessor system (λ=3, ρ = λ .E(t_s) / 4) are at Fig. 18., where x - axis contain values of parameter ρ (range of traffic intensity) and y - axis contain individual processing and waiting times per second.

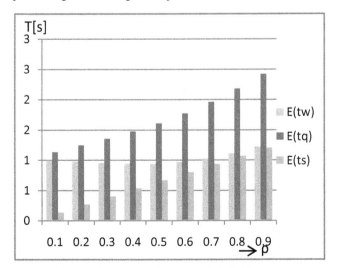

Figure 18. *Illustration of some values E(t_w), E(t_q), E(t_s) for M/D/4 system.*

Graphics illustration at Fig. 19 compare various queuing models (M/D/4, M/M/4) (λ=3, ρ = λ .E(t_s) / 4) for average time in system (queue + servicing), where x - axis contains values of parameter ρ (input load) and y - axis individual

processing times per second. From this comparison we can better results namely for higher values of traffic intensity parameter ρ.

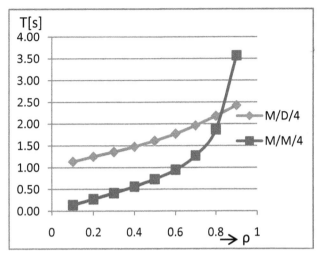

Figure 19. *Average waiting time in system (λ=3, T= E(t_q).*

8.2. Spec test ratio

We have been performed various tests (benchmarks) to verify derived analytical results. We illustrate some achieved results using Spec test ratio to compare performance of following processors

- AMD Athlon X2 6000+
- Intel Core2Duo E7300
- Intel i7-950.

Tab. 3 illustrates tested results for processor Intel i7-950 and at the same time description of used SPEC tests to evaluating performance. As we can see the used tests are really from various applications in order to come to more universal tested results.

Table 3. *Illustration of tested results for processor Intel i7-950*

Description	Name	Execution time [s]	SPEC ratio
String processing	Perl	445	21,9
Compression	bzip2	554	17,4
GNU C compiler	Gcc	321	25,1
Combinatorial optimization	Mcf	202	45,1
Artificial Intelligence	Go	460	22,8
Search gene sequence	Hmmer	516	18,1
Chess game (AI)	Sjeng	507	29,3
Quantum computer simulation	Libquantum	97,7	212
Video compression	h264avc	605	36,6
Discrete event simulation library	Omnetpp	269	23,3
Games/path finding	Astar	414	16,9
XML Processing	Xalancbmk	240	28,7
Geometric mean			29,1

To compare any computers using SPEC ratios test we prefer to use geometric mean value therefore it defines the same relative value regardless of used normalized reference computer. If we were evaluating normalized values using arithmetic mean value results would be depended from the type of used normalized computer. Graphical illustration of our tested computers is at Fig. 20 using standardized performance tests of SPEC consortium. According our expectations processor Intel i7-950 achieved the highest SPEC ratio value.

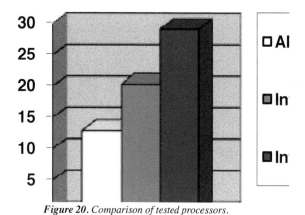

Figure 20. Comparison of tested processors.

9. Conclusions

Performance evaluation of computers generally used to be a very hard problem from birthday of computers. It was very hard to apply analytical methods based on queuing theory results to performance evaluation of sequential computers because of their high number of not predictable parameters. Secondly increasing of computer performance was done at first through technology improvements and processor's architecture changes.

From a point of user application of any analytical method (queuing theory, order analyze) is to be preferred in comparison with other possible methods because of transparent using of achieved results. Application of simulation method or Petri nets does not have such universal character as discussed analytical methods. Actually dominant using of multiprocessor or multicore computers opens more possibilities to apply a queuing theory results to their performance evaluation. This is based on a knowledge that outputs from more than one processor better approximate assumed Poisson distribution. Further the outputs from any computing node are going to another computing node in dominant parallel computers (NOW, Grid, meta computer). In relation to it we began to apply results of queuing theory to at first single computing node of parallel computer and then to dominant parallel computers based on NOW and their high integrated Grid (virtual parallel computer). To solve such coupled network of computing nodes appoints to couple network of queuing theory systems. We refer to it in another paper of this Special Issue [11]. The achieved results we can apply to performance modeling of multiprocessors or multicores) using as input parameter $\rho = \lambda .E(t_s) / m$ (for m=1 we can model computing node with one processor) as follows

- running of unbalanced parallel processes where λ is a parameter for incoming parallel processes with their

exponential service time distribution as $E(t_s) = 1/\mu$ (M/M/m model). In case of balanced parallel processes we could use the results as upper limits of exanimate parameters

- running of parallel processes (λ parameter for incoming parallel processes with their deterministic service time $E(t_s) = 1/\mu$ = constant). The same deterministic servicing time is a very good approximation for all optimal balanced parallel processes (M/D/m model)
- in case of using M/D/m model we can consider λ parameter also for incoming computer instructions with their average service time for instruction t_i, where $E(t_s) = 1/\mu = t_i$ = constant.

To consider incoming instructions at using M/M/m system it would be necessary to recalculate at entrance incoming parallel processes to instructions. To verify and to precise used analytical models we have been used SPEC ratio tests results (mean values of execution times for applied tests). To get comparable results we used the relation between throughput and execution time (latency) as

$$Throughput = \frac{1}{Latency}.$$

To model single computing nodes we can also use other more complicated single queuing theory systems than the analyzed ones (M/M/1, M/M/m, M/D/1, and M/D/m). We have choose the analyzed models from these causes

- to finish performance analysis of networks of queuing theory system we need results of chosen single queuing theory systems M/M/m and M/D/m
- we need their results to compute approximation relation for M/D/m
- M/M/1 and M/M/m models could be used to compare their results with other models M/D/1 and M/D/m respectively
- results of analyzed models M/M/m and M/D/m are necessary to finish coupled network of computing nodes [12].

Now according current trends in virtual parallel computers (SMP, NOW, Grid), based of powerful personal computers, we are looking for unified flexible models of any parallel computer that will be incorporated influences of

- other queue ordering discipline than FIFO
- various routing strategies
- various decomposition models etc.

In such flexible models we would like to study load balancing, inter process communication (IPC), transport protocols, performance prediction etc. We would refer to achieved results later.

Acknowledgements

This work was done within the project "Modeling, optimization and prediction of parallel computers and algorithms" at University of Zilina, Slovakia. The author gratefully acknowledges help of project supervisor Prof. Ing. Ivan Hanuliak, PhD.

References

[1] Abderazek A. B., Multicore systems on chip - Practical Software/Hardware design, Imperial college press, pp. 200, 2010

[2] Arora S., Barak B., Computational complexity - A modern Approach, Cambridge University Press, pp. 573, 2009

[3] Coulouris G., Dollimore J., Kindberg T., Distributed Systems – Concepts and Design (5 - th Edition), Addison Wesley, United Kingdom, pp. 800, 2011

[4] Dattatreya G. R., Performance analysis of queuing and computer network, University of Texas, Dallas, USA, pp. 472, 2008

[5] Dubois M., Annavaram M., Stenstrom P., Parallel Computer Organization and Design, Cambridge university press, United Kingdom, pp. 560, 2012

[6] Dubhash D.P., Panconesi A., Concentration of measure for the analysis of randomized algorithms, Cambridge University Press, United Kingdom, 2009

[7] Gautam Natarajan, Analysis of Queues: Methods and Applications, CRC Press, USA, pp. 802, 2012

[8] Gelenbe E., Analysis and synthesis of computer systems, Imperial College Press, United Kingdom, pp. 324, April 2010

[9] Giambene G., Queuing theory and telecommunications, Springer, Germany, pp. 585, 2005

[10] Hager G., Wellein G., Introduction to High Performance Computing for Scientists and Engineers, CRC Press, USA, pp. 356, 2010

[11] Hanuliak M., Modeling of dominant parallel computers based on NOW, American J. of Networks and Communication, Science PG, Vol. 3, USA, 2014

[12] Hanuliak M., Hanuliak P., Performance modeling of parallel computers NOW and Grid , AJNC (Am. J. of Networks and Communication), Science PG, USA, pp. 112-124, 2013

[13] Hanuliak J., Modeling of communication complexity in parallel computing, American J. of Networks and Communication, Science PG, Vol. 3, USA, 2014

[14] Hanuliak J., Hanuliak I., To performance evaluation of distributed parallel algorithms, Kybernetes, Volume 34, No. 9/10, United Kingdom, pp. 1633-1650, 2005

[15] HarcholBalterMor, Performance modeling and design of computer systems, Cambridge University Press, United Kingdom, pp. 576, 2013

[16] Hillston J., A Compositional Approach to Performance Modeling, University of Edinburg, Cambridge University Press, United Kingdom, pp. 172, 2005

[17] Hwang K. and coll., Distributed and Parallel Computing, Morgan Kaufmann, pp. 472, 2011

[18] John L. K., Eeckhout L., Performance evaluation and benchmarking, CRC Press, USA, 2005

[19] Kshemkalyani A. D., Singhal M., Distributed Computing, University of Illinois, Cambridge University Press, United Kingdom, pp. 756, 2011

[20] Kirk D. B., Hwu W. W., Programming massively parallel processors, Morgan Kaufmann, USA, pp. 280, 2010

[21] Kostin A., Ilushechkina L., Modeling and simulation of distributed systems, Imperial College Press,United Kingdom, pp. 440, 2010

[22] Kushilevitz E., Nissan N., Communication Complexity,Cambridge University Press, United Kingdom, pp. 208, 2006

[23] Kwiatkowska M., Norman G., and Parker D., PRISM 4.0: Verification of Probabilistic Real-time Systems, In Proc. 23rd Int. Conf. on CAV'11, Vol. 6806 of LNCS, Springer, Germany, pp. 585-591, 2011

[24] Le Boudec Jean-Yves, Performance evaluation of computer and communication systems, CRC Press, USA, pp. 300, 2011

[25] McCabe J., D., Network analysis, architecture, and design (3rd edition), Elsevier/Morgan Kaufmann, USA, pp. 496, 2010

[26] Miller S., Probability and Random Processes, 2nd edition, Academic Press, Elsevier Science, Netherland, pp. 552, 2012

[27] Misra Ch. S., Woungang I., Selected topics in communication network and distributed systems, Imperial college press, United Kingdom, pp. 808, April 2010

[28] Natarajan G., Analysis of Queues - Methods and Applications, CRC Press, USA, pp. 802, 2012

[29] Patterson D. A., Hennessy J. L., Computer Organization and Design (4th edition), Morgan Kaufmann, USA, pp. 914, 2011

[30] Peterson L. L., Davie B. C., Computer networks – a system approach, Morgan Kaufmann, USA, pp. 920, 2011

[31] Resch M. M., Supercomputers in Grids, Int. J. of Grid and HPC, No.1, Germany, pp. 1 - 9, 2009

[32] Riano l., McGinity T.M., Quantifying the role of complexity in a system's performance, Evolving Systems, Springer Verlag, Germany, pp. 189 – 198, 2011

[33] Ross S. M., Introduction to Probability Models, 10th edition, Academic Press, Elsevier Science, Netherland, pp. 800, 2010

[34] Wang L., Jie Wei., Chen J., Grid Computing: Infrastructure, Service, and Application, CRC Press, USA, 2009
www pages

[35] www.top500.org

[36] www. intel.com

[37] www.spec.org.

23

Decomposition models of parallel algorithms

Michal Hanuliak, Juraj Hanuliak

Dubnica Technical Institute, Sladkovicova 533/20, Dubnica nad Vahom, 018 41, Slovakia

Email address:

michal.hanuliak@gmail.com (M. Hanuliak)

Abstract: The article is devoted to the important role of decomposition strategy in parallel computing (parallel computers, parallel algorithms). The influence of decomposition model to performance in parallel computing we have illustrated on the chosen illustrative examples and that are parallel algorithms (PA) for numerical integration and matrix multiplication. On the basis of the done analysis of the used parallel computers in the world these are divided to the two basic groups which are from the programmer-developer point of view very different. They are also introduced the typical principal structures for both these groups of parallel computers and also their models. The paper then in an illustrative way describes the development of concrete parallel algorithm for matrix multiplication on various parallel systems. For each individual practical implementation of matrix multiplication there is introduced the derivation of its calculation complexity. The described individual ways of developing parallel matrix multiplication and their implementations are compared, analyzed and discussed from sight of programmer-developer and user in order to show the very important role of decomposition strategies mainly at the class of asynchronous parallel computers.

Keywords: Parallel Computer, Parallel Algorithms, Performance, Decomposition Model, Numerical Integration, Matrix Multiplication

1. Parallel Computing

Performance of actually computers (sequential, parallel) depends from a degree of embedded parallel principles on various levels of technical (hardware) and program support means (software). At the level of intern architecture of basic module CPU (Central processor unit) of PC they are implementations of scalar pipeline execution or multiple pipeline (superscalar, super pipeline) execution and capacity extension of cashes and their redundant using at various levels and that in a form of shared and local cashes (L1, L2, L3). On the level of motherboard there is a multiple using of cores and processors in building multicore or multiprocessors system as SMP (symmetrical multiprocessor system) as powerful computation node, where such computation node is SMP parallel computer too [1]. On the level of individual computers the dominant trend is to use multiple number of high performed workstations based on single personal computers (PC) or SMP, which are connected in the network of workstations (NOW) or in a high integrated way named as Grid systems [36]. A member of NOW or Grid could be any classic supercomputers [34].

1.1. Parallel Computers

From the point of system programmer we can divide all to this time realized parallel computers. The basic classification is from the point of realized memory as follows

- parallel computers with shared memory (multiprocessors, multicores)
- parallel computers with distributed memory (mainly based on computer networks)
- others.

1.1.1. Parallel Computers with Shared Memory

Basic common characteristics are as following [8, 13]

- shared memory
- using shared memory for communication
- supported developing standard
  - OpenMP
  - OpenMP Threads
  - Java
- typical architectures [37]
  - symmetrical multiprocessors (SMP)
  - supercomputers (massive SMP)

- Grid
- meta computers
- others.

1.1.2. Parallel Computers with Distributed Memory

Basic common characteristics are as following [22, 33]
- no shared memory (distributed memory)
- computing node could have some form of local memory where this memory in use only by connected computing node
- cooperation and control of parallel processes only using asynchronous message communication
- supported developing standard
  - MPI (Message passing interface)
  - PVM (Parallel virtual machine)
  - Java
- typical architectures
  - network of workstations (NOW)
  - Grid
  - meta computers
  - others.

2. Parallel Algorithms

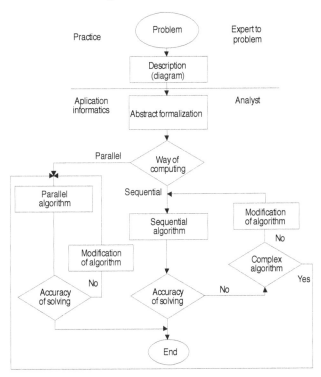

***Figure 1.** Deriving process of parallel algorithm.*

Users and programmers from a beginning of applied computer using request more powerful computers and more efficient applied algorithms. For a long time to effective technologies belong implementations of parallel principles so into computers as applied parallel algorithms. In this way term parallel programming could relate to every program, which contains more than one parallel process [21, 23]. This process represents single independent sequential part of program. Basic attribute of parallel algorithms is to achieve faster solution in comparison to quickest sequential solution. The role of programmer is for the given parallel computer and for the given application task to develop parallel algorithm (PA). Fig. 1 demonstrates how to derive parallel algorithm from existed sequential algorithms.

In last year's there is increased interest of scientific research into effective parallel algorithms. These trends to parallel algorithms also support actual trends in programming technologies to the development of modular applied algorithms based on object oriented programming (OOP). OOP algorithms are in their merits result of abstract thinking toward parallel solutions for existed complex problems.

2.1. General Classification of Parallel Algorithms

In general we supposed that potential effective parallel algorithms according defined algorithm classification (Fig. 2) should be in the group P as classified polynomial algorithms.

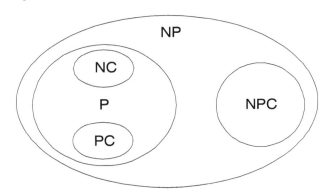

***Figure 2.** Parallel algorithm classification.*

Other used acronyms at Fig. 2 are as following [12]
- NP – general non polynomial group of all algorithms
- NC (Nick´s group). Group of effective polynomial algorithms
- PC – polynomial complete. Group of polynomial algorithms with high complexity
- NPC – non polynomial complete. This group consists of non-polynomial algorithms with their high solving complexity. The existence of any NPC algorithm in an effective way makes it available to solve effective also other NPC algorithms.

2.2. Parallel Processes

To derive PA we have to create conditions for potential parallel activities through dividing the input problem algorithm to its independent parts (decomposition model) according Fig. 3. These individual parts could be as following
- heavy parallel processes
- light parallel processes named as threads.

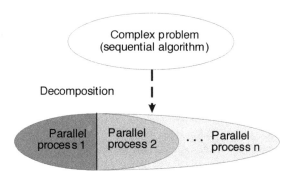

Figure 3. Illustration of decomposition process.

We will define standard process as developed sequential algorithm or its independent part. In detail standard process does not represent only some part of compiled program because to its characterization belongs also register status of processor. Illustration of such standard process is at Fig. 4.

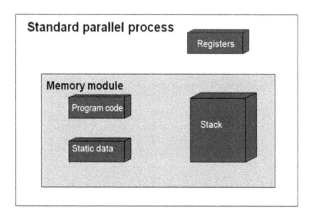

Figure 4. Illustration of standard process.

Every standard process has therefore own system stack, which contains process local data and in case of process interruption also actual register status of processor. It is obvious that we may have contemporary multiple numbers of standard processes, which used together some program part but their processes contexts (process local data) are different. Needed tools to manage processes (initialization, abort, synchronization, communication etc.) are within cores of multitask operation systems in form of services. Illustration of standard multi processes state is at Fig. 5.

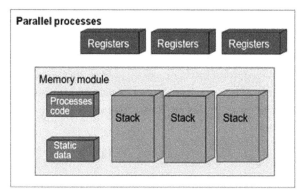

Figure 5. Parallel algorithm based on multiple parallel processes.

But concept of generating standard processes with individual address spaces is very time consuming. For example in operation system UNIX a new process is generating with operation fork(), which makes system call in order to create child process with new own address space. But in detail it means memory allocation, copying of data segment and descriptor of origin (parent) process and realization of child process stack. Therefore this concept we named as heavy-weighted process. It is obvious that heavy-weighted approach does not support effectiveness of applied parallel processing and needed scalability of parallel algorithms. In relation to it were necessary to develop another less time consuming concept of process generation named as light-weighted process. This lighten conception of generating new processes under another name as threads were implemented at various operation systems, supported threads libraries and parallel developing environments. Basic difference between standard process and thread is that we can within standard process generate additional new threads, which are together using the same address space including descriptor declaration of origin standard process.

2.3. Parallel Algorithms Classification

In principal parallel algorithms are dividing into two following basic classes
- parallel algorithms with shared memory (PA_{sm}). In this case parallel processes can communicate through shared variables using existed shared memory. For control of parallel processes are used typical synchronization tools as busy waiting, semaphores and monitors to guarantee exclusive using of shared resources only by single parallel process [10, 28]
- parallel algorithms with distributed memory (PA_{dm}). Distributed parallel algorithms have to synchronization and cooperation of parallel processes only network communication. The term distributed (asynchronous) parallel algorithms defines, that individual parallel processes are performed on independent computing nodes of used parallel computer with distributed memory [16, 31]
- mixed PA. Very perspective parallel algorithms which are to use advantages of dominant parallel computers based on NOW modules as following
  - using of parallel processes with shared memory in individual computing nodes of parallel computer
  - using of parallel processes based on distributed memory in parallel computers with distributed memory.

2.3.1. Parallel Algorithms with Shared Memory
Typical activity graph of parallel algorithms with shared memory PA_{sm} is at Fig. 6. To control decomposed parallel processes there is necessary synchronization mechanism as follows
- semaphors

- monitors
- busy waiting
- pathexpession
- critical region (CR)
  - conditional critical region (CCR).

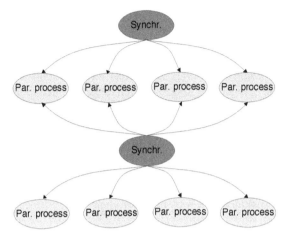

Figure 6. Illustration of parallel activities of PA$_{sm}$.

2.3.2. Parallel Algorithms with Distributed Memory

Parallel algorithms with distributed memory PA$_{dm}$ are parallel processes which are performing on asynchronous computing nodes of given parallel computer. Therefore for all needed cooperation of parallel processes we have available only inter process communication IPC. The principal illustration of parallel processes for PA$_{dm}$ is at Fig. 7.

Figure 7. Illustration of parallel activities of PA$_{dm}$.

3. The Role of Performance in Parallel Computing

Quantitative evaluation and modeling of hardware and software components of parallel systems are critical for the delivery of high performance. Performance studies apply to initial design phases as well as to procurement, tuning and capacity planning analysis. As performance cannot be expressed by quantities independent of the system workload, the quantitative characterization of resource demands of application and of their behavior is an important part of any performance evaluation study [6, 27]. Among the goals of parallel systems performance analysis are to assess the performance of a system or a system component or an application, to investigate the match between requirements and system architecture characteristics, to identify the features that have a significant impact on the application execution time, to predict the performance of a particular application on a given parallel system, to evaluate different structures of

parallel applications. In order to extend the applicability of analytical techniques to the parallel processing domain, various enhancements have been introduced to model phenomena such as simultaneous resource possession, fork and join mechanism, blocking and synchronization. Modeling techniques allow to model contention both at hardware and software levels by combining approximate solutions and analytical methods. However, the complexity of parallel systems and algorithms limit the applicability of these techniques. Therefore, in spite of its computation and time requirements, simulation is extensively used as it imposes no constraints on modeling.

3.1. The Role of Performance in Parallel Computing

To the performance evaluation in parallel computing we briefly review the techniques most commonly adopted for the evaluation in parallel computing as follows
- analytical
  - application of queuing theory results [11, 20]
  - order (asymptotic) analysis [3, 15]
  - Petri nets [7]
- simulation methods [24]
- experimental
  - benchmarks [32]
  - direct measuring [9, 29].

Analytical method is a very well developed set of techniques which can provide exact solutions very quickly, but only for a very restricted class of models. For more general models it is often possible to obtain approximate results significantly more quickly than when using simulation, although the accuracy of these results may be difficult to determine.

Simulation is the most general and versatile means of modeling systems for performance estimation. It has many uses, but its results are usually only approximations to the exact answer and the price of increased accuracy is much longer execution times. They are still only applicable to a restricted class of models (though not as restricted as analytic approaches.) Many approaches increase rapidly their memory and time requirements as the size of the model increases.

Evaluating system performance via experimental measurements is a very useful alternative for computer systems. Measurements can be gathered on existing systems by means of benchmark applications that aim at stressing specific aspects of computers systems. Even though benchmarks can be used in all types of performance studies, their main field of application is competitive procurement and performance assessment of existing systems and algorithms.

3.2. Performance Evaluation Metrics of Decomposition Models

To evaluating decomposition models in parallel algorithms we will be used defined complex basic concepts in [15] as follows

- parallel execution time T(s, p) as the execution time performed by p computing nodes (processors, cores, workstations) of given parallel computer and s defines input size (load) of given problem
- speed up factor S(s, p) as

$$S(s,p) = \frac{T(s,1)}{T(s,p)}$$

- efficiency E(s, p) as

$$E(s,p) = \frac{S(s,p)}{p} = \frac{T(s,1)}{p\,T(s,p)}$$

- the isoefficiency concept

$$w(s) = \frac{E}{1-E}\,h\,(s,p).$$

4. Developing Process of Parallel Algorithms

To exploit the parallel processing capability the application program must be parallelized. The effective way how to do it for a particular application problem (decomposition model) belongs to the most important step in developing an effective parallel algorithm [14, 18]. The development of the parallel network algorithm according Fig. 8 includes the developing activities as follow

- decomposition - the division of the application into a set of parallel processes
- mapping - the way how processes and data are distributed among the nodes
- inter process communication - the way of corresponding and synchronization among individual processes
- tuning - alternation of the working application to improve performance (performance optimization).

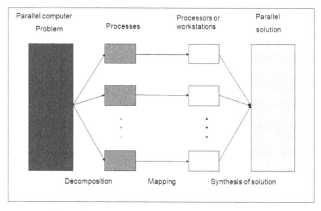

Figure 8. *Development steps in parallel algorithms.*

To do these steps there is necessary to understand the concrete application problem, the data domain, the used algorithm and the flow of control in given application. When designing a parallel program the description of the high level algorithm must include, in addition to design a sequential program, the method you intend to use to break the application into processes or threads (decomposition

model) and distribute data to different nodes (mapping). The chosen decomposition model drives the rest of program development.

4.1. Decomposition Models

Developing of sequential algorithms implicitly supposed existence of algorithm for given problem. Only later in stage of practical programming they are defined and used suitable data structures. In contrast to this classic developing method suggestion of parallel algorithm should include at beginning stage potential decomposition strategy including distribution of input data to perform decomposed parallel processes. Selection of suitable decomposition strategy has cardinal influence to further development of parallel algorithm.

Decomposition strategy defines potential dividing of given complex problem to their independent parts (Parallel processes) in such a way, that they could be performed in a parallel way through computing nodes of given parallel computer. Existence of some decomposition method is critical assumption to possible parallel algorithm. Potential decomposition degree of given complex problem is crucial for effectivity of parallel algorithm [4, 17]. To this time developed parallel algorithms and corresponding decomposition strategies were mainly related to available synchronous parallel computers based on classic massive parallel computers (supercomputers and their innovations). Developing parallel algorithms for actual dominant parallel computers NOW and Grid require at least modified decomposition strategies incorporating following priorities

- emphasis to functional parallelism of complex problems
- minimization of inter process communication IPC.

The most important step is to choose the best decomposition model for given complex problem. To do this it is necessary to understand concrete application problem, data domain, used algorithm and flow of control in given complex problem. When designing a parallel program the description of the high-level algorithm must include, in addition to design a sequential program, the method you intend to use to break the application into processes and distribute data to different computing nodes. The chosen decomposition models drive the rest of PA development. This is true is in case of developing new application as in porting serial code. The decomposition method tells us how to structure the code and data and defines the communication topology [25, 26].

Problem parallelization is very creative process, which creates potential degree of parallelism. This is a way how to divide complex problems to nondependent parts (Parallel processes) in such a way, to make possible to perform PA in parallel. The way of decomposition depends strongly from used task algorithm and from data structures. It has principal influences to performance and its communication consequences. To this time developed decomposition models and strategies seems to be close only to the in the world used supercomputers and their innovated types (classic parallel computers). On other way the realization of PA for in this time dominate parallel computers (SMP, NOW, Grid) demand modified decomposition models and strategies with respect to minimization of interposes communication intensity (NOW, Grid) and deriving waiting latency $T(s, p)_{wait}$ at using shared resources or at insufficient their capacities

- naturally parallel decomposition
- domain decomposition
- control decomposition
  - manager/workers
  - functional
- divide-and-conquer strategy for
- decomposition of big problems
- object oriented programming (OOP).

4.1.1. Natural Parallel Decomposition

Natural parallel decomposition allows simple creating of parallel processes whereby to their cooperation normally there is necessary low amount of inter process communication IPC. Also for parallel computation there is normally not important sequence of individual solutions. As a consequence there are not necessary any synchronization of performed parallel processes during parallel computation. Based on these attributes natural parallel algorithms allows to achieve practical ideal p - multiple speed - up using p - computation nodes of parallel computer (linear speed – up), and that with minimal additional efforts at developing parallel algorithms. Typical examples are numerical integration parallel algorithms. Based on this example we will illustrate in detail the role of decomposition models in developing steps of parallel algorithms.

4.1.1.1. Numerical Integration

Numerical integration algorithms are typical examples with implicitly latent decomposition strategy in which the parallelism is the integral part of own algorithm. Standard way of typical numerical integration algorithm (the computation of π number) assumes that we divide the interval <0, 1> to n identical subintervals whereby in each subinterval we approximate its part of curve with rectangle. Function values in middle of each subinterval determine height of the rectangle. Number of selected subintervals determines computation accuracy. The computed value of π will be given as sum of surface area of defined individual approximated rectangles. Illustration of numerical integration applied to π computation is at Fig. 9. Concretely for the calculation of the value π the following standard formula is used

$$\pi = \int_0^1 \frac{4}{1+x^2}\, dx$$

, where h = 1 / n is the width of the selected splitting interval, x_i = h (i - 0.5) are mid-ranges and n is the number of selected intervals (accuracy). For computation of π we can use an alternative following interpolating polynomial [93]

$$\pi = \int_0^1 f(x)\, dx = \int_0^1 \frac{4}{1+x^2}\, dx \cong \sum_{i=1}^n f(x_i)$$

or next possible relation

$$\pi \cong \sum_{0 \le i \le N} \frac{4}{\left(1 + \frac{(i+0,5)}{N}\right)^2}\left(\frac{1}{N}\right)$$

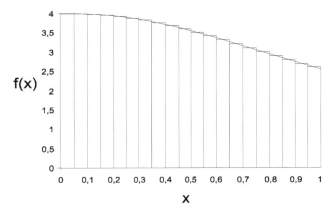

Figure 9. Principle of numerical integration.

4.1.1.2. Decomposition Model

For the parallel way of numerical integration computation we are used the property of latent decomposition strategy in all natural parallel algorithms. We divide the whole needed computation to its individual parallel processes according to the Fig. 10 where are for simplicity illustrated four parallel processes. For the parallel computation of number π we then use these created parallel processes.

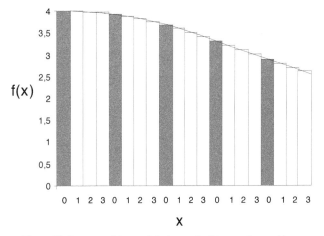

Figure 10. Decomposition model of numerical integration problems.

enter the desired number of subintervals n
compute the width w of each subinterval
for each subinterval
   find its centre x
   compute f(x) and the sum
end of cycle
multiply the sum with width to obtain π
return π

The prospective parallel implementations on dominant parallel computers (SMP, NOW, Grid) allow analysis of communication load depending on input computation load because input load is proportional to changes in load communication.

4.1.1.3. Mapping of Parallel Processes

The individual independent processes we distribute forcomputation in such way that every created parallel process will be executed on different computing node of parallel computer (mapping). After the parallel computation in individual nodes of a network of workstations was performed we need only to sum the partial results to get final result. To manage this task we have to choose one of the computing nodes to handle it. As well at the computation begin the chosen node (let it be node 0) must know the value of n (number of the strips in every process) and then selected node 0 has to let it know to all other computing nodes. Example of the parallel computation algorithm (manager process) is then as following

```
if my node is 0
    read the number n of strips desired and send it to all
    other nodes
else
    receive n from node 0
end if
for each strip assigned to this node
    compute the height of rectangle (at midpoint) and sum
    result
end for
if my node is not 0
    send sum of result to node 0
else
    receive results from all nodes and sum
    multiple the sum by the width of the strips to get π
return
```

Sending values of n / p could be done in case of parallel algorithm with distributed memory PA_{dm} using MPI API explained collective communication command Broadcast in case of the same size of parallel processes or MPI command Scatter in case of variable size of parallel processes.

We illustrate on this relative simple example parallel algorithm in parallel language FORTRAN for parallel computer with distributed memory. The algorithm extends and modifies the starting serial algorithm for the specific parallel implementation. To characterize shortly this parallel algorithm, we note that number of computing nodes of a parallel computer and identification of each computing node performed procedures numnodes() (number of computing nodes), mynode() (my computing node) and my pid() (number of my parallel process). These procedures allow the activation of any number of computing nodes of a parallel system. Then if we allow designated node, for example computing node 0, perform

management functions and certain partial computations, while other nodes perform a substantial part of the rest of computations for assigned subintervals, and communicated obtained partial amounts specified computing node 0. Procedures silence and crecv serve to ensure the required collective data communications and the final summation of achieved partial results.

```
f(x) = 4.0 /(1.0 + x*x)
    integer n, i, p, me, mpid
    real w, x, sum, pi
    p = numnodes ()              return number of nodes
    me = mynode ()               return number of my node
    mpid = mypid ()              return id of my process
    msglen = 4                   estimate message length
    allnds = -1                  message name for all nodes
    msgtp0 = 0                   name for message 0
    msgtp1 = 1                   name for message 1
    if (me .eg. 0) then          if i am node 0
        read *, n                read number of subintervals n
        callcsend (msgtp0,n,msglen,allnds,mpid) and send it to
all other nodes
    else                             if I am any other node
        callcrecv (msgtp0, n, msglen)    receive value n
    end if
    w = 1.0/n
    sum = 0.0
    do 10 i = me+1, n    dividing subintervals among nodes
        x =w*(i-0.5)
        sum = sum + f(x)         10 continue
        if (me .ne. 0) then      if I am not node  0
        callcsend (msgtp1, sum, 4, 0, mpid)      send partial
result to node 0
        else                         if I am node 0
        do 20 i = 1, p-1         for every other used node
            callcrecv (msgtp1, temp, 4)   receive partial result
to temp
            sum = sum + temp       and add it to sum
        20 continue
        pi = w*sum
        compute final result
        print *,pi                        and print it
    endif
end
```

The disadvantage is that the implementation of the central necessary routing communications through the designated node 0 (manager node), and as a result there may be a bottleneck which could negatively affect the efficiency of the parallel algorithm implementation.

4.1.1.4. Performance Optimization

In the above example of numerical integration this requirement leads to reduce the bottleneck, which is inter-process communication (IPC) latency. This latency should be minimized, since it could be used more effective in useful computation of parallel algorithm. It is therefore very important to minimize the number of communicating data messages proportionally to number of computational operations, thereby minimizing also overall execution time of a parallel algorithm. In computation of number π demanded centralization of needed communication through manager computing node 0 may come to computation bottleneck for two following reasons

- manager computing node 0 node could simultaneously receive data message only from one other computation node
- summary of the partial results at manager node 0 is done sequentially, in a sequential way, which is just a prerequisite to come to bottleneck.

The outgoing point in mentioned both cases is to consider using of collective communication commands of standardized development environments as MPI API, and that collective command Reduce or Gather. For some parallel computers are available alternative global summarization operation gssum(), just to eliminate such bottlenecks. This operation in iterative manner always sums partial results of two computing nodes, which both computing nodes are exchanged. Each partial sum, which receives one of computing node pair, is added to the obtained sum in a given node, and this result is transmitted to next computing node of defined communication chain. In this way there is gradually obtained the total accumulated amount of the computing nodes whereby manager computing node 0 can perform the last sum and print the final result. This procedure of global summarization can be also programmed, but the sequence of implemented appropriate procedures simplifies implementation of parallel algorithms and contributes to its effectiveness too.

In other applied tasks using a larger direct inter process communication implementation of communication used, for example form of asynchronous communication using direct support of multitasking in a given node, thereby achieving parallelism implementation of communication activities other node. Of course, an example of reducing the ratio of communication computing activity is not the only task in optimizing the performance of parallel algorithms. Available methods for optimizing performance are virtually the same colorful as diverse mere parallel application tasks. Inspired examples and procedures will therefore be included in the illustrative application examples in the following sections. Commonly used method and procedure for the decomposition of the application tasks indirectly implies the possibilities of optimizing its performance, i.e., optimization can also cause re-evaluation of strategies used decomposition.

Then if we allow designated node, for example node 0, perform management functions and certain partial calculations while other nodes perform a substantial part of the calculations for assigned subintervals, and communicate obtained partial amounts to specified node 0. Procedures silence and crecv serve to ensure the required communication.

It is also important to note that for mentioned module is required direct communication of every computing node with other computing nodes as assumption of parallel communication between multiple pairs of computing nodes. In this approach the final sum can be obtained after performing the second, third or even fourth cycle of the communication chain. In fact we need only $\log_2 p$, where p is the number of computing nodes of parallel computer, cycles of communication chains compared to n data messages at initial implementation. Used parallelism of data message exchange will therefore increase efficiency of parallel algorithms. For the implementation of an improved approach for data messages communications it's necessary to replace following part

if (me .ne. 0) then
  callcsend (msgtp1, sum, msglen, 0, mpid)
else
  do 20 i = 1, p-1
    callcrecv (msgtp1, temp, msglen)
    sum = sum + temp
  20 continue
  pi = w*sum
  print *, pi
endif
end
with next part
  callgssum (sum, 1, temp)
  if (me .eg. 0) then
    pi = w*sum
    print *, pi
  endif

In other applied parallel algorithms there is possible at using a larger direct inter process communication to perform communication for example in form of asynchronous communication using multitasking support in a given node, thereby it is achieved parallelism of performing communication with other computing node activity. Of course described example of reducing ratio of communication/ computing activity is not the only task in optimizing the performance of parallel algorithms. Available methods for performance optimizing are practical so various as are diverse parallel application problems.

4.1.2. Domain Decomposition

Typical characteristic of many complex problems is some regularity in sequential algorithms or in their data structures (computational or data modularity). Existence of such computational or data modules then represent domain of computation or data. Decomposition strategy based on such domain is substantial part of these complex problems to generate parallel processes. Mostly such domain is characterized by massive, discrete or static data structure. Typical examples of computational domains are matrix parallel multiplication and matrix parallel algorithms representing for example with system of linear equations.

These both matrix parallel algorithms we have chosen to represent typical examples of matrix data decomposition models.

4.1.2.1. Decomposition Methods for Matrix Multiplication

We will illustrate the role of optimal selection of decomposition model on matrix multiplication. The principle of matrix multiplication we illustrate for simplicity for the matrixes A, B with numbers of rows and columns k=2. The result matrix C = A x B is

$$\begin{bmatrix} a_{11} & a_{12} \\ a_{21} & a_{22} \end{bmatrix} \times \begin{bmatrix} b_{11} & b_{12} \\ b_{21} & b_{22} \end{bmatrix} = \begin{bmatrix} a_{11} \cdot b_{11} + a_{12} \cdot b_{21} & a_{11} \cdot b_{12} + a_{12} \cdot b_{22} \\ a_{21} \cdot b_{11} + a_{22} \cdot b_{21} & a_{21} \cdot b_{12} + a_{22} \cdot b_{22} \end{bmatrix} = \begin{bmatrix} c_{11} & c_{12} \\ c_{21} & c_{22} \end{bmatrix}$$

The way of sequential calculation is following
Step 1:
  compute all the values of the result matrix C for the first row of matrix A and for all columns of the matrix B
Step 2 :
  take the next row of matrix A and repeat step 1.

In this procedure we can see the potential possibility of parallel computation that is the repetition of activities in the step 1 always with another row of matrix A. Let consider the calculation example of matrix multiplication on parallel system. The basic idea of the possible decomposition procedure illustrates Fig. 11.

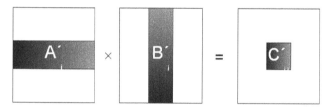

Figure 11. *Standard decomposition of matrix multiplication.*

The procedure is as following
Step 1:
  give to the i-th node horizontal column of the matrix A with their names A'_i and i-th vertical column of the matrix B named as B'_i
Step 2:
  compute all values of the result matrix C for A'_i and B'_i and named them as C'_{ii}
Step 3:
  give to i-th computing node its value B'_i to the node i-1 and get B'_{i+1} value from computing node i+1.
Repeat the steps 2 and 3 to the time till i-th node does not computed $C'_{i,i-1}$ values with B'_{i-1} columns and row A'_i. Then i-th node computed i-th row of the matrix C (Fig. 12.) for the matrix B with the number k-columns. The advantage of such chosen decomposition is the minimal consume of memory cells. Every node has only three values (rows and columns) from every matrix.

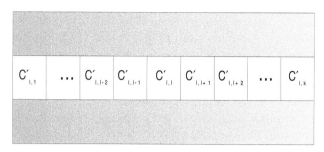

Figure 12. *Illustration of the gradually calculation of matrix C.*

This method is also faster as second possible way of decomposition according Fig. 13.

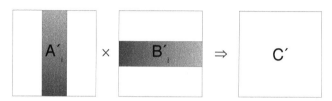

Figure 13. *Matrix decomposition model with columns of first matrix.*

The procedure is following
Step 1:
  give to the i-th node vertical column of the matrix A (A'_i) and the horizontal row of the matrix B (B'_i).
Step 2:
  perform the ordinary matrix computation A'_i and B'_i. The result is the matrix C'_i of type n x n. Every element from C'_i is the particular element of the total sum which corresponds to the result matrix C.
Step 3:
  call the function of parallel addition GSSUM for the creating of the result matrix C through   corresponded elements C'_i. This added function causes increasing of the calculation time, which strong depends on the magnitude of the input matrixes (Fig. 14.).

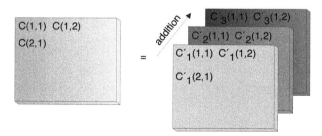

Figure 14. *Illustration of the gradually calculation of the elements $C'_{i,j}$.*

Let k be the magnitude of the rows or columns A, B and U defines the total number of nodes. Then

$$C'_1(1,1) = a_{1,1} \cdot b_{1,1} + a_{1,2} \cdot b_{1,2} + \ldots + a_{1,k} \cdot b_{1,k}$$
$$C'_2(1,1) = a_{1,k+1} \cdot b_{1,k+1} + a_{1,k+2} \cdot b_{1,k+2} + \ldots + a_{1,2k} \cdot b_{1,2k}$$

.

.

.

$$C'_U(1,1) = a_{1,(U-1)k+1} \cdot b_{1,(U-1)k+1} + \ldots + a_{1,Mk} \cdot b_{1,Mk}$$

and the finally element of matrix C

$$C(1,1) = \sum_{i=1}^{U} C'_i(1,1)$$

4.1.2.2. Decomposition Matrix Models

Any matrix is a regular data structure (domain) for which we can create parallel processes dividing matrix into strips (rows, columns). In generation process of matrix parallel processes there is necessary to tray allocate domain parts in such a way that every computing node has roughly the same number of strips (rows, columns). If the total number of strips is divisible by the number of processors without rest then all computing node will have the same number of decomposed parts (load balance). Otherwise some computing node has some strips extra.

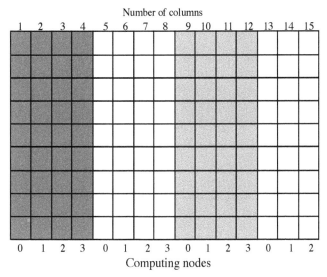

Number of columns

Computing nodes

Figure 15. Matching of data blocks.

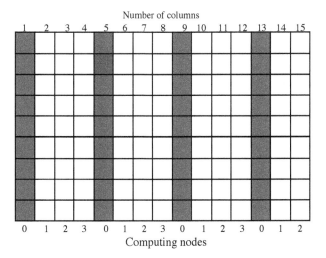

Number of columns

Computing nodes

Figure 16. Assigns columns.

Matrix allocation methods of decomposed strips (rows, columns) for solving system of linear equations by Gauss

eliminated method (GEM) are as follows [5]
- allocation of block strips
- gradually allocation of strips.

In the first allocation method strips are divided to set of strips and to every computing node is assigned one block. Illustration of these allocation methods is at Fig. 15. At another method with gradual allocation of columns they are allocated columns to individual computing nodes like the card are gradually passing out at games to game participants. Illustration of gradual assignment of columns is at Fig. 16.

4.1.3. Functional Decomposition

Functional decomposition strategies concentrate their attention to find parallelism in distribution of sequential computation stream in order to create independent parallel processes. In comparison to domain decomposition we are concerned to create potential alternative control streams of concrete complex problem. In this way we are streaming in functional decomposition to create so much as possible of parallel threads. Illustration of functional decomposition is at Fig. 17.

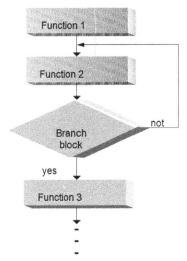

Figure 17. Illustration of functional decomposition.

The widely distributed functional strategies are as following
- controlled decomposition
- manager / workers (server / clients).

Typical parallel algorithms are complex optimization problems, which are connected with consecutive searching of massive data structures.

4.1.3.1. Control Decomposition

Control decomposition as an alternative of functional decomposition concentrates to given complex problems as sequence of individual activities (operations, computing steps, control activities etc.), from which we are able to derive multiple control processes. For example we can consider searching tree, which respond to game moves where branch factor is changed from node to node. Any static allocation of tree is not possible or activates unbalanced load.

So in this way this decomposition method supposed irregular structure controlled decomposition which is soft connected with complex problems in artificial intelligence and similar non numerical applications. Secondary it's very natural to look at any complex problem as collection of modules, which represent needed functional parts of given algorithm.

4.1.3.2. Decomposition Strategy Manager / Workers

Another alternative of functional decomposition is the strategy manager / workers. In this case there is used one parallel process as control one (manager). Manager process then sequentially and continuously generates needed parallel processes (workers) to their performing in controlled computing nodes. The illustration of parallel structure of decomposition model manager / workers is at Fig. 18. One of the possible applied solutions illustrates Fig. 19.

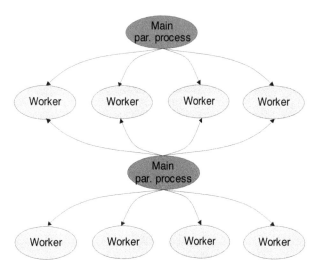

Figure 18. *Manager / worker parallel structure.*

Manager process controls computation sequence in relation to sequential finishing allocated parallel processes from individual workers. This decomposition strategy is suitable mainly in such cases in which given problem does not content static data or known fixed number of computations. In such cases there is necessary to concentrate to control aspects of individual parts of complex problems. From performed analysis then comes out needed communication sequence to achieve demanded time sequences of created parallel processes. Dividing degree of given complex problem coincide with number of computing nodes of parallel computer, with parallel computer architecture and with performance knowledge of computing nodes. One of important element of previous steps is allocation algorithm. It is more effective to allocate parallel process to first free computing node (worker) in comparison to some defined sequence order of allocation.

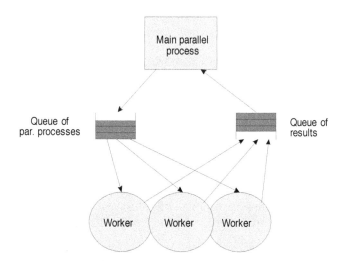

Figure 19. *Illustration of manager / workers strategy.*

4.1.4. Divide and Conquer Decomposition Model

Divide and conquer strategy decomposed complex problem into subtasks of the same size but it iteratively keeps repeating this process to obtain yet smaller parts of given complex problems. In that sense this decomposition model iteratively applies the problem partitioning technique as we can see it at Fig. 20. Divide and conquer is sometimes called recursive partitioning. Typically complex problem size is an integer power of 2 and the divide and conquer strategy halves complex problem into two equal parts at each iteration step.

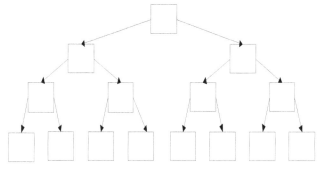

Figure 20. *Illustration of divide and conquer strategy (n=8).*

Every from N / 2 – point DFT we can again divide to next parts, that is to two N / 4 – point DFT. Applied decomposition strategy could follows till to exhausting dividing possibility for given N (one point value). Dividing factor is named as radix - q, and that for dividing number higher than two. We have applied decomposition model of divide-and-conquer strategy for parallel solution of fast discrete Fourier transform (DFFT) in [15].

4.1.5. Decomposition Models for Complex Problems

To decompose complex problem there is in many cases necessary to use more than one decomposition strategy. This is true mainly in hierarchical structure of concrete complex problem. The hierarchical character of complex problem means that we are looking to such complex problem as set of various hierarchical levels whereby it would be useful to apply at every existed level another decomposition model. This approach we are naming as multilayer decomposition.

To effective using of multilayer decomposition contributes new generation of common parallel computers based on implementation of the order more than thousand computing nodes (processors, cores). Secondly unifying trends of high performance parallel computing (HPC) based on massive parallel computers (massive SMP, supercomputers) and distributed computing (NOW, Grid) open to programmers new horizons.

To the typical complex problems belong weather prognosis, fluid flow, structural analysis of substance building, nanotechnologies, high physics energies, artificial intelligence, symbolic processing, knowledge economic etc. Multilayer decomposition model makes it available to decompose complex problem at first to simpler modules and then in second phase to apply suitable decomposition model only to given decomposed part.

4.1.6. Object Oriented Decomposition

Object oriented decomposition is integral part of object oriented programming (OOP). Actually it presents modern way of parallel program development. OOP beside increased demand to abstract thinking of programmer contents decomposition of complex problem to independent parallel modules named as objects [35]. In this way object oriented approach looks at complex problem as collection of abstract data structures (objects), where integral parts of objects are also build object functions as other form of parallel processes. In the same way OOP creates bridge between sequential computers and modern parallel computers based on SMP, NOW and Grid. Illustration of object structure is at Fig. 21.

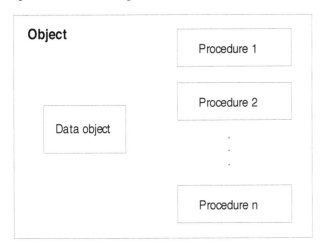

Figure 21. Object structure.

4.1.7. Mapping

This step allocates created parallel processes to computing nodes of parallel computer for their parallel execution. There is necessary to achieve that every computing node should perform allocated parallel processes (one or more) with at least approximate input

loads (load balancing) on real assumption of equal powerful computing nodes. Fulfillment of this condition contributes to optimal parallel solution latency.

4.1.8. Inter Process Communication

In general we can say that dominated elements of parallel algorithms are their sequential parts (Parallel processes) and analyzed inter process communication (IPC) among performed parallel processes using high speed communication networks [2, 37].

4.1.8.1. Inter Process Communication in Shared Memory

Any concrete communication mechanism make use existence of shared memory which allows every parallel process to story communicating data at some addressed memory place and then another parallel process to read stored data (shared variable). It looks very simple but there is necessary to guarantee that in the same time can use the addressed memory place only one parallel process. These needed control mechanism are named as synchronization tools. The typical synchronization tools are as following

- busy waiting
- semaphores
- conditional critical regions (CCR)
- monitors
- path expressions.

4.1.8.2. Inter Process Communication in Distributed Memory

Inter process communication (IPC) for parallel algorithm with distributed memory (PA_{dm}) is defined within supporting developing standards as following

- MPI (Message passing interface)
    - point to point (PTP) communication commands
        - send commands
        - receive commands
    - collective communication commands
        - data distribution commands
        - data gathering commands
- PVM (Parallel virtual machine)
- Java (Network communication support)
- other.

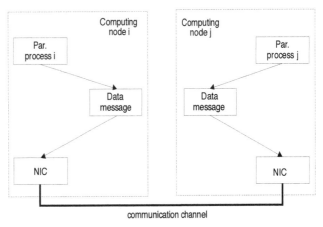

Figure 22. Illustration of MPI network communication.

To create needed synchronization tools in MPI we have available only existed network communication of connected computing nodes. Typical MPI network communication is at Fig. 22. Based on existed communication links MPI contains synchronization command Barrier.

4.1.9. Performance Tuning

After verifying developed parallel algorithm on concrete parallel system the further step is performance modeling and optimization (effective PA). This step contents analysis of previous steps in such a way to minimize whole latency of parallel computing T(s, p). Performed optimization of T(s, p) for given parallel algorithm depends mainly from following factors

- allocation of balanced input load to used computing nodes of parallel computer (load balancing) [19]
- minimization of accompanying overheads amounts (parallelization, inter process communication IPC, control of PA) [30].

To do load balancing we need in case of obvious using of equally powerful computing nodes of PC results of load allocation for given developed PA. In dominated asynchronous parallel computers (NOW, Grid) there are necessary to reduce (optimize) mainly number of inter process communications IPC (communication load) for example through using of alternative existing decomposition model.

5. Chosen Illustration Results

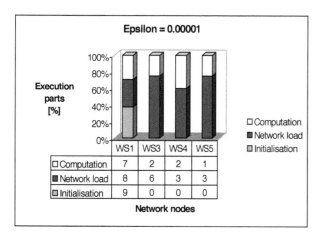

Figure 23. Computation and communication latencies for epsilon=10^-5.

We illustrate some of chosen performed tested results. For experimental testing we have used workstations of NOW parallel computer as follows

- WS 1 – Pentium IV (f = 2,26 G Hz)
- WS 2 - Pentium IV Xeon (2 proc., f = 2,2 G Hz)
- WS 3 - Intel Core 2 Duo T 7400 (2 cores, f=2,16 GHz)
- WS 4 - Intel Core 2 Quad (4 cores, 2.5 GHz)
- WS 5 - Intel Sandy Bridge i5 2500S (4 cores, f=2,7 GHz).

5.1. Numerical Integration

We have measured defined performance evaluation metrics in NOW parallel computer with Ethernet communication network and previous defined specification of used workstations. The individual latencies of complex execution time for accuracy epsilon=10^{-5} are illustrated at Fig. 23.

From this figure we can see that for used computation loads communication latencies (initialization, network load) are dominant.

From Fig. 24 we can see that with decreasing order of epsilon (higher computation accuracy) growths needed computing time in a linear way caused with linear raise of needed computations.

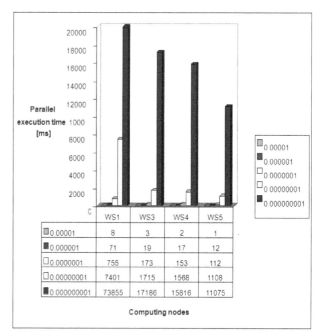

Figure 24. Computation latency for various accuracy (epsilon =10^{-5} - 10^{-9}).

From Fig. 25 we can see that increased input computation loads could result in dominating influence of computation time. This is caused by low increased communication load, which remains nearly constant.

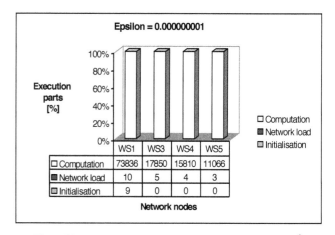

Figure 25. Dominancy of computation latency for epsilon = 10^{-9}.

Based on these results we know that dominate influence to the whole complexity of analyzed parallel algorithm has computation latency $T(s, p)_{comp}$ in comparison to communication latency $T(s, p)_{comm}$. To map mentioned assumption to the relation for asymptotic isoefficiency $w(s)$ means that

$$w(s)=\max\left[T(s,p)_{comp},T(s,p)_{comm}<T(s,p)_{comp}\right]=\max\left[T(s,p)_{comp}\right]$$

Such parallel algorithms are very effective also at using dominant parallel computers with distributed memory (NOW, Grid).

5.2. Parallel Multiplication

The comparison on analyzed decomposition models for parallel multiplication in part 4.1.2.1. for various number of computing nodes of parallel computer illustrates Fig. 26. The first chosen decomposition model (decomposition 1) goes straightforward to the calculation of the individual elements of the result matrix C through multiplication of the corresponding matrix elements A and B. The second decomposition model (decomposition 2) for the getting the final elements of the matrix C besides multiplication of the corresponded matrix elements A and B demand the additional addition of the particular results, which causes the additional time complexity in comparison to the first used method. This additional time complexity depends strong on the magnitude of the input matrixes. On this example of matrix multiplication we can see potential crucial influence of decomposition model to complexity of parallel algorithms.

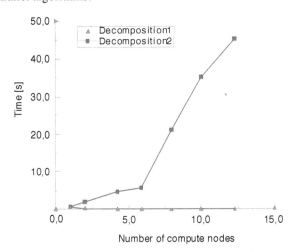

Figure 26. Influence of decomposition models in parallel multiplication.

6. Conclusions and Perspectives

Performance modeling in parallel computing as a discipline has repeatedly proved to be critical for design and successful use of parallel computers and parallel algorithms too. At the early stage of design, performance models can be used to project the system scalability and evaluate design alternatives. At the production stage, performance evaluation methodologies can be used to detect bottlenecks and subsequently suggests ways to

alleviate them. Analytical methods (order analysis, queuing theory systems, Petri-nets), simulation and experimental measurements have been successfully used for the evaluation of parallel computers and parallel algorithms too.

On the given illustrative examples of chosen parallel algorithms (numerical integration, matrix multiplication) we have been demonstrated great influence of selection of decomposition model to their potential effectiveness. This is very important mainly in using dominant parallel computers (NOW, Grid) where the dominant influence used to have communication complexity of parallel algorithms. Therefore according the latest trends in parallel computing based on developing of mixed parallel algorithms (shared memory, distributed memory) there are important optimal decisions in developing PA which parts would be executed on SMP computing nodes and which on NOW modules.

Via the extended form of complex isoefficiency concept we have illustrated its concrete using to predicate the performance in applied matrix parallel algorithms. To derive complex isoefficiency function in analytical way it is necessary to derive al typical used criterion for performance evaluation of parallel algorithms including their overhead function $h(s, p)$. Based on these relations we are able to derive complex issoefficiency function as real criterion to evaluate and predict performance of parallel algorithms also for theoretical (not existed) parallel computers. So in this way we can say that this process includes complex performance evaluation including performance prediction.

Acknowledgements

This work was done within the project "Complex modeling, optimization and prediction of parallel computers and algorithms" at University of Zilina, Slovakia. The authors gratefully acknowledge help of project supervisor Prof. Ing. Ivan Hanuliak, PhD.

References

[1]   Abderazek A. B., Multicore systems on chip - Practical Software/Hardware design, Imperial college press, pp. 200, 2010

[2]   Arie M.C.A. KosterArie M.C.A., Munoz Xavier, Graphs and Algorithms in Communication Networks, Springer-Verlag, Germany, pp. 426, 2010

[3]   Arora S., Barak B., Computational complexity - A modern Approach, Cambridge University Press, United Kingdom, pp. 573, 2009

[4]   Bahi J. H., Contasst-Vivier S., Couturier R., Parallel Iterative algorithms: From Seguential to Grid Computing, CRC Press, USA, 2007

[5]   Bronson R., Costa G. B., Saccoman J. T., Linear Algebra - Algorithms, Applications, and Techniques, 3rd Edition, Elsevier Science & Technology, Netherland, pp. 536, 2014

[6]   Dattatreya G. R., Performance analysis of queuing and computer network, University of Texas, Dallas, USA, pp. 472, 2008

[7]   Desel J., Esperza J., Free Choise Petri Nets, Cambridge University Press, United Kingdom, pp. 256, 2005

[8]   Dubois M., Annavaram M., Stenstrom P., Parallel Computer Organization and Design, Cambridge university press, United Kingdom,pp. 560, 2012

[9]   Dubhash D.P., Panconesi A., Concentration of measure for the analysis of randomized algorithms, Cambridge University Press, United Kingdom, 2009

[10] Edmonds J., How to think about algorithms, Cambridge University Press, United Kingdom, pp. 472, 2010

[11] Gelenbe E., Analysis and synthesis of computer systems, Imperial College Press, United Kingdom, pp. 324, 2010

[12] GoldreichOded, P, NP, and NP - Completeness, Cambridge University Press, United Kingdom, pp. 214, 2010

[13] Hager G., Wellein G., Introduction to High Performance Computing for Scientists and Engineers, CRC Press, USA, pp. 356, 2010

[14] Hanuliak P., Complex modeling of matrix parallel algorithms, American J. of Networks and Communication, Science PG, Vol. 3, USA, 2014

[15] Hanuliak P., Hanuliak J., Complex performance modeling of parallel algorithms, American J. of Networks and Communication, Science PG, Vol. 3, USA, 2014

[16] Hanuliak P., Hanuliak I., Performance evaluation of iterative parallel algorithms, Kybernetes, Volume 39, No.1/ 2010, United Kingdom, pp. 107- 126, 2010

[17] Hanuliak P., Analytical method of performance prediction in parallel algorithms, The Open Cybernetics and Systemics Journal, Vol. 6, Bentham,United Kingdom, pp. 38-47, 2012

[18] Hanuliak P., Complex performance evaluation of parallel Laplace equation, AD ALTA – Vol. 2, issue 2, Magnanimitas, Hradec Kralove, Czech republic, pp. 104-107, 2012

[19] Harchol-BalterMor, Performance modeling and design of computer systems, Cambridge University Press, United Kingdom, pp. 576, 2013

[20] Hillston J., A Compositional Approach to Performance Modeling, University of Edinburg, Cambridge University Press,United Kingdom, pp.172, 2005

[21] Hwang K. and coll., Distributed and Parallel Computing, Morgan Kaufmann, USA, pp. 472, 2011

[22] Kshemkalyani A. D., Singhal M., Distributed Computing, University of Illinois, Cambridge University Press, United Kingdom, pp. 756, 2011

[23] Kirk D. B., Hwu W. W., Programming massively parallel processors, Morgan Kaufmann, USA, pp. 280, 2010

[24] Kostin A., Ilushechkina L., Modeling and simulation of distributed systems, Imperial College Press, United Kingdom, pp. 440, 2010

[25] Kumar A., Manjunath D., Kuri J., Communication Networking, Morgan Kaufmann, USA, pp. 750, 2004

[26] Kushilevitz E., Nissan N., Communication Complexity, Cambridge University Press, United Kingdom, pp. 208, 2006

[27] Le Boudec Jean-Yves, Performance evaluation of computer and communication systems, CRC Press, USA, pp. 300, 2011

[28] Levesque John, High Performance Computing: Programming and applications, CRC Press, USA, pp. 244, 2010

[29] Lilja D. J., Measuring Computer Performance, Cambridge University Press, United Kingdom, pp. 280, 2005

[30] McCabe J., D., Network analysis, architecture, and design (3rd edition), Elsevier/ Morgan Kaufmann, USA, pp. 496, 2010

[31] Misra Ch. S.,Woungang I., Selected topics in communication network and distributed systems, Imperial college press, United Kingdom, pp. 808, 2010

[32] Paterson D. A., Hennessy J. L., Computer Organisation and Design, Morgan Kaufmann, USA, pp. 912, 2009

[33] Peterson L. L., Davie B. C., Computer networks – a system approach, Morgan Kaufmann, USA, pp. 920, 2011

[34] Resch M. M., Supercomputers in Grids, Int. J. of Grid and HPC, No.1, pp. 1 - 9, 2009

[35] Shapira Y., Solving PDEs in C++ - Numerical Methods in a Unified Object-Oriented Approach (2nd edition), Cambridge University Press, United Kingdom, pp. 800, 2012

[36] Wang L., Jie Wei., Chen J., Grid Computing: Infrastructure, Service, and Application, CRC Press, USA, 2009 www pages

[37] www.top500.org.

Permissions

All chapters in this book were first published in AJNC, by Science Publishing Group; hereby published with permission under the Creative Commons Attribution License or equivalent. Every chapter published in this book has been scrutinized by our experts. Their significance has been extensively debated. The topics covered herein carry significant findings which will fuel the growth of the discipline. They may even be implemented as practical applications or may be referred to as a beginning point for another development.

The contributors of this book come from diverse backgrounds, making this book a truly international effort. This book will bring forth new frontiers with its revolutionizing research information and detailed analysis of the nascent developments around the world.

We would like to thank all the contributing authors for lending their expertise to make the book truly unique. They have played a crucial role in the development of this book. Without their invaluable contributions this book wouldn't have been possible. They have made vital efforts to compile up to date information on the varied aspects of this subject to make this book a valuable addition to the collection of many professionals and students.

This book was conceptualized with the vision of imparting up-to-date information and advanced data in this field. To ensure the same, a matchless editorial board was set up. Every individual on the board went through rigorous rounds of assessment to prove their worth. After which they invested a large part of their time researching and compiling the most relevant data for our readers.

The editorial board has been involved in producing this book since its inception. They have spent rigorous hours researching and exploring the diverse topics which have resulted in the successful publishing of this book. They have passed on their knowledge of decades through this book. To expedite this challenging task, the publisher supported the team at every step. A small team of assistant editors was also appointed to further simplify the editing procedure and attain best results for the readers.

Apart from the editorial board, the designing team has also invested a significant amount of their time in understanding the subject and creating the most relevant covers. They scrutinized every image to scout for the most suitable representation of the subject and create an appropriate cover for the book.

The publishing team has been an ardent support to the editorial, designing and production team. Their endless efforts to recruit the best for this project, has resulted in the accomplishment of this book. They are a veteran in the field of academics and their pool of knowledge is as vast as their experience in printing. Their expertise and guidance has proved useful at every step. Their uncompromising quality standards have made this book an exceptional effort. Their encouragement from time to time has been an inspiration for everyone.

The publisher and the editorial board hope that this book will prove to be a valuable piece of knowledge for researchers, students, practitioners and scholars across the globe.

List of Contributors

Tarmizi Amani Izzah and Syed Sahal Nazli Alhady
School of Electrical Electronics Engineering, Universiti Sains Malaysia (Engineering Campus), Nibong Tebal, SPS, Penang

Wan Pauzi Ibrahim
School of Medical Sciences, Universiti Sains Malaysia (Health Campus), 16150, Kubang Kerian, Kelantan

Michal Hanuliak
Dubnica Technical Institute, Dubnica Nad Vahom, Slovakia

Anmar Hamid Hameed and Salama A. Mostafa
College of Graduate Studies, Selangor, Malaysia

Universiti Tenaga Nasional, Selangor, Malaysia

Mazin Abed Mohammed
Dept. of Planning and Follow up, Anbar, Iraq

University of Anbar, Anbar, Iraq

Peter Hanuliak and Michal Hanuliak
Dubnica Technical Institute, Sladkovicova 533/20, Dubnica nad Vahom, 018 41, Slovakia

Sharada Santosh Patil
MCA, Deptt. SIBAR Kondhwa, Pune, Maharshtra, INDIA

Director MCA, SIBAR, Kondhwa,Pune, Maharashtra INDIA

Arpita N. Gopal
Director MCA, SIBAR, Kondhwa,Pune, Maharashtra INDIA

A. A. Ojugo
Department of Mathematics/Computer, Federal University of Petroleum Resources Effurun, Delta State

A. O. Eboka, M. O. Yerokun and I. J. B. Iyawa
Department of Computer Sci. Education, Federal College of Education (Technical) Asaba, Delta State

R. E. Yoro
Department of Computer Science, Delta State Polytechnic Ogwashi-Uku, Delta State

Oladosu Olakunle Abimbola, Okikiola Folasade Mercy, Alakiri Harrison Osarenren and Oladiboye Olasunkanmi Esther
Department of Computer Technology, Yaba College of Technology, Yaba, Lagos

A. A. Ojugo. and D. Oyemade.
Department of Mathematics/Computer, Federal University of Petroleum Resources Effurun, Delta State

R. E. Yoro.
Department of Computer Science, Delta State Polytechnic Ogwashi-Uku

M. O. Yerokun., A. O. Eboka, E. Ugboh and F. O. Aghware
Department of Computer Science, Federal College of Education Technical Asaba

Nosiri Onyebuchi Chikezie and Chukwudebe Gloria Azogini
Department of Electrical & Electronic Engineering, Federal University of Technology Owerri, Nigeria

Onoh Gregory Nwachukwu
Department of Electrical & Electronic Engineering, Enugu State University of Science & Technology, Nigeria

Azubogu Austin Chukwuemeka
Department of Electronic and Computer Engineering, Nnamdi Azikiwe University Awka, Nigeria

Muhammad Siraj Rathore, Markus Hidell and Peter Sjödin
Network Systems Laboratory, School of ICT, KTH Royal Institute of Technology, Stockholm, Sweden

Abdulaziz Rashid Alazemi
Computer Engineering Department, Kuwait University, Kuwait

Luo Jin, Ma Qihua and Luo Yiping
Faculty of Automotive Engineering, Shanghai University of Engineering and Science, Shanghai, China

Jyothilal Nayak Bharothu and D Sunder Singh
Associate professor of Electrical & Electronics Engineering, Sri Vasavi Institute of Engineering & Technology, Nandamuru, A.P.; India

Basma A. Mahmoud, Esam A. A. Hagras and Mohamed A. Abo El-Dhab
Department of Electronics & Communications, Arab Academy for Science, Technology & Maritime Transport, Cairo, Egypt

Diponkar Paul
Department of Electrical and Electronic Engineering, Prime University, Dhaka, Bangladesh

Shamsuddin Majamder
Department of Electrical and Electronic Engineering, World University of Bangladesh, Dhaka, Bangladesh

S. Uvaraj
Arulmigu Meenakshi Amman College of Engineering, Kanchipuram

N. Kannaiya Raja
Defence Engineering College, Ethiopia

Nguyen Minh Quy and Ho Khanh Lam
Faculty of Information Technology, Hung Yen University of Technology and Education, Hung Yen, Vietnam

Huynh Quyet Thang
School of Information Communication and Technology, Hanoi University of Science and Technology, Ha Noi, Vietnam

Peter Hanuliak
Dubnica Technical Institute, Sladkovicova 533/20, Dubnica nad Vahom, 018 41, Slovakia

Peter Hanuliak and Juraj Hanuliak
Dubnica Technical Institute, Sladkovicova 533/20, Dubnica nad Vahom, 018 41, Slovakia

Michal Hanuliak
Dubnica Technical Institute, Sladkovicova 533/20, Dubnica nad Vahom, 018 41, Slovakia

Peter Hanuliak and Michal Hanuliak
Dubnica Technical Institute, Sladkovicova 533/20, Dubnica nad Vahom, 018 41, Slovakia

Michal Hanuliak and Juraj Hanuliak
Dubnica Technical Institute, Sladkovicova 533/20, Dubnica nad Vahom, 018 41, Slovakia

Index

Printed in the USA
CPSIA information can be obtained
at www.ICGtesting.com
JSHW051432221024
72173JS00006B/1446